Making Sustainability Work

Best Practices In Managing And Measuring Corporate Social, Environmental, And Economic Impacts

Marc J. Epstein
Adriana Rejc Buhovac
With Forewords By John Elkington
Herman B. "Dutch" Leonard

16pt

Read How You Want
LARGE PRINT BOOKS, BRAILLE & DAISY

Copyright Page from the Original Book

Making Sustainability Work

Berrett-Koehler Publishers, Inc.
235 Montgomery Street, Suite 650
San Francisco, California 94104-2916
Tel: (415) 288-0260, Fax: (415) 362-2512
www.bkconnection.com

Ordering information for print editions
Quantity sales. Special discounts are available on quantity purchases by corporations, associations, and others. For details, contact the "Special Sales Department" at the Berrett-Koehler address above.
Individual sales. Berrett-Koehler publications are available through most bookstores. They can also be ordered directly from Berrett-Koehler: Tel: (800) 929-2929; Fax: (802) 864-7626; www.bkconnection.com
Orders for college textbook/course adoption use. Please contact Berrett-Koehler: Tel: (800) 929-2929; Fax: (802) 864-7626.
Orders by U.S. trade bookstores and wholesalers. Please contact Ingram Publisher Services, Tel: (800) 509-4887; Fax: (800) 838-1149; E-mail: customer.service@ingrampublisherservices.com; or visit www.ingrampublisherservices.com/Ordering for details about electronic ordering.

Published simultaneously in the UK in print by:

Greenleaf Publishing Limited
Aizlewood's Mill, Nursery Street
Sheffield S3 8GG, UK
Tel: +44 (0)114 282 3475
Fax: +44 (0)114 282 3476
www.greenleaf-publishing.com
Hardcover print UK edition ISBN 978-1-90764-393-4

Second Edition
Hardcover print edition ISBN 978-1-60994-993-8
PDF e-book ISBN 978-1-60994-994-5
IDPF e-book ISBN 978-1-60994-995-2

2014-1

Cover by LaliAbril.com.

TABLE OF CONTENTS

About the authors

Marc J. Epstein is a Distinguished Research Professor of Management at Jones Graduate School of Business at Rice University in Houston, Texas. Prior to joining Rice, Dr. Epstein was a professor at Stanford Business School, Harvard Business School, and INSEAD (European Institute of Business Administration). In both academic research and managerial practice, Dr. Epstein is considered one of the global leaders in the areas of innovation, sustainability, governance, performance measurement and accountability in both corporations and not-for-profit organizations. His 20 authored or co-authored books and well over 200 professional papers have won numerous top academic, professional, and business awards.

Adriana Rejc Buhovac is an Associate Professor of Management at the Faculty of Economics at the University of Ljubljana. She is a lecturer in several undergraduate and graduate courses and a renowned speaker and lecturer of executive program courses. Her primary areas of expertise are strategic management, performance measurement and management control, and corporate sustainability. In these areas, Dr. Rejc Buhovac co-authored a number of academic articles and several *Management Accounting Guidelines* for CMA Canada and the AICPA.

List of cases, figures, and tables

Cases

Figures

Tables

Foreword from the First Edition

John Elkington

On the face of it, few books seem further removed from Making Sustainability Work than Robert Pirsig's bestseller Zen and the Art of Motorcycle Maintenance. But, at least to my mind, there are interesting links. Pirsig's book, which first appeared in 1974 and sold many millions of copies in 27 languages, turned into a global phenomenon, as has the sustainability agenda which Marc Epstein presents in the following pages. What sticks in my mind over 30 years after reading Zen is the way Pirsig spotlighted two very different personality types. The first type is mostly interested in what the Germans call the Gestalt, focusing on big-picture trends and the configuration of elements, rather than the elements themselves. This 'Romantic' personality enjoys the experience of bike-riding, but is none too keen on the messy business of engineering, let alone maintenance. The second, 'Classic' type enjoys the experiences, but is much more interested in the details, the inner workings, the mechanics.

Reading through Marc Epstein's latest book, I was tempted to conclude that he falls into the second, Classic type. His title as Distinguished Research Professor, coupled with the Rice and Harvard affiliations, seemed ample proof. And, adding further circumstantial evidence, the titles of his books over the years underscore his intense analytical focus on the inner workings of what he and his co-editors dubbed 'The Accountable Corporation' in a series of books of the same name. His other works include Counting What Counts: Turning Corporate Accountability to Competitive Advantage and Measuring Corporate Environmental Performance: Best Practices for Costing and Managing an Effective Environmental Strategy.

Case proven? Well, not entirely. Because if there is one thing that really influenced me in Pirsig's book, which was subtitled An Inquiry into Values, it was the notion that the most successful people in any field combine elements of both the Classic and the Romantic world-views. They are vigorous hybrids. My sense is that Marc Epstein is such a hybrid. Making Sustainability Work, which in the context of mounting challenges in areas like climate change, pandemic risks, and poverty must be a central political and business priority in the coming decades, is now throwing up some very unlikely champions. In the United States, which has been on something of an excursion away from sustainability in recent years, we now have companies like GE and Wal-Mart, combinations of corporations and NGOs like the U.S. Climate Section Partnership, and politicians like Arnold Schwarzenegger embracing issues once seen as almost un-American—and, more to the point, driving forward with imaginative, entrepreneurial market solutions.

In short, the timing could not be better for such a book as you hold in your hands. As Epstein notes in his opening lines, "With growing sensitivity toward social and environmental issues and shareholder concerns, companies are increasingly striving to become better corporate citizens. Executives recognize that long-term economic growth is not possible unless that growth is socially and environmentally sustainable. A balance between economic progress, social responsibility, and environmental protection, sometimes referred to as the triple bottom line, can lead to competitive advantage."

The time has come to kick the tires and look under the hoods of our most powerful institutions, most particularly our corporations, to test their capacity to help drive the sustainability transformation of our politics, governance, economies, corporations, communities, and, ultimately, societies. Those who lead the way will be able to see the big picture, mapping the future and engaging a wide range of decision-makers and other stakeholders in the process, while simultaneously being able to drill down to the detail, to the critical points where the rubber hits the road. Marc Epstein is a successful, proven navigator in these complex new risk and opportunity spaces. Fasten your safety belts—and make sure your CEO and board have copies of this invaluable guide ready to hand.

***

In 1987, John Elkington co-founded SustainAbility (www.sustainability.com) and blogs at www.johnelkington.com. In 2004, *BusinessWeek* described him as "a dean of the corporate-responsibility movement for three decades."

Foreword from the First Edition

Herman B. "Dutch" Leonard

If you are a corporate leader who is seriously interested in getting your organization to find, develop, and actually carry out successful programs in the domain of social responsibility—programs that actually improve social and environmental outcomes while building business value for your firm—then you have long needed this book.

There are two forms of corporate social responsibility (CSR) programs: the kind where corporate leaders talk a lot about what their firms are doing (but don't actually do very much or generate much impact), and the kind where socially responsible activities are being carried out on a material scale and significant results are actually being achieved. Sadly, at this stage in our history, there is still far too much of the former—and not nearly enough of the latter.

The reasons for this are not far to seek. First, there are some payoffs from just talking about CSR or running low-impact CSR programs—critics can sometimes be mollified and stakeholders reassured if a firm develops and describes a small collection of well-intentioned and plausible-sounding "citizenship" initiatives. Second, going beyond a few simple, nice-sounding initiatives to develop significant programs that build both business and social value is much more difficult than it might appear.

Ultimately, the real policies of an organization are not what its leaders say they are; the real policies are what the people in the organization are actually doing. It is easy for corporate leaders to talk about the "business case for social responsibility"—the idea that doing things in a way that improves social and environmental outcomes will also build greater business value (often with the caveat "...in the long run")—but talking about it is a far cry from making it be what is actually happening throughout the firm.

So, if you do actually want to make social responsibility be what your firm is doing, what do you need to do? You will need to articulate a combination of business, social, and environmental goals and then build structures, systems, and procedures within your firm that will focus attention on the combined goals—and enact your stated policy by embedding it in the ongoing actions and decisions of the firm. Unless and until the wide range of consequences of business activities—impacts on customers, revenue, markets, cost, social conditions, and environmental outcomes—are viewed at the same time and within the same discussions and analyzed and examined with the same rigor, CSR-related programs will remain sideline, non-strategic, secondary activities.

And that is where this book comes in. In this work, Marc Epstein presents a wide range of tools, methods, and approaches to bring social and environmental results into focus in the same ongoing business processes that drive the mainline business activities of the firm. He begins with an overview of the leadership necessary to animate and organize a serious corporate effort to build social and economic value through social responsibility, and lays out the elements necessary to make such an effort an integral aspect of an overall, comprehensive business strategy. He then examines the organizational structure issues that need to be addressed to create and maintain alignment among the activities designed to address the broader array of corporate goals that result from pursuing an integrated strategy.

The heart of this book—and the centerpiece of its contributions to corporate performance—is the series of sections on how to build and operate the organizational processes that will determine whether the firm is paying lip service to CSR or, instead, is enacting it in its daily operations and work. How can the costs of meeting social goals—and the risks of not meeting them—be factored into capital investment and allocation decisions? How can performance evaluation and reward systems be reconstructed to reflect the broadened set of goals? And how can organizational information systems be constructed to help managers achieve the high performance those personnel systems seek to reward? Both the evaluation systems and the management and learning systems will require metrics to inform them: How can we construct organizational processes that will define, collect, track, and analyze relevant data to provide managerial incentives, drive organizational learning, and guide strategic action across the full integrated panoply of firm objectives? How can the standard corporate processes associated with important business decisions—budgeting, personnel assignments and career tracking, and so on—be modified to include the full array of consequences from financial to social, that the firm now seeks jointly to manage? And, finally, how should firms organize the development of and carry out the internal and external communication of its goals and accomplishments across the full domain of consequences for which it is now taking responsibility?

On all of these subjects, this book provides practical advice grounded in examples drawn from a wide array of businesses. Epstein engages the issues at the frontier of CSR today: the practical questions of how to make it work in practice, in detail, day in and day out, so that what the firm wants its CSR policies to achieve actually turns out to be what the firm is accomplishing.

Many books have been written about why corporations should redefine their intentions and accept greater responsibility for the wide array of consequences that flow from business action. Many others have been written about what CSR strategies should look like in the abstract. This book transcends that rather stilted (and often moralistic) discussion. It assumes that there are good business reasons to pursue social and environmental goals, and then helps business leaders build the organizational processes necessary to discover and develop those opportunities—and to deliver on them.

***

Herman B. "Dutch" Leonard is Co-Chair of the Initiative on Social Enterprise and Eliot I. Snider and Family Professor of Business Administration, Harvard Business School; and George F. Baker, Jr. Professor of Public Management, John F. Kennedy School of Government, Harvard University.

Preface

Marc J. Epstein's work with CEOs and other senior executives of global corporations over the last 30 years has often begun with a query that goes something like: "Marc, I have spoken publicly about the importance of stakeholder engagement and sustainability [or corporate social responsibility], but do you realize how great the challenge is to implement this in my company of 50,000–100,000 employees? How can we get this done?" One can fully appreciate how daunting the challenge can seem; and much of our work has focused on answering these questions. This book will steer companies through this process.

Making Sustainability Work: Second Edition is not so much about what, whether, or why to focus on the "triple bottom line" of social, environmental, and economic impacts—but *how.* The study and practice of sustainability has matured. It is no longer just about risk and compliance, but also about innovation and opportunity and how to simultaneously achieve excellence in both sustainability and financial performance. With corporations facing more risks, with greater potential impacts, from a larger number of sources, the issues are more critical than ever. And, with the opportunities for innovation and growth in these areas more pronounced, this topic has come to the forefront of senior management discussions in most large organizations.

The book is grounded in extensive academic research and the best practices of corporations throughout the world. The research is the best in the field today, augmented by our own work, usually with academic colleagues, which includes field research and interviews, surveys of corporate practices, and conceptual development on approaches to improve the identification, measurement, and management of corporate social, environmental, and economic impacts. It also builds on research work that we have undertaken in the field, and the practices of the 100 or so leading companies that are discussed throughout the book (see list of companies cited). But of course this book could not have been completed without the hard work of managers worldwide who are even now implementing sustainability in their organizations. We are indebted to them for educating us, guiding us, and allowing us access to their corporations and their work.

We have been working with companies and conducting research in CSR (corporate social responsibility) for most of our professional careers. In his doctoral studies, Marc was fascinated by the development of CSR and worked to develop approaches to measuring companies' impacts on society. He continued this work as Director of Social Measurement Services at Abt Associates Inc. in Cambridge, Massachusetts. During his time there he saw the inception of the "corporate social audit" with both for-profit and nonprofit entities focusing on measurement and reporting of social impacts. Over the years his articles in academic and managerial publications on social and environmental responsibility—along with governance, accountability, and related topics—have run into dozens. He has also completed quite a number of books on this topic, including *Corporate Social Performance: The Measurement of Product and Service Contributions* (with Eric Flamholtz and Jack McDonough; New York: National Association of Accountants, 1977); *Counting What Counts: Turning Corporate Accountability to Competitive Advantage* (with Bill Birchard; Reading, MA: Perseus Books, 1999); and *The Accountable Corporation* (Westport, CT: Praeger, 2006), a four-volume edited series with Kirk Hanson. In many senses, though, this book is a follow-up to *Measuring Corporate Environmental Performance: Best Practices for Costing and Managing an Effective Environmental Strategy* (Burr Ridge, IL: Institute of Management Accountants/Irwin Professional Publishing, 1996).

Adriana has been concerned about this area and has been involved in research projects with company visits in North America and Europe on this topic.

Since Marc wrote *Measuring Corporate Environmental Performance,* he has been affiliated with four business schools: Stanford Business School, Harvard Business School, INSEAD (European Institute of Business Administration in France), and the Jones Graduate School of Management at Rice University. This book could not have been completed without the intellectual contributions, discussions, research assistance, collaboration, and other support by many, many friends and associates. Colleagues at each of those schools (and other schools and organizations besides) have provided intellectual stimulation and lively discussions around these issues and have significantly impacted his thinking. We thank all of them for their contributions. Though we don't have space to list them all individually, we do acknowledge that many of the thoughts on this topic were due to substantial learning from smart and dedicated colleagues, many of whom also collaborated with us on research projects. At these schools this includes Srikant Datar, Dutch Leonard, Kash Rangan, Jim Austin, Bob Kaplan, Bob Simons, Krishna Palepu, Greg Dees, Kirk Hanson, Jean-François Manzoni, Henri-Claude DeBettignies, Michael Brimm, Steve Currall, Sally Widener, Karen Schneitz, and Rick Bagozzi.

We also cannot say enough about our research colleagues at other schools who have engaged in many research projects with us on these and related topics. Their contribution to the research was often far greater than ours and their dedication and discussions always provided great stimulus. We have learned much from each of them, especially Marie-Josée Roy, Tony Davila, Priscilla Wisner, Wendy Smith, Tamara Bekefi, Melissa Tritter, Bill Birchard, Jed Emerson, and Kristi Yuthas. Our collaboration with these colleagues has produced much of the work upon which this book is based.

Numerous bright and diligent researchers have provided extraordinary assistance, including Nicolas

Lacouture, Rachel Gelman, Alicia Yancy, and Tammy Knotts. Tammy's work with Marc over the last two years has been extremely important in finalizing all of the details of the manuscript. Karen Lavelle has been working with Marc for many years providing valuable assistance in all facets of this research. He couldn't have done it without their dedicated work.

We also want to thank our editors John Stuart at Greenleaf Publishing and Johanna Vondeling and Neal Maillet at Berrett-Koehler Publishers who provided guidance and enthusiastic support for this project.

Making Sustainability Work: Second Edition is dedicated to corporate managers throughout the world who face the challenge of integrating sustainability considerations into their daily decision-making. And it is dedicated to those managers who are not just thinking about corporate social responsibilities and risks—but also about corporate social *opportunities.* The goal of simultaneously improving both corporate and societal performance is certainly a noble one.

Marc J. Epstein and Adriana Rejc Buhovac
July 2013

INTRODUCTION

Improving sustainability and financial performance in global corporations

On April 24, 2013, the catastrophic collapse of the Rana Plaza building on the outskirts of Dhaka, Bangladesh, killed more than 1,100 garment workers who had been producing apparel for some of the world's largest retailers.[1] The deadliest disaster in the history of the garment industry triggered a chain of actions that may result in a major shift in corporate liability in terms of global supply chain safety. Up until today, retailers would be using factory inspections and audits to claim their global supply chains are safe and in line with their business codes. From now on, where their contractors have factories, retailers may also become accountable for building safety. The world's largest apparel companies, including H&M (Hennes & Mauritz), Marks & Spencer, Inditex (the parent of Zara), and others, have already signed a far-reaching and legally-binding safety agreement. The plan requires retailers to have rigorous independent inspections and to help pay for fire safety upgrades and some other building improvements. While global climate change, local air and water pollution, child labor, and workers' rights, remain among the key challenges that face corporate executives on a daily basis, the scope of corporate responsibility is changing.

The issue of whether companies should consider their sustainability or the impacts of their activities on their stakeholders is thus no longer up for discussion. On the contrary, these issues, and many many more like them, have become a central part of the creation of shareholder value and the management of both global and local enterprises. The challenge has moved from "whether" to "how" to integrate corporate social, environmental, and economic impacts—corporate sustainability—into day-to-day management decisions when managers at all levels have significant incentive pressures to increase short-term earnings. It is now about how to be more socially responsible or sustainable, and how to engage corporate stakeholders more effectively. It is about the specific actions that managers can take to effectively deal with the paradox of trying simultaneously to improve corporate sustainability and financial performance.

Sustainability has been defined as economic development that meets the needs of the present generation without compromising the ability of future generations to meet their own needs.[2] For businesses, this includes issues of corporate social responsibility and citizenship along with improved management of corporate social and environmental impacts and improved stakeholder engagement. In this book, the terms "(corporate) sustainability" or "(corporate) sustainability performance" will imply that a company is contributing to sustainable development of society, which includes economic growth, environmental protection, and social progress. Further, "sustainability strategy" will relate to a set of strategic activities by which companies are following sustainability principles and contributing to sustainable development.[3]

Developing innovative sustainability strategies is often an important challenge for senior executives, but implementation is usually the larger challenge. In most of the successful implementations, CEOs are involved and are the drivers of corporate concern to implement sustainability. But these senior managers are often challenged as to how to manage the paradox of simultaneously improving social, environmental and economic, on the one hand, and financial performance, on the other. Business unit and facility managers are pressured to deliver profits and their performance is typically measured primarily on how successfully they deliver. So, there is often difficulty in obtaining an alignment of strategy, structure, systems, performance measures, and rewards to facilitate effective implementations. It is also often difficult to obtain the resources to manage the various drivers of social, environmental, and economic performance effectively.

Leading companies have increasingly recognized the critical importance of managing and controlling corporate social, environmental, and economic performance. The impetus for implementing a corporate strategy to integrate social, environmental, and economic impacts may be driven by internal factors, such as a management commitment to sustainability as a core value or by management recognition that sustainability can create financial value for the corporation through enhanced revenues and lower costs. Often, however, the leading impetus for a sustainability strategy is from external pressures such as government regulation, marketplace demands, competitors' actions, or pressure from NGOs (nongovernmental organizations). For example, controls on GHG (greenhouse gas) emissions will affect the price of energy and the products, services, and sectors that rely on that energy. Executives should assess their business strategies to determine whether there are opportunities in new worldwide markets in carbon, capital, advanced technologies, and products and services that emit lower amounts of GHGs.[4]

Managers have now recognized the importance of stakeholder input and engagement and the potential impact on long-term corporate profitability. The consequences for businesses when they do not effectively consider the impacts of their activities on society are often substantial. Thus, effective management of stakeholder impacts and relationships is critical.

Some companies have not developed any coherent sustainability strategy or even any systematic way of thinking about or managing their social, environmental, and economic (sustainability) impacts. Negative sustainability impacts have tarnished the reputation of many corporations. However, some have recognized the social, environmental, and economic effects of their actions, developed a corporate sustainability statement, and made progress toward defining a policy that confronts the problems. These companies have developed partial systems to deal with social, environmental, and economic problems and may have transferred technologies from other parts of the company to use in implementing sustainability. They may have set up systems for improved costing, capital budgeting, performance evaluations, or product design but have not developed an integrated program that includes sustainability in day-today decision-making. Some companies have developed effective reactive systems to address these issues and others have been more aggressively proactive.

It is unlikely that any company has fully integrated or achieved sustainability—this is a huge task—but numerous companies have taken important steps toward improving their sustainability performance and reducing their negative social, environmental, and economic impacts. Many of these companies are included in this book as exemplars of best practice. Rather than searching for one best company example to model, those companies and managers that want to improve their sustainability performance should instead look to adapt and adopt the various best practices of individual sustainability elements illustrated in this book. Through the detailed model, measures, and guidance to implementation presented here and the extensive best practice company examples from around the world, companies can select those practices that can be used to better implement sustainability in their own organizations to simultaneously improve corporate social, environmental, economic, *and* financial performance.

Leading companies are examining the impacts of their products, services, processes, and other activities more broadly. They are looking at a more comprehensive set of social, environmental, and economic impacts on a broader set of stakeholders. Managers recognize that stakeholders have numerous impacts on company profits—employees in their desire to work for the company, customers in their desire to buy from the company, the community in its desire to permit the company a license to operate. But they have faced difficulty in managing competing stakeholder interests and simultaneously improving both sustainability and financial performance. Business leaders who want to respond sensibly to activist calls for corporate responsibility should think about the issue in the same way they would about any other business problem.

But stakeholder management has to be more than identifying the squeakiest wheels and greasing them. Sustainability cannot be managed as just a public relations strategy to pacify stakeholder concerns. Doing so can be quite risky as stakeholders expect actions and results to be consistent with rhetoric. Furthermore, it is only through the identification, measurement, and management of sustainability impacts that social, environmental, economic, and financial performance can be improved and value created. For sustainability to be valuable to both the organization and its stakeholders, it must be integrated into the way a company does business.

The size of corporate sustainability expenditures is increasing rapidly and the necessity of improved identification and management of these impacts has become critical. Business leaders need to make an independent assessment of their social, economic, and environmental impacts to see where pressure is most likely to come and also to see where the company is providing unpriced social, environmental, and economic benefits for which it is not receiving credit. Firms should not underestimate their ability to turn sustainability into a competitive advantage. Patrick Cescau, the former group chief executive of Unilever, once said: “We have come to a point now where this agenda of sustainability and corporate responsibility is not only central to business strategy but will increasingly become a critical driver of business growth.... how well and how quickly businesses respond to this agenda will determine which companies succeed and which will fail in the next few decades.”[5]

Why it's important

Although this book focuses on *implementation,* here are the four main reasons why sustainability now demands our urgent attention:

1. Regulations. Government regulations and industry codes of conduct require that companies must increasingly address sustainability. Noncompliance with regulations was (and still is) costly, as regulatory noncompliance costs to companies include:

- Penalties and fines
- Legal costs
- Lost productivity due to additional inspections
- Potential closure of operations
- The related effects on corporate reputation

2. Community relations. The general public and activist NGOs are becoming increasingly aware of sustainability and the impacts that corporations have on society, the environment, and economy. Identifying the social, environmental, and economic issues that are important to key stakeholders and improving stakeholder relationships can foster loyalty and trust. Gaining a license to operate from governments, communities, and other stakeholders is of critical importance for corporations to be able to conduct business on an ongoing basis. Good performance on sustainability can garner a positive reputation with stakeholders and improve community relations and business performance. Alternatively, the consequences of mismanaging sustainability and stakeholder relationships can be significant and costly in terms of reputational damage and potential impacts on the bottom line. Coca-Cola, for example, has been working hard to provide data that show it uses water responsibly, and to make the case that it would be contrary to its own business interests to damage water sources and harm residents in the process. In doing so, it has gained ground in the courts and among regulators. Coca-Cola has been successful in securing the "technical" and "regulatory" licenses for using the water it needs. But where it has sometimes been challenged is in securing the "social license" for its operations among the communities that host them. To secure the trust and goodwill of the people in the communities where these operations are located, candid, continual dialogue with residents is needed, as well as showing residents how the company is managing water resources and what it is doing to improve them.[6]

3. Cost and revenue imperatives. Sustainability can also create financial value for the corporation through enhanced revenues and lower costs. In other words, managing sustainability is a good business decision. Revenues can be increased through increased sales due to improved corporate reputation. Costs can be lowered due to using resources more efficiently, product and process improvements, and a decrease in regulatory fines. The key is to identify the areas where good for society, good for the environment, and good for the company intersect.

4. Societal and moral obligations. Because of their impact on environment, society, and economy, companies have a responsibility to manage sustainability. A personal concern for social, environmental, and economic impacts and their social and moral obligations has led some executives and corporations to include sustainability in their strategies.

These four reasons may also be interrelated. Leadership organizations recognize the relationship between business and society and are redefining their economic, environmental, and social responsibilities around the concept of sustainability. Some corporate leaders have adopted sustainability for each of the reasons listed above. Yvon Chouinard, founder of Patagonia, an outdoor clothing and equipment company, always wanted to put the environment first in his business. Patagonia was one of the first companies to reuse materials and it used its mail-order catalog as a platform to speak out on environmental issues such as genetically modified foods and overfishing.[7]

In contrast, it is clear that the focus of GE (General Electric) on sustainability is driven by its goal of improving the bottom line. GE's CEO, Jeffrey Immelt, has publicly stated that his company must focus on innovation and the environment in order to increase revenues and stay competitive. However, he has made it clear that this is about business first. The social and environmental strategies developed at GE to reduce social and environmental impacts must also achieve financial goals. For example, Ecomagination, announced in 2005, is a major GE program to dramatically increase the company's business in environmental technologies. The company has pledged to increase investment in environmental technologies to US$1.5 billion and sales of environmental technologies to US$20 billion by 2010. In 2012, Ecomagination met this objective, with revenue totaling US$25 billion. Ecomagination R&D investments in 2012 totaled US$1.4 billion and overall R&D investment totaled more than US$5 billion between 2010 and 2012. It has also pledged to reduce GHG emissions by 1% and improve energy efficiency by 30% by 2012. Indeed, GHG emissions were lowered to 4.88 million metric tons of CO_2 equivalents, a reduction of 32% from GE's adjusted 2004 baseline. Products included in the Ecomagination initiative, for example, include a fluorescent light bulb that saves 70–80% of energy compared with an ordinary light bulb and a wire coating for cars and electronics that does not include any pollutants in its production.[8]

Managing corporate sustainability

Corporations have become more sensitive to social, environmental, and economic issues and stakeholder concerns and are striving to become better corporate citizens. Whether the motivation is concern for society and the environment, government regulation, stakeholder pressures, or economic profit, the result is that managers must make significant changes to more effectively manage their social, economic, and environmental impacts. The best practices in corporate sustainability performance are no longer primarily focused on companies like Ben & Jerry's or The Body Shop, as they were ten or 20 years ago. It is now also some of the world's largest corporations such as GE and Walmart (along with many others) that are leading the way with significant financial and organizational commitments to social, environmental, and economic issues.

As companies search for ways to improve their performance, determining the best ways to thoroughly integrate these improvements into all parts of the organization still presents challenges. These challenges are because implementing sustainability is fundamentally different than implementing other strategies in the organization. For operating goals, the direct link to profit is usually clear. For innovation, though long-term and often difficult to predict and measure, the intermediate goal is new products and the ultimate goal is increased profit. However, for sustainability, the goal is to achieve excellence in social, environmental, economic, *and* financial performance. Managing and measuring this paradox creates challenges since financial initiatives are associated with clear, measurable, short-term metrics, whereas sustainability measurements are often uncertain and long-term.

A particular challenge is how to integrate sustainability impacts and financial performance into day-to-day management decision-making. Such decision-making is related to the various tensions between these goals. The relationship between social, environmental, and economic, on the one hand, and financial goals, on the other hand, is typically characterized with short-term competition and inconsistencies and long-term benefits.[9] More specifically, while these initiatives may benefit one another in the longterm, they often conflict in their need for resources.[10] Managers must make resource allocation trade-offs between these multiple goals, which is difficult because the longterm financial gains of sustainability initiatives may not fit well into a traditional capital budgeting format, unless the risks and reputation-related impacts are measured and integrated into decision-making.

There are also other tensions. Pursuing social goals demands cooperation to achieve public benefits, while financial goals encourage competition for individual gains.[11] Through decentralization and employee empowerment, typical of some large organizations, cooperation is even more difficult to achieve. The tensions evolve further as business unit and facility managers have significant incentive pressures to increase short-term earnings. Their performance is typically measured and rewarded primarily based on profits while they are accountable for excellent performance in all areas (social, environmental, economic, and financial performance).

In addition, it is often unclear how stakeholders will respond. For companies with leading positions in their respective industries, every major action taken is visible and is picked up by the media on some level. However, when there is a significant financial cost in improving social, environmental, and economic performance, managers are faced with a dilemma of how to make the choices and which actions to take.[12]

Often, it is unclear how trade-offs between financial and environmental or social performance should be made. Moreover, the trade-offs keep changing—at certain times, shareholders may want the company to place substantial weight on social performance and the environment, whereas at other times they may want the company to place more weight on short-term profits.

The costs of implementing sustainability are also constantly changing. For example, potential technology improvements may make it far cheaper to implement pollution reduction later rather than earlier. Even when sustainability is thought to provide financial benefits, the benefits can, at best, only be measured over long time horizons. This makes it difficult to measure the impact of social, environmental, and economic performance and to quantify the resulting benefits. The constant uncertainty about how far to move toward sustainability, the constantly changing emphasis on and costs of implementing sustainability, and the long time horizons therefore make it difficult to implement sustainability in the same way that other strategic initiatives are implemented.

For these reasons, the standard implementation approaches often fail. In order to improve the integration of social, environmental, and economic impacts into day-to-day management decisions, companies must tie the measurement and reporting of these impacts into decision-making processes. Further, these impacts must be measured and reported in financial terms and then integrated into the traditional investment models. So, how can companies integrate sustainability into day-to-day decision-making? Through the combination of a clear and well-articulated and well-communicated sustainability strategy, senior management commitment to a broader set of objectives than profit alone, and utilizing appropriate structures and systems to drive sustainability through the organization.

The importance of vision and communicated core values are well accepted. But these commitments to social, environmental, and economic concerns must be consistently communicated both in words and actions. Top managers must exercise leadership to decide how much integration of sustainability concerns they want and how they want to do it. They must articulate the trade-offs to managers, help them deal with the trade-offs by leading by example, and continually reinforce these objectives throughout the organization. As one senior executive at Nike, the world's leading designer, marketer, and distributor of athletic products and clothing, stated: "Consistent support of leaders is more important than refined measures on environmental impact and compliance."[13] They must also choose a strategy that is consistent with mission, culture, and aligned with geography, customer, product, community, and other stakeholder requirements. Strategy and leadership are minimum enablers to successful sustainability implementation.

Just as the formulation of sustainability strategy is critical, so is the execution. Management must also make choices about how to implement the sustainability strategy and integrate economic, social, and environmental impacts into their organizations. These impacts are sometimes managed using "soft" leadership elements, such as managing people and culture, along with a variety of informal systems. For example, in their recruitment and development practices, companies may seek to create in their employees a passion and commitment to sustainability. They in effect create a culture to support sustainability decisions. This culture is firmly embedded in the beliefs, values, and mission and vision statements of companies that serve to inspire and motivate employees to take sustainability obligations seriously.

Sustainability impacts can also be managed through "hard" or formal implementation systems such as performance measurement and evaluation, compensation, and incentives. Many companies have created performance measurement and management systems that include social, environmental, and economic indicators in addition to financial performance measures. Some are also including rewards and incentives that are based on social, environmental, and economic performance. Companies can also change their organizational design or structure to signal a commitment to sustainability. The right mix of soft and hard systems depends on the nature of the impacts: the potential magnitude, the degree of uncertainty, and the time horizons involved. It also depends on customer, product, geographic, and other characteristics.

Managing sustainability may not be paradoxical after all

In some companies, the trade-offs between sustainability and financial goals are not seen as difficult or paradoxical. These companies creatively use technology and innovation to overcome "win-lose" scenarios. For example, Nike, the leading designer, marketer and distributor of athletic products and clothing, created predictive tools for designers (Considered Index®) that help them focus on environmental issues rather than rely on measuring tools at the end of the innovation process. At Nike, the use of environmentally preferred materials is a "win" for the environment, but that use also increases the costs ("lose" for the company). But, by innovatively reducing waste the company also reduces costs ("win-win"). By using innovation, they try to balance the costs and increase longterm financial performance. When P&G (Procter & Gamble), one of the world's leading branded consumer products companies, designs a technology system, it also has to deliver on service at a low cost and be environmentally friendly. Innovation is a critical driver of these processes. This includes, for example, the Purchasing Asset Recovery Materials program, looking specifically at finding value in waste ("win-win"). At Nissan North America, a unit of Nissan Motor Co., a leading global auto manufacturer, even though expensive, some decisions are mandated, such as the achievement of the 95% recycling in plants. Plant groups are tasked with this environmental challenge. They use creativity and innovation to achieve this global goal.

These companies may hold on to social, environmental, economic as well as financial performance simultaneously, because they are using the tensions as a source of new ideas, innovation and creativity rather than seeing them as impediments to effective decision-making. In addition, they have keen awareness of anticipated stakeholder reactions to corporate social, environmental, and economic performance.[14]

The Corporate Sustainability Model

So what can companies do to improve their sustainability performance? More specifically, how can executives identify, measure, and manage the drivers of improved sustainability performance and create systems and structures that improve it? How does sustainability performance impact overall long-term corporate profitability, and how should executives communicate these impacts to general managers, financial managers, employees throughout their companies, and to external audiences?

For organizations, a sustainability framework or model of social, environmental, and economic performance creates a powerful opportunity to create enduring value for multiple stakeholders. At the same time, it challenges managers to understand the complex interrelationships between social, environmental, and economic performance. This book presents a model or framework to aid companies in identifying, measuring, and integrating social, environmental, and economic impacts into corporate strategy and into management decisions to successfully manage those impacts and increase profitability. It explains how various inputs and processes affect sustainability performance and stakeholder reactions, and how they drive long-term corporate financial performance.

The Corporate Sustainability Model describes the inputs, processes, outputs, and outcomes necessary to implement a successful sustainability strategy. The inputs include:

- The broader external context
- The internal context
- The business context
- Human and financial resources

Though the inputs sometimes act as constraints to improved corporate sustainability, managers have significant ability through leadership and the formulation and implementation of various processes including sustainability strategy, structure, actions, and systems to effect corporate sustainability performance. The output of these processes is the sustainability performance—that is, the effect of corporate activity on the social, environmental, and economic fabric of society. In addition to having an effect on society, these activities often affect corporate financial performance.

This typically occurs through various positive and negative stakeholder (such as customers, employees, regulators, and consumer activists) reactions such as additional purchases, consumer protests, employee loyalty or resistance, and government regulations. These stakeholder reactions affect corporate profits and are a part of the business case for sustainability that has been widely discussed in both academic and managerial circles.[15] The model of the drivers, actions, and measures that managers can use to implement corporate sustainability can provide guidance for future research and managerial practice. It can help executives better manage the pressure to simultaneously achieve excellence in social, environmental, economic, *and* financial performance and create sustainability programs that maximize sustainability and financial outcomes.

Background to this book

In 1996, Marc J. Epstein wrote *Measuring Corporate Environmental Performance: Best Practices for Costing and Managing an Effective Environmental Strategy.* The book has been used extensively by managers in business and government, researchers, and students. It was widely used by corporate executives in both small and large companies and in general management functions at the senior and middle levels of organizations. It was also widely used by functional managers in the social and environmental management functions and the finance function. *Making Sustainability Work* builds on this earlier work and numerous other articles and books, and develops an entirely new framework for the measurement and management of corporate social, environmental, and economic impacts. It is written to be accessible to corporate managers but is built on a solid academic research foundation.

Relying on the best practices of major corporations and the latest academic research, this book covers the broad dimensions of sustainability along with the specificity of how to execute it within companies. The academic research relies on:

- Our own extensive field studies with dozens of companies
- An extensive review of the many academic and managerial articles and books on various aspects of implementing sustainability
- A large body of empirical work including surveys of company practices
- Archival data from various sources
- Other academic and company research, analysis, and discussions

It also includes best practice examples and models from dozens of global companies that are listed below. The examples include companies that have primary activities across the globe: in Europe, Asia, North America, South America, Australia, and Africa. Companies in different industries with different challenges are used to examine how to formulate and execute a sustainability strategy.

Companies cited in this book

- ABN AMRO
- adidas Group
- Advanced Micro Devices (AMD)
- Alcatel-Lucent

- Alcoa
- Allied Waste Industries
- Allstate Insurance
- Alpha Natural Resources
- Amanco Guatemala
- AMP Ltd
- Anglo-American
- Apple
- Avon Products
- Banco Real
- Bank of America
- Barclays
- Baxter
- Bayer
- Ben & Jerry's
- BG Group
- BHP Billiton
- The Body Shop
- The Boeing Company
- BP
- Bristol-Myers Squibb
- British American Tobacco
- British Telecom
- Browning-Ferris Industries
- Canadian Pacific
- Canon
- Cargill
- CEMEX
- China National Petroleum Corporation
- Chiquita Brands International
- Citigroup

- Coca-Cola
- Colgate
- Compañía de Minas Buenaventura
- The Co-operative Bank
- Danone
- Dean Foods
- De Beers
- Deutsche Bank
- Dell
- The Dow Chemical Company
- DuPont
- DyeCoo Textile Systems
- Eaton Corporation
- EGYPTAIR
- Emirates
- Ethiopian Airlines
- FleetBoston Financial
- Ford Motor Company
- Foxconn Technology Group
- Fujitsu Group
- Gazprom Bank
- Gazprom Group
- General Electric
- General Mills
- General Motors
- Georgia-Pacific
- Gillette
- GlaxoSmithKline
- Grameen Telecom
- Heineken
- Heinz

- Henkel International
- Hennes & Mauritz
- Herman Miller
- Hewlett-Packard
- The Home Depot
- Honda
- Honda North America
- HSBC
- ICICI Bank
- Inditex
- ING
- Intel
- Interface
- Johnson & Johnson
- Kingfisher
- L’Oréal
- Lucent Technologies
- Marks & Spencer
- Massey Energy
- Mattel
- McDonald’s
- Microsoft
- MillerCoors
- Mitsubishi Corporation
- National Australia Bank
- Nestlé
- Newmont Mining
- Niagara Mohawk Power
- Nike
- Nissan Motor Company
- Nissan North America

- Novartis
- Novo Nordisk
- Ontario Hydro
- Patagonia
- Perrier
- Pfizer
- Philips
- Procter & Gamble
- PUMA
- Reebok
- Rio Tinto
- Roche
- Royal Dutch Shell
- Samsung
- Santander
- Sberbank
- ScottishPower
- Seiko
- Shaw Industries
- Siemens
- Sony Corporation
- Starbucks
- Star-Kist
- Steelcase
- Stonyfield Farms
- Suncor Energy
- Sun MicroSystems
- Teck
- Texaco
- Timberland
- Toyota Motor Corporation

- Unilever
- Union Carbide
- United Technologies Corporation
- UPS
- Uzbekistan Airways
- Verizon Communications
- Visa
- Vodafone
- Volkswagen Group
- Volvo Car Corporation
- Walmart
- West LB
- XcelEnergy
- Zara

The academic research and the examination of best-practice companies have all been integrated into a model (the Corporate Sustainability Model, described in Chapter 1) and guide to best practice. The subsequent chapters offer guidance to help translate sustainability strategies into specific policies, programs, systems, and measures that will provide direction and boundaries for decision-making and move the entire company toward its sustainability and financial performance goals.

Sustainability at CEMEX

CEMEX, a leading global cement company headquartered in Mexico, has a rich history of improving the wellbeing of those it serves through its efforts to pursue innovative industry solutions, efficiency advancements, and efforts to promote a sustainable future. Since launching its Eco-efficiency Program in 1994, CEMEX is aggressively pursuing a leadership role in the development of products and building solutions that reduce the environmental impacts of construction projects while fostering social and economic growth. CEMEX product innovations help to improve the sustainability of buildings and other structures (such as with self-compacting concrete, which has a dense formulation, improves the strength, durability, and life of a structure, while reducing labor costs, energy use, and health risks during construction, and maintenance costs throughout its life). The company is evaluating and improving the LCA (life-cycle analysis) of its products to enhance the understanding of, and ultimately reduce, its environmental footprint. A specific example is the LCA of concrete pavements versus asphalt pavements that demonstrate the reduction of CO_2 emissions during the life utility of the pavement by several factors, through, for example, less maintenance, less power consumption in illumination during its use, and less fuel consumption due to reduced rolling resistance.

Through partnerships with private enterprises, governments, and academic institutions, CEMEX supports programs that help people in emerging markets gain the knowledge to save money for housing, start a business, build homes and infrastructure, and help improve community services. Over 450,000 Latin American families benefited from Patrimonio Hoy, Productive Centers of Self-Employment, and ConstruApoyo. Combining the global presence of CEMEX distribution with the power of microcredit, Patrimonio Hoy offers integral solutions to low-income families by providing financial and technical assistance in the construction of their homes. Productive Centers of Self-Employment are community spaces where individuals manufacture concrete blocks and other precast forms, keeping half of their production for personal construction purposes while selling the other half to state and municipal governments. Today, there are 76 such centers in Mexico and Columbia. Through its ConstruApoyo program, CEMEX facilitates the distribution of funds for the construction, repair, or extension of homes with a prepaid debit card system, creating a transparent system through which aid recipients are able to purchase the building materials they need.[16]

Identifying the impacts created by an industry can aid in the development and implementation of a sustainability strategy. The cement industry embarked on a collaborative research project to identify the challenges and opportunities in achieving sustainability. The environmental issues include:

• Depletion of nonrenewable resources (i.e., fossil fuels)

• Impacts of resource extraction on landscape and environmental quality

• Dust emissions

• Other emissions including nitrogen oxides, sulfur dioxide, and carbon monoxide

The industry has positive and negative social impacts. Communities are concerned about health effects, worker safety, noise, and dust. On the other hand, in many developing countries, cement companies are contributing to improved roads and sewers, and training workers. The economic issues include job creation and economic growth due to the development of cement facilities and financial prosperity for the company.[17] The cement industry will continue to face challenges. To succeed, companies in this industry must monitor changes in the industry, be proactive in responding to challenges, and realize the opportunities that effective management of these challenges can have for the company and for society.

Making sustainability work: an overview of the revised book

We look at the important role of leadership, organizational culture, and strategy in achieving success in corporate sustainability in Chapter 2, examining the role of senior managers and corporate boards in leading and governing the sustainability activities and developing the sustainability strategy, along with the importance of senior management commitment and the various choices of strategy.

We also show how organizational design impacts the success of corporate sustainability, looking at the choices of organizational structures and the applicability to different organizational types. This includes centralized and decentralized functions, outsourced activities, and approaches to integration. One of the major challenges to successful sustainability implementation is to fit this new strategy into existing organizational structures simultaneously to improve social, environmental, economic, *and* financial performance. Chapter 3 discusses various organizational designs issues that can improve sustainability.

The various management systems that can be used to execute a sustainability strategy are critical elements in any successful implementation. This includes the variety of information that is needed to improve both operational and capital investment decisions. It includes improving the financial analysis needed for better management decision-making throughout the organization, along with a more formal integration of social, environmental, and political risks into the analysis. These systems provide the levers that managers can use to increase sustainability and financial performance. Chapter 4 looks at capital investment, costing, and risk management systems.

We take an in-depth look at specific ways to measure and reward sustainability performance. In this book, the emphasis is on measuring the performance of the *process* of sustainability along with measuring sustainability performance *results* as an ultimate goal and also as an intermediate goal to achieving financial success. We discuss each of these along with the role of incentives and rewards in improving sustainability performance, which are the focus of Chapter 5. Just as effective leadership and strategy are minimum enablers for sustainability success, some of the various formal and informal organizational systems must be used to effectively implement sustainability.

The measurement of social, environmental, and economic impacts of products, services, processes, and other corporate activities is critical. Chapter 6 gives an overview of the approaches that can be used to measure these impacts effectively, along with more detailed and applied examples of how to do this for inputs, processes, outputs, and outcomes.

Chapter 7 gives specific guidance on how to implement social, environmental, and economic impact measurement systems, including an extensive list of useful measures that can be used or adapted to measure the inputs, processes, outputs, and outcomes of sustainability investments. One of the biggest challenges for managers is to determine how to measure progress in sustainability. This requires process measures (which typically do not exist), in addition to results measures. Guidance is provided on the development of high-performance sustainability metrics to measure sustainability success and improve performance, as well as a framework and set of measures that can be used to measure performance and payoffs of sustainability investments. The extensive discussion of the foundations of sustainability measurement, along with the list of sample measures, is one of the unique features of this book.

Feedback and internal reporting are also needed to improve sustainability. This includes the design, content, audience, frequency, distribution, and communication of sustainability information. Chapter 8 describes how organizations can use this information to improve organizational learning and change products, processes, services, and other activities to be more sensitive to sustainability issues. It also includes a discussion of the feedback loops in the Corporate Sustainability Model and the importance for both learning and organizational performance.

External reporting is also important for communicating sustainability performance to stakeholders. Chapter 9 provides an overview of the existing regulations and guidance for social, environmental, and economic reporting and describes best practices. This includes a discussion of the reporting related to the Global Reporting Initiative, the choices for reporting in corporate annual reports, sustainability reports, and the web, and the choices for verification and auditing of the sustainability reports.

Chapter 10 summarizes the book's main points and provides guidance for managers with additional examples of best practices. It describes the opportunities available for innovation when companies proactively manage sustainability. And it focuses on the significant benefits that can accrue to both corporations and society by making sustainability work.

The development and implementation of a sustainability strategy is important for companies with either high or low social, environmental, and economic impact, companies small and large, manufacturers and service companies, with large community affairs or environmental, health, and safety (EH&S) staffs, and with no full-time EH&S staff at all. The numerous examples and approaches suggested in this book are at this very moment being introduced and used successfully in a variety of companies and can be readily adapted to companies of different sizes and complexities, in different industries, and with different environmental, economic, and social sensitivities.

The concepts discussed in this book are especially relevant to corporate general managers, sustainability managers, and financial managers who take a proactive role in creating systems to measure and manage corporate performance. It is also imperative that financial executives understand the relationships between social, environmental, and economic performance, as these complexities are increasingly key components of corporate valuations, analyses, and reporting. Most organizations now have sustainability managers who need to have the knowledge and tools to help create a strategic sustainability management system that links to corporate value.

The approach presented here also provides an opportunity to make better resource allocation decisions throughout the organization. It also provides an opportunity for sustainability managers to more effectively measure and report the value created through more effective management of stakeholder impacts and improvement of sustainability performance. Through more careful analysis and measure of the payoff of sustainability investments, general managers, financial managers, and sustainability managers can more effectively justify investments. In this way, sustainability investments can be integrated into the same capital investment process as other investments and the value of these investments to improving shareholder and other stakeholder value can be seen more clearly.

Operational managers, who are on the frontline of managing operations, need an understanding of the potential synergies and conflicts between operational, environmental, economic, and social performance, so that they can make informed decisions that create value for the organization. Many of the concepts and practices discussed in this book also have relevance for marketing managers, distribution managers, and legal managers, for the complexities involved in managing the impacts of an organization's products, services, processes, and other activities touch on all aspects of an organization and its constituents. R&D leaders, and product and process design engineers, will be interested in how analysis and management of social, environmental, and economic impacts present opportunities for innovation.

As well as senior and middle managers, academics and others interested in the field will benefit from reading this book. Nonprofit and governmental organizations, alike, will continue to be very interested in this topic as they have become increasingly sensitive to their social, environmental, and economic impacts and the evaluation of the costs and benefits of their activities.

Sustainability at Chiquita Brands International

Developing and initiating a sustainability strategy involves many steps. Chiquita Brands International, a leading producer and distributor of bananas, began its sustainability program by creating a Corporate Responsibility Steering Committee consisting of senior and middle managers. The goal of the committee was to determine a way Chiquita could introduce values management into the organization. The result was Chiquita's Code of Conduct ... Living by Our Core Values, which established standards including social responsibility. The code includes the requirements of Social Accountability 8000 (SA8000) and a goal to have third-party certification to SA8000 of all facilities located in Latin America. It also details its reporting guidelines, which includes an identification of measures and indicators. These reports are distributed to employees and disclosed publicly. Chiquita also established the position of Corporate Responsibility Officer. The officer reports directly to the CEO and board of directors. Prior to the creation of this position, sustainability was the responsibility of operating managers who did not receive much oversight.[18]

Despite these developments, Chiquita continued to face the difficulties that plague many multinational corporations. In many countries, government security of employees is not effective. Chiquita, for example, was fined for financially supporting a terrorist organization to protect its employees in Colombia. Additional lawsuits were filed against Chiquita seeking compensation for the deaths of people allegedly killed by the terrorist group.[19]

Many companies are globalizing into countries where current social and environmental regulations are lax. These companies are faced with severe competitive pressures that question whether global standards are too costly or unsafe for the operations in many countries. Deciding whether to follow a global standard or to follow common country practices or locally adapted standards is just one of the many challenges that multinational corporations encounter when trying to set a sustainability strategy.

And finally...

Companies know that it is critical to formulate a sustainability strategy, but how to formulate and execute it remains a challenge. This book provides a framework and model for implementing sustainability in large, complex, global organizations. But, for this to happen:

- Sustainability must be an integral component of corporate strategy
- Leadership must be committed to sustainability and build additional organizational capacity
- Sustainability strategies should be supported with management control, performance measurement, and reward systems, as appropriate
- Sustainability strategies should be supported with mission, culture, and people as appropriate
- Managers must integrate sustainability into all strategic and operational decisions. Then, additional systems and rewards can be introduced to formalize and support decision-making
- Managing sustainability performance should be viewed not only as risk avoidance and compliance,

but also as an opportunity for innovation and competitive advantage

CHAPTER 1

A new framework for implementing corporate sustainability

With growing sensitivity toward social, environmental, and economic issues and shareholder concerns, companies are increasingly striving to become better corporate citizens. Executives recognize that long-term economic growth is not possible unless that growth is socially and environmentally sustainable. A balance between economic progress, social responsibility, and environmental protection, sometimes referred to as the triple bottom line, can lead to competitive advantage.[1] Through an examination of processes and products, companies can more broadly assess their impact on the environment, society, and economy, and find the intersection between improving sustainability impacts and increased long-term financial performance. To aid executives in achieving sustainability, this chapter will:

- Define the principles of sustainability
- Identify important stakeholder relationships

- Introduce a framework—the Corporate Sustainability Model—to guide managers in measuring and managing sustainability performance. This framework will be the basis for the remainder of the book and provides a tool for the implementation of corporate sustainability and the evaluation of corporate impacts

The evaluation of social, economic, and environmental impacts of organizational actions is necessary to make effective operational and capital investment decisions that positively impact organizational objectives and satisfy the objectives of multiple stakeholders. In many cases, reducing these impacts increases long-term corporate profitability through higher production yields and improved product quality. Novo Nordisk, the global Danish-based healthcare company specializing in diabetes care, strives to conduct its business in a financially responsible (profitable for the long-term), socially responsible (patients first), and environmentally responsible (doing more with less) way. The aim is to ensure long-term profitability by minimizing any negative impacts from business activities and maximizing the positive footprint from its global operations: improved health, employment, economic prosperity, and social equity (see Fig.1.1).

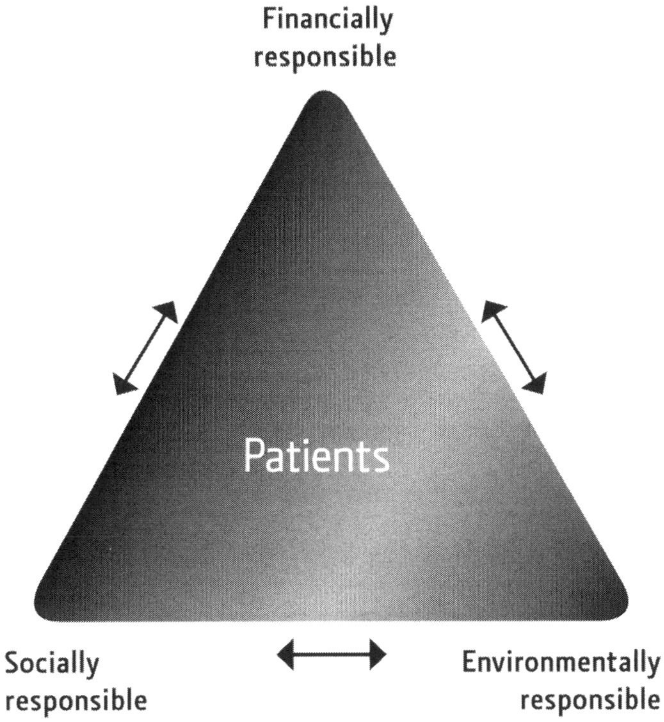

FIGURE 1.1 **Novo Nordisk's triple bottom line business principle**

Source: Novo Nordisk (2012) *Annual Report*

There is growing interest among the business community in the development and implementation of sound, proactive sustainability strategies, including significantly increased stakeholder engagement. The financial payoff of a proactive sustainability strategy can be substantial.[2] By addressing the nonfinancial aspects of business, companies can improve the bottom line and earn superior returns. The Dow Chemical Company, a global diversified chemical company, focuses on manufacturing efficiency inside the company while maximizing the contributions of Dow products to improve efficiency and expand affordable alternatives. Dow's manufacturing energy intensity has improved more than 40% since 1990, saving the company a cumulative US$24 billion. Dow is committed to bringing solutions to the challenge of climate change by producing products that help others reduce greenhouse gas emissions, such as lightweight plastics for automobiles and insulation for energy efficient homes and appliances.[3]

Henkel International, a German-based manufacturer of laundry and homecare products, beauty care, and adhesive technologies, has developed a sustainability strategy to create more value for its customers and consumers, for the communities it operates in, and for the company—at a reduced ecological footprint. Henkel concentrates its activities along the value chain on six focal areas that reflect the challenges of sustainable development as they relate to Henkel's operations. Figure 1.2 presents Henkel's six focal areas with five-year targets for 2015. Focal areas are subdivided into two dimensions—"more value" and "reduced footprint". To accomplish these, the company uses innovations, products, and technologies, but recognizes that these dimensions must be ever-present in the minds and day-to-day actions of around 47,000 employees.

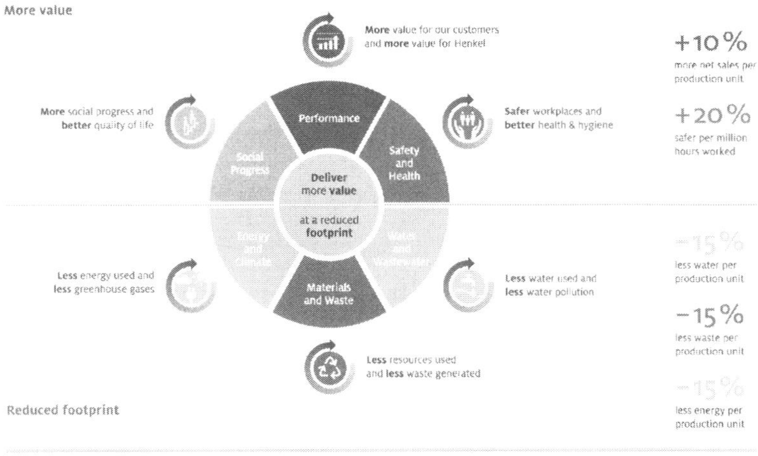

FIGURE 1.2 **Henkel's six focal areas in sustainability performance**

Source: Henkel (2012) *Sustainability Report*

To become a leader in sustainability, it is important to articulate what sustainability is, develop processes to promote sustainability throughout the corporation, measure performance on sustainability, and ultimately link this to corporate financial performance. Corporate citizenship is an important driver for building trust, attracting and retaining employees, and obtaining a "license to operate" within communities. However, corporate citizenship is much more than charitable donations and public relations—it's the way the company integrates sustainability principles with everyday business operations and policies, and then translates it all into bottom-line results.

For sustainability to be long-lasting and useful, it must be representative of and integrated into day-to-day corporate activities and corporate performance. If it is seen only as an attempt to provide effective public relations, it does not create long-term value and can even be a value destroyer. The key is integrating sustainability into business decisions, and identifying, measuring, and reporting (both internally and externally) the present and future impacts of products, services, processes, and activities. In fact, this book is all about the integration of sustainability into corporate operations to simultaneously achieve increases in social, environmental, economic, *and* financial performance.

What is sustainability?

To help understand what sustainability is in the context of corporate responsibility, we have broken it down into nine principles (see Table 1.1).[4] These principles have three attributes:

1. They make the definition of sustainability more precise
2. They can be integrated into day-to-day management decision processes and into operational and capital investment decision-making
3. They can be quantified and monetized

These nine principles of sustainability will be used as a foundation throughout this book. They highlight what is important in managing stakeholder impacts (i.e. the impact of company products, services, processes, and other activities on corporate stakeholders).

Although we are presenting in Table 1.1 a broad definition of sustainability, this book focuses on the criteria that are usually included in sustainability discussions, analyses, measurements, and reports—social, environmental, and economic. So, though the principles of ethics and governance, for example, are important aspects of sustainability, they are not the focus of most corporate applications of corporate social responsibility or sustainability. But the discussion of systems, structures, performance measures, culture, and so forth necessary for implementation can be easily adapted to improve performance on all nine principles.[5] Further, the formal and informal organizational processes described in this book should be applied to all of these principles.

TABLE 1.1 The broad definition of sustainability performance—nine principles

1. Ethics	The company establishes, promotes, monitors, and maintains ethical standards and practices in dealings with all of the company stakeholders
2. Governance	The company manages all of its resources conscientiously and effectively, recognizing the fiduciary duty of corporate boards and managers to focus on the interests of all company stakeholders

3. Transparency	The company provides timely disclosure of information about its products, services, and activities, thus permitting stakeholders to make informed decisions
4. Business relationships	The company engages in fair-trading practices with suppliers, distributors, and partners
5. Financial return	The company compensates providers of capital with a competitive return on investment and the protection of company assets
6. Community involvement/economic development	The company fosters a mutually beneficial relationship between the corporation and community in which it is sensitive to the culture, context, and needs of the community
7. Value of products and services	The company respects the needs, desires, and rights of its customers and strives to provide the highest levels of product and service values
8. Employment practices	The company engages in human-resource management practices that promote personal and professional employee development, diversity, and empowerment
9. Protection of the environment	The company strives to protect and restore the environment and promote sustainable development with products, processes, services, and other activities.

Source: Epstein and Roy (2003) "Improving Sustainability Performance"

1. Ethics

Ethical companies establish, promote, monitor, and maintain fair and honest standards and practices in dealings with all of the company stakeholders and encourage the same from all other stakeholders, including business partners, distributors, and suppliers. To follow this principle, a company needs to place particular emphasis on human rights and diversity to ensure that workers are treated fairly. This means that, although a company has to adhere to local laws, its ethical practices will often necessitate standards far in excess of industry, international, national, and local guidelines or regulations.

Ethical companies set high standards of behavior for all employees and agents, and have in place effective systems for monitoring, evaluating, and reporting on how the company does business. The reporting of ethical violations to appropriate authorities is also actively promoted.

Ethical companies create codes of conduct, develop ethics education programs, and honor internationally recognized human rights programs.

2. Governance

The governance principle is a commitment to manage all resources conscientiously and effectively, recognizing the fiduciary duty of corporate boards and managers to focus on the interests of all company stakeholders.

This duty is of primary importance and is superior to the interests of management. The company follows practices of fair process and seeks to enhance both financial and human capital while balancing the interests of all of its stakeholders.

The company encourages the achievement of its mission while being sensitive to the needs of its various stakeholders. Its mission must be clearly stated and widely understood, and must recognize the interests of multiple stakeholders. The company must have a strategy and performance metrics that are consistent with its mission. The mission, strategy, policies, practices, and procedures are communicated openly and clearly to employees. Decision-making processes are engrained within this principle as performance is directly related to a particular course of action taken by the company.

Companies that value governance evaluate the CEO and senior management on financial and nonfinancial performance and have a board structure that represents a wide range of stakeholder views.

3. Transparency

While the governance principle relates to internal management issues, the transparency principle is about disclosure of information to company stakeholders. Transparent companies provide full disclosure to existing and potential investors and lenders of fair and open communication related to the past, present, and likely future financial performance of the company.

Transparent companies broadly identify their stakeholders. Indeed, companies embracing this principle recognize that they are accountable to internal and external stakeholders, understanding both their informational needs and their concerns about the company's effects on their lives.

4. Business relationships

Companies must encourage reciprocity in their relationships with suppliers, by treating them as valued long-term partners in enterprise, enlisting their talents, loyalty, and ideas. Companies endorse long-term stable relationships with suppliers in return for quality, performance, and competitiveness. Companies select their suppliers, distributors, joint-venture partners, licensees, and other business partners not only on the basis of price and quality but also on social, ethical, and environmental performance.

Companies that embrace this principle set specific targets for utilizing indigenous, disadvantaged, or minority-owned businesses and use their purchasing power to encourage suppliers to improve their own social, environmental, and economic practices.

5. Financial returns to investors and lenders

The company compensates providers of capital with a competitive return on investment and the protection of company assets. Company strategies promote growth and enhance long-term shareholder value. The interests of investors and lenders must be explicitly recognized and companies must develop formal mechanisms to foster an ongoing dialogue with their investors. However, though improved financial results are a natural product of creating value for customers, employees, and other stakeholders, the company is committed to balancing the interests of all stakeholders.

6. Community involvement and economic development

Increasingly, companies recognize that it is in the best long-term interest of both the company and the community to improve the community, community resources, and the lives of its members. Thus, the company fosters a mutually beneficial relationship between the corporation and the community in which it is sensitive to the culture, context, and needs of the community. The company plays a proactive and cooperative role in making the community a better place to live and conduct business.

Companies that value community involvement and economic development collaborate with community members who promote rigorous standards of health, education, safety, and economic development.

7. Value of products and services

This principle requires companies to specify their relation and obligations to their customers. A proactive stance on this principle requires the company to respect the needs, desires, and rights of its customers and ultimate consumers, and to provide the highest levels of product and service values, including a strong commitment to integrity, customer satisfaction, and safety.

Companies create explicit programs to assess the impacts on their stakeholders of the products and services they provide.

8. Employment practices

Companies must decide on the type of management practices they want to engage in. Adopting this principle means that companies engage in management practices that promote personal and professional employee development, diversity, and empowerment. Companies regard employees as valued partners in the business, respecting their right to fair labor practices, competitive wages and benefits, and a safe, family-friendly work environment.

Indeed, companies adopting this principle recognize that concern for and investing in employees is in the best long-term interests of the employees, the community, and the company. Thus, companies strive to increase and maintain high levels of employee satisfaction and respect international and industry standards for human rights. To do this they offer programs such as tuition reimbursement, family leave time, day care, and career development opportunities.

9. Protection of the environment

To follow this principle, companies must define their commitment to the natural environment. For proactive companies, it means striving to protect and restore the environment and promoting sustainable development with products, processes, services, and other activities. Companies must be committed to minimizing the use of energy and natural resources, and decreasing waste and emissions. At a minimum, the company fully complies with all existing international, national, and local regulations and industry standards regarding emissions and waste. It strives for continuous improvement in the efficiency with which it uses all forms of energy, in reducing its consumption of water and other natural resources, and its emissions into air, water, and land of hazardous substances. It also entails a commitment to maximize the use and production of recycled and recyclable materials, the durability of products, and to minimize packaging.

Increasingly, companies have recognized that sustainability values and principles are important for long-term corporate profitability and are using them to define their sustainability strategies. Alcatel-Lucent, a French-based global communications industry leader, focuses on three core sustainability priorities: extreme green innovation; employees; and digital inclusion. A core element of Alcatel-Lucent sustainability strategy and day-to-day businesses are the following three values:

- "We take a zero tolerance stance on compliance violations and reinforce full integrity in every business action from every employee, treating each other with respect and empathy
- We collaborate and do business only with partners, suppliers, contractors and subcontractors who share and support our values. We commit to regularly and thoroughly assessing their performance and partnering to ensure improvement
- "We commit to engaging with pride and passion as citizens of the communities where we do business around the globe"[6]

However, identifying the values or dimensions of sustainability in a theoretical way is only the first step in improving corporate accountability and long-term profitability. The values of the company need to align with the values of its important stakeholders; so stakeholder identification is the next step.

Identify your stakeholders

In managing sustainability, stakeholder value (a significantly broader concept than shareholder value) is critical. How an organization chooses to define its stakeholders is an important determinant of how stakeholder relations are considered in sustainability decision-making and how stakeholder reactions are managed. Some definitions cover those individuals who can either be affected by or affect the organization, while others require that a stakeholder be in a position to both influence and be influenced by organizational activities.[7] Additionally, there are core stakeholders and fringe stakeholders. Core stakeholders are those that are visible and are able to impact corporate decisions due to their power or legitimacy. Fringe stakeholders, on the other hand, are disconnected from the company because they are remote, weak, or currently disinterested.[8] Typical stakeholders include shareholders, customers, suppliers, employees, regulators, and communities.

Few companies have advanced their stakeholder engagement on sustainability as quickly and effectively as Dell, an American multinational computer technology corporation. The company realized long ago that engaging with stakeholders adds enormous value to their sustainability efforts. In addition to the usual groups such as suppliers, NGOs (nongovernmental organizations) and industry consortiums, Dell has benefited enormously from the perspectives of peers and competitors, government agencies, investment groups, academics, faith-based groups, and customers. However, the company has found that about 10–12 different organizations is the optimum number of participating stakeholders on any particular issue. Beyond that, some stakeholders feel their opinions are not being heard or considered.[9]

Relationships with stakeholders should evolve over time. These relationships often go through the following four stages:

1. **Awareness.** At this stage stakeholders know that the company exists. Companies will want to communicate with these stakeholders by providing them with more information about the company so that they can begin to appreciate the company's mission and values
2. **Knowledge.** Stakeholders have begun to understand what the company does, its values, strategy, and mission. During this stage, companies want to provide stakeholders with information to make decisions. Customers want to know how the organization's products meet their needs, employees need to understand organization structure and systems, and suppliers want to understand what the company needs from them
3. **Admiration.** Once stakeholders have gained knowledge about the company, trust needs to be

developed. This is the stage where stakeholders will develop commitment toward the company

4. **Action.** Companies are now taking action to collaborate further with stakeholders. Customers refer business, investors recommend the stock, and employees are willing to take on greater responsibility[10]

To move toward a more complete understanding of sustainability and a further integration of social, environmental, and economic issues into core business strategy and operational decisions, sustainability values and organizational stakeholders must be identified and specified. (Stakeholder engagement and the measurement of stakeholder reactions are discussed in more depth in Chapter 7.) Many constituents have a legitimate stake in company activities and, therefore, a variety of interests and opinions are important in developing sustainability strategies. The long-term value of a company is influenced by the knowledge and commitment of its employees and its relationships with investors, customers, and other stakeholders.[11] Additionally, corporate stakeholders are demanding increased information about corporate governance and the impact of corporate activities on society. This call for corporate transparency requires companies to account for their social, environmental, and economic impacts.

This is particularly critical since the aftermath of Enron, when ethical obligations and accountability, including social, environmental, and economic responsibility, transparency, and proactive engagement with stakeholders became a higher priority for top executives. In many corporations, the operational and reputational effects of negative social, environmental, and economic impacts, along with financial analysts' concerns of increased risk leading to future liabilities, have caused stock prices to be lower and costs of capital to be higher than in comparable more socially, environmentally, and economically responsible companies. Because of the increased scrutiny and effects that it can have on the bottom line, many corporations are focusing more on improving their reputation for effective management of social, environmental, and economic impacts.

Be accountable

To better integrate a broader set of stakeholder concerns into management decisions, consideration of impacts and recognition of the importance of accountability is necessary. In *Counting What Counts: Turning Corporate Accountability to Competitive Advantage,* along with Bill Birchard, Marc Epstein developed an accountability cycle (Fig.1.3) which defines four approaches to becoming an accountable organization.[12] The four primary elements are:

1. Improved **corporate governance** centered around two essential conditions: director independence and enhanced board performance. Both are

necessary to enable the board to make better decisions and to stimulate continuous improvement in company performance

2. Improved measurements that include operational and social **measures of performance** along with a broadened set of financial metrics that include both leading and lagging indicators
3. Improved **reporting** to a broad set of internal and external stakeholders of information relevant to decisions. This begins with internal reporting to managers and the selection of various voluntary disclosures to supplement the mandatory external disclosures that are currently the primary content of corporate reports
4. Improved **management systems** to drive these improvements through corporate culture and change the way managers make decisions to improve both corporate accountability and corporate performance

The model integrates both internal and external reporting, along with a broader set of measures, and provides a mechanism to link social, environmental, and ethical concerns to financial performance. It provides broad guidance for the integration of social concerns into day-to-day management decisions and does so in a format that examines the relevance of social and other leadership issues to overall corporate performance. It also provides a framework of corporate accountability that can be used as a foundation for the implementation of a sustainability strategy.

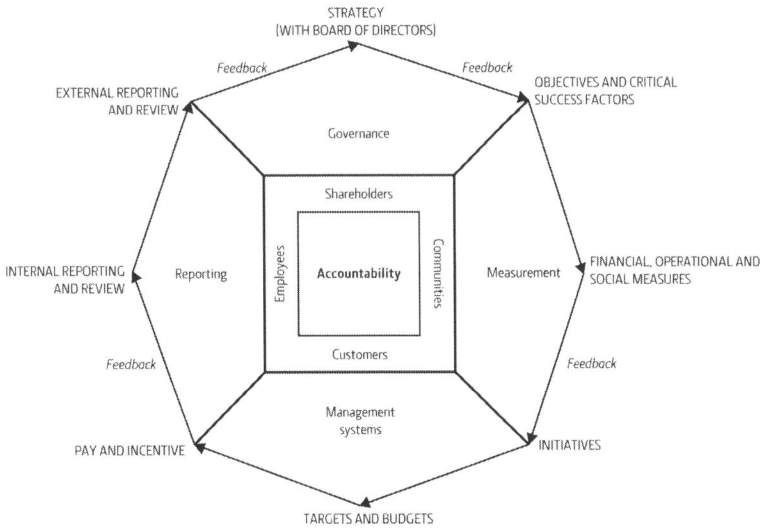

FIGURE 1.3 **The accountability cycle**

Source: Epstein and Birchard (1999) *Counting What Counts*

Corporate Sustainability Model

So, to have an effective sustainability strategy, it is critical that managers understand:

- The causal relationships between the various alternative actions that can be taken
- The impact of these actions on sustainability performance
- The likely reactions of the corporation's various stakeholders to sustainability performance

- The potential and actual impacts on financial performance

By carefully identifying these interrelationships and by establishing relevant performance metrics to measure success, a company can improve operational decision-making and make the "business case" for a sustainability strategy by better linking it with the impacts of the strategy on both the company, society, and the environment.

However, effective implementation and measurement of success are significant challenges. Companies must find ways to motivate employees to focus on sustainability issues while managing sustainability and financial outcomes simultaneously. Additionally, to get adequate resources for the strategy, senior sustainability managers need better ways to measure the payoffs of these actions and programs. General corporate and business unit managers often request an analysis of the payoffs of proposed expenditures so that the resource allocation decisions can be made using the same ROI (return on investment) model that is used throughout the organization. Therefore, we need more guidance in understanding the drivers and measures of success.

The Corporate Sustainability Model (see Fig.1.4) uses the social, environmental, and economic dimensions of sustainability as its foundation.[13] The model describes the drivers of corporate sustainability performance, the actions that managers can take to affect that performance, and the consequences of those actions on corporate environmental, social, economic, and financial performance. By more carefully understanding both the drivers of sustainability performance and the impacts of that performance on the various corporate stakeholders it is easier to integrate the information into day-today operational decisions.

The inputs of the model include the broader external context (regulatory, geographical), internal context (mission, vision, strategy, structure, systems), the business context (industry sector, customers, products), and the human and financial resources. These guide the decisions of leaders and the processes that the organization undertakes to improve its sustainability. They provide a foundation for understanding the complex factors leaders should consider and often take the form of constraints that must be addressed.

After evaluating the inputs and their likely effects on sustainability and financial performance, leaders can develop the appropriate processes to improve sustainability. The sustainability strategy, structure, systems, programs, and actions have three major ultimate sets of impacts: corporate financial costs and benefits of actions; social, environmental, and economic impacts (sustainability performance); and long-term financial impacts through sustainability performance.

The managerial actions taken lead to sustainability performance (positive or negative) and stakeholder reactions, ultimately affecting long-term corporate financial performance. Also included in the model are continual feedback loops that leaders can use to evaluate and improve corporate strategies. Managers should customize this general framework to reflect their particular industry, geographical, internal, or external business context. They must map a corporate performance framework that reflects their specific concerns and interests in sustainability performance and provides rewards for supportive managerial actions.

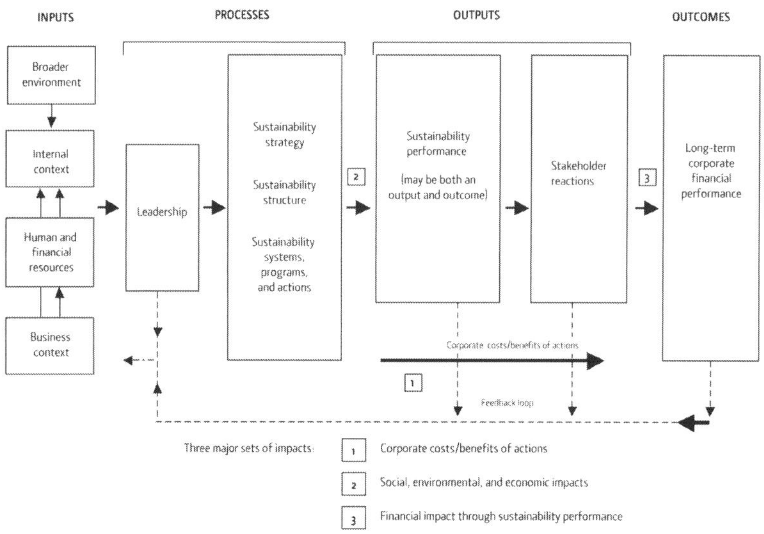

FIGURE 1.4 Corporate Sustainability Model

A fundamental aspect of this framework is the distinction between intermediate results (outputs) and financial outcomes. In Figure 1.4:

- Arrow 1 portrays processes that have immediate and identifiable costs and benefits that affect long-term corporate financial performance
- Arrow 2 shows the impact of the various inputs and processes on sustainability performance
- Arrow 3 shows how corporate financial performance is impacted by stakeholder reactions to corporate sustainability performance

Therefore, intermediate outputs, such as environmental, social, and economic performance (but also public image, employee hiring, customer reactions, and market share) must be monitored to determine the effectiveness of sustainability management practices.

Arrow 3 depicts what is often termed the "business case" for sustainability or corporate social responsibility. Whereas arrow 2 portrays the effect of sustainability actions on social, environmental, and economic performance, arrow 3 reflects how, through stakeholder reactions, social, environmental, and economic performance affects financial performance. Thus, sustainability performance should be seen as both an intermediate output and as an outcome. That is, it is important to understand, measure, monitor, and manage sustainability performance because of concern for societal, environmental, and economic impacts and for long-term corporate financial performance.

So, the inputs, processes, and outputs are all critical elements of the process to drive the outcome of corporate profitability. In the discussion below, the details of these inputs, processes, outputs, and outcomes are further explored. They are then discussed in greater detail in the chapters that follow.

Inputs

Broader external context

The local and global broader external context significantly affects the choices a corporation makes regarding the formulation and implementation of sustainability actions. Pressure is exerted by government regulations for corporations to follow minimum standards of sustainability performance: for example, hazardous and other waste disposal regulations, pollution standards, nondiscrimination laws, and regulations governing working conditions. Regulatory pressure may vary by geographic region, with regulatory pressures typically stronger in some European and Asian countries. If these types of regulation are required by government, a corporation must respond effectively by developing a thorough sustainability plan. Additionally, the appropriate level of wages (living, minimum, or prevailing) and the desirability of the employment of children are issues that have caused significant dismay to many widely recognized companies.

Another external influence on a corporation's choice of sustainability strategy is the marketplace for the corporation's products and services. In studies of corporations operating in China and in Mexico, it was shown that corporations selling to customers in economies with a relatively stronger culture of sustainability performance outperformed their peer companies in terms of environmental performance.[14] Additionally, some locations are more tolerant of pollution due to their topography and weather patterns, in addition to public reactions and the regulatory environment; and so a corporation must consider whether it wants to adapt sustainability strategies to locational differences.

Conflict Minerals Rule

In 2012, the US Securities and Exchange Commission (SEC) enacted a rule that represents an attempt to curtail human rights abuses in Africa through regulation of US public companies. The Conflict Minerals Rule requires companies to trace the conflict minerals (gold, tantalum, tin, and tungsten) in their supply chains. The SEC estimated that 6,000 US issuers will be directly affected by the rule. It also estimated the initial compliance costs of US$3 billion to US$4 billion as end users of the four conflict minerals attempt to find out whether their raw materials originated at mines run by warlords in the Democratic Republic of the Congo or its nine adjoining neighbors (Angola, Burundi, Central African

Republic, the Republic of Congo, Rwanda, South Sudan, Tanzania, Uganda, and Zambia). The complexity and far-reaching effect of the new rule can be demonstrated by an estimate made by Hewlett-Packard (HP), an American multinational information technology corporation, that about 1,000 suppliers in its chain ultimately provide a product to HP that may contain one of the conflict minerals. Each supplier will be asked to do its part in the due diligence process required by the new rule.

Companies making public anti-conflict minerals statements include Intel, Philips, and Samsung. Intel, a multinational semiconductor chip maker corporation, set a goal for 2013 to manufacture the world's first verified, conflict-free microprocessor. Philips, a Dutch multinational engineering and electronics conglomerate, has committed not to purchase materials it knows finance armed groups in the affected countries. Samsung, a multi-faceted family of businesses, including high-tech electronics manufacturing and digital media, prohibits the use in its business units of conflict minerals identified as sourced from conflict mines in the affected countries. Some companies want to go beyond the minimum of complying and reporting and make a perhaps costly effort to make sure none of their materials come from mines run by warlords in the affected countries.[15]

Internal context

This comprises corporate and business unit missions, visions, strategies, structures, and systems; it is through the development and implementation of these that sustainability performance occurs. Thus, companies that are striving toward improved sustainability performance must examine the various sustainability elements that relate to their current strategies, and assess whether and how their corporate and business unit strategies will probably impact issues such as human rights, employee rights, and environmental protection.

Business context

Additional important considerations are the industry sector of the business, and the characteristics of customers and products. Companies that operate in high social and environmental impact industries, such as chemicals, oil, paper, and mining, may exhibit relatively poor performance in terms of sustainability elements such as consumption of natural resources, emissions, and health risk of their products or services compared to companies operating in other industries. The industry also impacts where companies focus their sustainability efforts. For example, oil and mining companies may focus more on environmental and health issues, while service-oriented companies may emphasize the social aspects of sustainability. Although all companies can improve their social, environmental, and economic impacts, some industries have greater opportunities and risks. These include companies with:

- High brand exposure (consumer products companies)
- Big environmental impact (oil companies, manufacturing)
- Natural-resource dependence (fish, food, forest products)
- Current exposure to regulations (hazardous materials, utilities)
- Increasing potential for regulation (automobiles, electronics)
- Competitive markets for talent (service sectors)
- Low market value (small-to-medium B2B [business to business] companies)[16]

Further, companies in different industries are exposed to widely different pressures from political institutions, customers, and community activists. These various pressures become important external drivers of corporate sustainability. Issues such as labor practices and environmental management exist in many industries and have been of increasing community concern. Company and industry codes of conduct are widely and rapidly being established in many industries, including those in the apparel, toy, and footwear industries. Many companies are now working together to establish global labor standards and common factory inspection systems.

Human and financial resources

Another important input is the resources constraint of the corporation. The corporation needs financial resources to implement the various sustainability programs and to pay and train sustainability staff. In addition, organizations need educated and trained individuals throughout the organization who can be sensitized to sustainability issues, along with staff who can be specifically dedicated to sustainability programs. The amount of financial and human resources allocated to sustainability will significantly impact the ability to implement sustainability programs.

Processes

Leadership

It is important for corporate leaders to consider all of these inputs if they want to formulate effective sustainability strategies. Research has shown that sustainability strategies are typically top-down and that the most effective ones are when top management is clearly committed to the strategy.[17] Signals of this commitment are given through the way the strategy is communicated throughout the organization. Senior executives must be knowledgeable, support the organization, and effectively communicate the mission, vision, and strategy to the other members of the organization. The commitment of the board of directors and management encourages employees to act in ways that are compliant and consistent with company strategy. If leaders are not knowledgeable enough about sustainability to motivate their subordinates or institute the proper strategy, structure, or systems, then sustainability actions are unlikely to be successful. It is the responsibility of top leaders to create an environment that encourages sustainability. Verizon Communications, the large telecommunications company, has created a Corporate Responsibility and Workplace Culture Council to foster a culture that encourages sustainability. The council is chaired by the senior vice president for public policy development and corporate responsibility and by the vice president for workplace culture, diversity, and compliance, and consists of managers from each major business unit. It is responsible for identifying and addressing challenges associated with corporate citizenship in key areas, including accessible product design, broadband deployment, and supply chain and environmental management. Increasingly, companies also have committees of the board and other senior management committees devoted to issues of sustainability, and chief sustainability officers as members of the top management team. The importance of board and CEO leadership on sustainability is discussed in Chapter 2.

Sustainability strategy

Top management of some companies have neither developed a strategy for addressing environmental, social, and economic concerns nor developed any systematic way of evaluating or managing their sustainability impacts. In many cases, this lack of corporate responsiveness is evidence of companies that:

- Are crisis-prone rather than crisis-prepared
- May produce social, environmental, and economic impacts that have substantial future consequences involving increased costs, increased community concerns, increased legal claims, and damaged corporate reputation
- May decrease current and future corporate profitability through decreased potential revenues related to sustainability issues

Best practice companies pursue coherent sustainability strategies.

Guidance in the development of a sustainability strategy sometimes comes from governments and industries that have established minimum compliance standards or best practices for corporations. However, many companies go beyond a minimum compliance strategy. For example, prior to any industry standards, toy manufacturer Mattel established its own Global Manufacturing Principles for company-owned, -contracted, and -licensed facilities. These Global Manufacturing Principles provide a framework for its worldwide manufacturing practices requiring fair treatment of employees and protection of the environment.

Companies operating globally also have to choose whether to implement a global sustainability strategy or adapt it locally. As well as regulatory issues, cultural and environmental issues can complicate this decision. There are also significant implications here for corporate and sustainability structures and systems. The process of formulating a sustainability strategy is discussed more fully in Chapter 2.

Sustainability structure

Some companies narrowly view sustainability as an operations function responsibility for environmental performance, as a human resource function responsibility for labor performance, as a community affairs function for community interaction, and perhaps place responsibility in the legal department to ensure that the company is doing things "right," which through a legalistic lens means according to the law or extant regulation. Companies that define sustainability as a legal issue, or as solely an operations, community affairs, or human resources issue, often find themselves in a reactive position regarding sustainability issues and are missing significant opportunities to more fully integrate sustainability into their business practices.

Companies need to leverage sustainability concerns throughout the organization. A study of Mexican firms found that sustainability outcomes were significantly improved when more than two departments had functional responsibility for sustainability performance.[18] For example, at UPS, a global shipping company, health and safety managers are placed in each business unit to implement strategic safety initiatives. How to improve sustainability through organizational design is covered in Chapter 3.

Sustainability systems, programs, and actions

To drive a sustainability strategy through an organization, various management systems, such as product costing, capital budgeting, information, and performance evaluation, must be designed and aligned. Many companies have revised their performance measurement and evaluation systems to help gauge the sustainability performance of business units and company facilities. For example, Sony uses an intranet-based data system to collect sustainability information from its sites worldwide. Managers at each site input data on energy, water, waste, and other environmental costs, which allows Sony to track its impact on the environment. An effective performance evaluation system should integrate economic, environmental, and social objectives and reward the contributions of individuals, facilities, and business units toward meeting those corporate goals.

Many companies have been using the ISO 14001 environmental management system (EMS) for guidance on their environmental strategy. Indeed, a strong EMS is essential in helping companies systematically identify, measure, and appropriately manage their environmental obligations and risk. Without appropriate management systems, corporations may not reap all the benefits associated with sustainability performance. The alignment of strategy, structure, and management systems is essential in both coordinating activities and motivating employees.

The actions taken by the organization toward sustainability should be both internally and externally focused. Internally focused actions include:

- Labor practices and benefits programs
- Life-cycle analysis and design for environment
- Plant certifications
- Audits for social and environmental standards and practices
- Employee volunteer programs
- Training of employees—both sustainability training and also training to improve employee capabilities, integration of sustainability throughout the organization, and effective monitoring and reporting of results

Externally focused actions include:

- Philanthropy
- Community outreach programs
- Supplier certification requirements
- Supplier audits for workplace practices
- Public reporting of sustainability performance

Some actions are proactive, designed to impact sustainability performance (for example, life-cycle analysis), while others are implemented reactively to respond to the performance indicators and to stakeholder concerns. There is a growing body of research that reports that the most effective sustainability initiatives, in terms of impacting organizational performance, are those that are proactive rather than reactive.[19] Many different plans and programs can be devised to improve sustainability performance. These can be minor changes of existing routines or radical new ways of doing business. They may include capital investments in new technologies, product or process redesign, or R&D spending. They may also include programs to promote ethical sourcing, workforce diversity, or more stringent codes of conduct in terms of labor practices.

Other plans and programs are directed at promoting a company's sustainability performance to stakeholders. This requires both responsible actions and communication with the stakeholders. These stakeholder initiatives may include marketing efforts to promote social, environmental, and economic product features and lobbying efforts to governmental agencies related to social, environmental, and economic issues. The various systems, programs, and actions that can be used to promote sustainability are discussed thoroughly in Chapters 4 and 5.

Figure 1.5 lists some of the various organization processes that lead to success—including leadership, strategy, structure, and systems.

Leadership	Strategy
1. Show commitment from the top 2. Scan business environment for potential risks and opportunities 3. Lead a cultural transformation	1. Develop a mission statement 2. Consider global and local regulations, as well as voluntary standards 3. Consider the impact of social investors
Structure	**Systems**
1. Integrated throughout organization 2. Effective use of human resources 3. Manager access to top leadership 4. Aligned with strategy	1. Costing and capital investment systems 2. Risk management systems 3. Performance evaluation and reward systems 4. Measurement systems 5. Feedback systems 6. Reporting and verification systems

Figure 1.5 Sustainability actions leading to financial and sustainability success

Outputs

Sustainability performance

Companies, through their actions, can either improve or impair their sustainability performance. Sustainability performance is the social, environmental, and economic performance of a company and relates to the objectives that are important to the internal and external stakeholders of the organization. Performance goals and objectives are typically determined only after the organization has a clear understanding of its strategy, who its stakeholders are, and its relevant objectives. Social, environmental, and economic performance objectives typically relate to a broad set of company stakeholders and often address impacts that are at times broader and less tangible than financial performance objectives. This performance includes impacts as diverse as child labor, environmental emissions, product packaging, workplace practices, product quality, and so forth. It includes all of those impacts, both positive and negative, on the company's various stakeholders. Because sustainability goals are often broad, organizations must focus on specific issues or areas of priority when assessing performance.

As mentioned earlier, sustainability performance can be both an intermediate output and a final outcome. In the development and evaluation of corporate sustainability strategies, companies typically attempt to improve their contributions by reducing negative corporate social, environmental, and economic impacts, increasing positive impacts, or both. Companies can view the social, environmental, and economic impacts as ultimate outcomes developed expressly for improving society with no explicit goal of improving profitability, or companies can attempt to improve their social, environmental, and economic impacts as an intermediate output to improving corporate profitability (often called the "business case"). In both considerations, it is important to:

- Measure sustainability performance and evaluate the effectiveness of programs

- Recognize the corporate impacts on society, the environment, and the economy
- Determine how the company can improve its contribution to society, the environment, the economy, and the corporation

The scope must be wide, with an extensive analysis of a broad set of stakeholders and impacts. Where current impacts or stakeholder reactions are low, companies must consider potential changes in impacts or likely future stakeholder reactions to current and future impacts. Chapters 6 and 7 deal with measurement of sustainability impacts.

Stakeholders' reactions

Sustainability performance is converted to having an effect on corporate financial performance, through stakeholder reactions (again, see Fig.1.4). Though critical, integrating consideration of all major stakeholder interests into day-to-day management decisions is a complex undertaking. Companies wishing to do so must broadly identify their stakeholders and the impacts of their products, services, and activities on those stakeholders. They must communicate openly to both internal and external stakeholders and implement the proper mechanisms to listen to their specific concerns through broad stakeholder identification and engagement.

Stakeholder reactions are an important component of the framework as they may significantly affect short-term revenues and costs and long-term corporate performance on many levels. Because gaining advantage through stakeholders has been recognized as a driver of strategic success, companies must identify the key stakeholder groups that are the primary drivers of their strategy, including shareholders, customers, suppliers, employees, and communities. Companies are now gaining lasting advantage through stakeholder relationships uniquely structured to provide strategic competitive advantage.

- Customers provide this advantage through loyalty and long-term purchasing. They can choose to buy more sustainability-positioned products or they can boycott products that are deemed to have negative social, environmental, or economic impacts
- Employees do the same when they commit to great service, innovation, and reliability. Potential employees can choose to work (or not work) for the company based on sustainability performance
- Regulators and communities can increase or decrease regulation, monitoring, and enforcement based on company performance
- Shareholders provide a lasting advantage when they provide long-term capital, and potential investors use sustainability performance as an important component in their investment decisions

Thus, companies must carefully consider likely stakeholder reactions in developing and implementing their sustainability strategy. The framework acknowledges that a company's stakeholders can react to both sustainability performance and the actions taken to promote that performance. Methods of engaging stakeholders will be discussed in Chapter 7, while reporting sustainability performance to internal and external stakeholders is examined in Chapters 8 and 9.

Outcomes

Corporate financial performance

For most companies, the ultimate focus of sustainability strategies and programs must be short-term or long-term corporate financial performance. To effectively capture the impact on organizational performance, the outputs of the sustainability processes must be ultimately converted to monetary measures. The impacts of sustainability actions should include present and future benefits and costs, represented through additional revenues to the organization or a reduction in costs.

Extensive research has shown that improved corporate sustainability performance impacts financial results through both enhanced revenues and lower costs.[20] Numerous studies have shown that consumers have a more favorable image of corporations that support causes that the consumers care about, and that many consumers report that they would switch brands based on social reputation.[21] Revenues related to sustainability management initiatives can be positively impacted through reputational effects as well as through "green" marketing initiatives. Social initiatives undertaken by corporations also impact revenue streams and the level of annual expenditures for cause-related marketing is steadily increasing.

Costs are also positively influenced by sustainable management initiatives. Process improvements may lower costs of energy and water usage and decrease costs of waste handling and recycling. In an effort to reduce its environmental footprint while simultaneously boosting its bottom line, Alcoa's Poços de Caldas refinery and smelter complex in Brazil reused or sold more than 30,000 metric tons of industrial byproducts for a combined income and cost savings of almost US$6.5 million. Alcoa is the world's leading producer and fabricator of aluminum with over 61,000 employees.[22]

Siemens, a German-based integrated technology company, launched an environmental program—Serve the Environment—for industrial environmental protection that contributed to cost reduction by, among other things, improving energy efficiency. In 2012, for example, Siemens Real Estate realized savings potential at selected locations of more than €4.5 million in reduced energy costs and around 17,000 tons of CO_2 emissions.[23]

Similarly, Unilever's sustainability plan resulted in many immediate payoffs. On the one hand, reducing packaging saved the company money; for example, the Vaseline Petroleum Jelly jar uses 3% less plastic than it used to, saving 113 tons of resin a year as well as energy costs in its production. On the other hand, as a result of eliminating waste, several of the company's factories in the US send no waste to landfill. In addition, Unilever operates in more than 190 countries and runs up big travel expenses; its offices in over 50 of those countries are equipped with videoconferencing facilities. A program called Agile Workplace enables people to work at home, which reduces the company's real estate needs.[24]

Globally, McDonald's restaurants are adopting energy-efficient behaviors and installing state-of-the-art equipment to further improve efficiency. McDonald's continues working closely with equipment suppliers to identify and implement these new technologies in its restaurants. Between 2011 and 2012, nearly 60 suppliers introduced new, more energy-efficient equipment to restaurants, helping them avoid approximately US$5 million in energy costs.[25]

Companies with stronger environmental performance also tend to have lower costs attributed to fines, penalties, and legal fees related to environmental activities. Other companies report lower packaging and distribution costs through environmental management initiatives.

Organizations are also able to reduce costs through attention to processes related to social issues, such as absenteeism, lateness, presenteeism, worker turnover, loss of productivity, and healthcare costs. Companies that broaden their employee benefits to include stress management practices or take steps to reduce employee stress often see significant benefits. Organizations also have found that employees involved in company-sponsored volunteer programs report, on average, higher levels of satisfaction, increased enthusiasm for their jobs, and lower turnover rates. Corporate sponsorship through cause-related marketing has also been shown to increase employee loyalty to the organization, and 72% of Americans report that they would prefer to work for a corporation that supports charitable causes over one that does not.[26] The costs related to locating and operating in a community are also impacted by a corporation's sustainability strategies.

So, for many companies, excellence in sustainability performance is a desired final outcome—the sustainability actions taken to reduce the organization's "footprint" on society, the environment, and the economy. However, for many organizations, few actions are pursued without a clear linkage of these actions and their related outputs to the financial health of the organization. Managers are encouraged to allocate resources in a manner that leads directly to improved corporate value. By understanding how sustainability performance impacts stakeholders' actions, and how stakeholders' actions impact organizational revenue and cost streams, the "business case" for sustainability becomes much clearer. This can lead to improved resource allocation decisions, more resources for sustainability programs, and improved sustainability and corporate financial performance. Figure 1.6 summarizes the Corporate Sustainability Model (Fig.1.4) and the factors leading to sustainability success.

Inputs 1. Broader external context 2. Internal context 3. Business context 4. Human and financial resources	**Processes** 1. Leadership 2. Sustainability strategy 3. Sustainability structure 4. Sustainability systems, programs, and actions
Outputs 1. Sustainability performance 2. Stakeholder reactions	**Outcomes** 1. Sustainability performance 2. Long-term corporate financial performance

FIGURE 1.6 Factors leading to sustainability success

Feedback

The feedback process is an important aspect of the proposed framework (as noted by the dashed arrows in Figure 1.4). It is likely that this process will challenge and change strategies and assumptions. Various mechanisms must be in place so that the feedback process does not rely exclusively on the data relating to the financial performance. Instead, sustainability performance, stakeholder reactions, and the effect on financial performance all must be reported and used to modify future sustainability strategy formulation and implementation. Indeed, appropriate management control systems should feedback information on sustainability initiatives, potential environmental, social, and economic impacts, actual sustainability performance (at all organizational levels), stakeholder reactions, and corporate financial performance.

Furthermore, the potential for learning associated with appropriate information is significant and should not be ignored by corporations implementing sustainability actions. Companies must develop mechanisms to access and share best practices and initiatives across the organization. Feedback mechanisms and continuous learning are important parts of any learning organization and in the implementation of systems to improve corporate sustainability. Managers must constantly use feedback to challenge their assumptions as to the viability of various decisions and their long-term implications for the company, society, the environment, and the economy. We discuss the feedback mechanism in more detail in Chapter 8.

Implementing the model

The framework presented in the model is only a starting point for understanding relationships among the key factors for success. Customizing and implementing the framework carefully throughout the company is critical and challenging. As the focus of the framework relates to the simultaneous improvement in both sustainability and financial performance, senior managers throughout the organization must be engaged.

An appropriate set of measures should be developed to test the foundation of the customized corporate framework. Managers must quantify how one variable drives another until the link to profit is clear. This argues for explicitly linking corporate strategy and sustainability actions to sustainability and financial performance. For example, one company wanted to better control costs related to employee absence, including the costs of hiring temporary workers and overtime pay. The managers identified four primary drivers of absenteeism: substance abuse; accidents; carpal tunnel syndrome; and eyestrain/headaches. Management determined that an investment in employee health, safety education, and workstation design would pay off in terms of decreased employee absence. The company's model also showed that employee absenteeism negatively impacted productivity and service quality, so a decrease in absenteeism would positively impact internal business processes and customer value.[27]

Summary

Successful strategies require a better understanding of the implications of management decisions. This includes a careful analysis of the key drivers of performance and measurement of both the drivers and the linkages between them. It also requires a clear understanding of the broad set of impacts that are caused by corporate activities and to understand this impact on a broad set of stakeholders.

We now have several decades of evidence as to what differentiates the successes and failures in this area. In addition, advances in IT have significantly improved our potential to put corporate sustainability into action.

In the next chapter we look at the importance of leadership in kick-starting the whole process.

CHAPTER 2

Leadership, organizational culture, and strategy for corporate sustainability

Identifying, measuring, and reporting social, environmental, and economic impacts cannot begin until the board of directors and CEO are committed to improved sustainability management. Often it is through mission and vision statements and values, or the development and articulation of a corporate sustainability strategy that the board and CEO set the tone at the top. It is then necessary to drive this commitment through the organization by implementing the various systems for identifying and measuring impacts, stakeholder engagement, product design, product costing, capital budgeting, information management, and performance evaluation.

The CEO communicates the values of the organization, the behaviors expected, and the results ultimately achieved. The CEO is responsible for inspiring, insisting on, and implementing action plans for boosting performance. Effective and consistent leadership provides an alignment between sustainability activities and corporate goals and provides internal credibility to promote progress toward improved sustainability management within business units and organizational functions. Management support is particularly important when companies are implementing global sustainability standards across their business units.

Leaders make important choices regarding the formulation and implementation of corporate sustainability strategies in relation to the inputs discussed in Chapter 1 (broader external context, internal context, business context, and human and financial resources) (see Fig.1.4). These input factors also influence the structure, systems, programs, and actions that leaders design to promote the strategy. As we saw, these processes should encourage sustainability performance, improve stakeholder reactions, and produce financial benefits for the organization.

This chapter will discuss:

- The importance of leadership in communicating corporate commitment to sustainability
- The critical role of leadership in developing and implementing sustainability strategy
- Corporate mission and vision statements
- The role of organizational culture and people in instilling sustainability
- The challenges multinational corporations face in implementing sustainability when operating globally

- Important industry standards that should be considered
- How government regulations can affect sustainability decisions
- How socially, environmentally, and economically responsible investment and rating systems can influence sustainability strategies

Board commitment to sustainability

The development of a strong corporate sustainability strategy is critical to changing corporate culture and reducing potential negative impacts. The commitment of the board and management to the enforcement of sustainability principles and development of organizational systems can encourage all employees to comply with company strategy. A high-performance board should achieve three core objectives. It should:

1. Provide superior strategic guidance to ensure the company's growth and prosperity
2. Ensure accountability of the company to its stakeholders, including shareholders, employees, customers, suppliers, regulators, and the community
3. Ensure that a highly qualified executive team is managing the company

These objectives guide not only overall company strategy but also the sustainability strategies undertaken by the organization. Any sustainability strategy should work within the boundaries of the general strategies and frameworks of the organization. These board objectives are critical in establishing a strong foundation for corporate sustainability. In Nike, for example, power is concentrated in the board chaired by founder Phil Knight, who retains controlling ownership in the firm. This allows the company greater flexibility in the options available for balancing sustainability and financial objectives, and enables the company to make sustainability investments with a very long payback period.[1]

Numerous decisions concerning board operations will have significant impact on the board's ability to achieve these objectives. These include decisions about board composition, structure, and the supporting systems. To create effective boards, corporations must provide them with all the means and resources necessary to enable them to fulfill their responsibilities.[2] Based in Spain, Santander, one of the world's largest banks, like most other companies, views its director responsibility as critical. The company has contracted an outside consulting firm to assess the board of directors on a regular basis and all directors receive training in the following areas: financial markets, corporate governance, supervision and regulation, and financial information. These training programs help to equip directors with the knowledge they need to make very difficult decisions.[3]

The following six core principles can help boards in formulating strategies in general and to improve sustainability in particular:

- **Leadership.** Provide a framework for checks and balances; identify and build skills to address sustainability issues
- **Engagement.** Support engagement as a corporate value through dialog and consultation with stakeholders
- **Alignment.** Establish operational practices and incentives that align with sustainability policies and performance goals
- **Diversity.** Include a diversity of races, skills, experiences, genders, and ages in executive and director positions
- **Evaluation.** Evaluate the performance of the board and the company in progressing toward a higher level of accountability and sustainability performance
- **Responsibility.** Ensure that the board responds to and maintains trust with company stakeholders[4]

Numerous corporate collapses and scandals have spurred recent changes and boards are being required to take a more active role in monitoring, evaluating, and improving the performance of the CEO and the company. Many companies are creating board committees specifically focused on governance, accountability, and sustainability. The board of BP, a British multinational oil and gas company, delegates some of its oversight and monitoring activities to its various committees, composed entirely of nonexecutive directors. The Safety, Ethics and Environment Assurance Committee (SEEAC), for example, monitors the management of nonfinancial risk (see Fig.2.1).

Boards are also being asked to focus more on evaluating and improving their own performance as a means of improving corporate governance and transparency. Boards can significantly improve both the evaluation and management of the fundamental elements of corporate governance and sustainability through better measurement systems that will provide relevant information on the performance of the board, the company, and the CEO. (We have more to say about these performance evaluation systems in Chapter 5.)

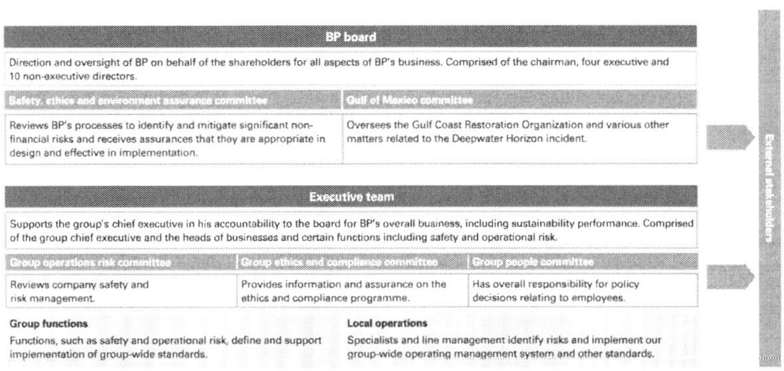

FIGURE 2.1 **BP's committees providing nonexecutive oversight**

Source: BP (2012) *Sustainability Review: Building a Stronger, Safer BP*

CEO commitment to sustainability

The CEO is in a key position to convince the company's constituencies that achieving sustainability is a corporate goal. Research has clearly shown that sustainability strategies are typically top-down, and that the most effective implementation occurs when top management is clearly committed to the strategy.[5] The CEO leads the company in setting sustainability policies and making key decisions for sustainability strategies. But instilling values also takes participation and two-way communication, not just a decree from above. Some of the leading companies engage multiple levels of employees in the institutional task of identifying and communicating values, creating and reinforcing principles through active recitation and search for interpretation, such as Banco Real, a Brazilian bank owned by Santander.[6]

To deliver positive sustainability outputs and outcomes, leaders should:

- Know their company's current sustainability activities and impacts
- Set the organization's sustainability strategy and goals and gather information on sustainability performance through benchmarking with peers and competitors

- Understand and engage with stakeholders
- Implement sustainability policies that support the overall business and sustainability strategies[7]

The CEO, working with the board of directors and other constituencies, conveys the company's position on social, environmental, and economic issues to employees, shareholders, and other stakeholders. The letter from the chairman or CEO to the shareholders in corporate annual reports should express the goals, missions, and strategy of the company. Here are some examples of how some company CEOs and chairmen communicate a commitment to sustainability through their letters to shareholders.

> **Fujitsu Group President Masami Yamamoto:** "No company, no matter how spectacular, can continue to conduct business without the Earth that we all inhabit. We believe that the essence of CSR is the transformation of business activities to suit the surrounding environment so as to leave a beautiful planet for our children."[8]
>
> **Unilever CEO Paul Polman:** "We remain convinced that businesses that address both the direct concerns of citizens and the needs of the environment will prosper over the long term. We need to build new business models that enable responsible, equitable growth that is decoupled from environmental impact."[9]

BHP Billiton Chief Executive Officer Marius Kloppers: "Our approach to sustainability is about ensuring our organization contributes lasting benefits through the consideration of social, ethical, and environmental aspects in all that we do. We will only be successful when our workforce returns home safe and healthy every day."[10]

Starbucks Chairman Howard Schultz: "In business, success is most often measured by numbers. For Starbucks, these metrics are important indications of how we are growing our brand and returning value to our shareholders. But equally important to the value we create are the values we live by. We've been building a company with a conscience for more than four decades, intent on the fair and humane treatment of our people as well as the communities where we do business, and the global environment we all share. We are proud of our heritage. Yet never before have we seen the marketplace and today's consumers have such a deep interest in and knowledge about what companies stand for and how they are living up to their promises."[11]

Nestlé Chairman Peter Brabeck-Letmatheand and CEO Paul Bulcke: "Our position as the world's leading nutrition, health and wellness company brings both opportunities and responsibilities: to do business in compliance with national laws, international standards and our own

> Corporate Business Principles; and in ways that help protect the environment for future generations."[12]

A primary goal of leadership for sustainability is setting principles and practices that will help institutionalize the concept of sustainability in the organization. The board and senior management establish and protect these principles, and are responsible for driving them into the culture of the organization.[13] Social, environmental, and economic responsibility must be seen as a core corporate value.

A leader needs to demonstrate a combination of humility and ambition toward achieving social, environmental, and economic goals. This combination is a key factor in creating legitimacy and wielding influence over employees.[14] Tony Trahar, former CEO of Anglo American, said, "We [Chairman and CEO] devote time and are part of the sustainability initiatives. You can no longer delegate that responsibility to some functional department. If you did you simply wouldn't achieve the buy-in you need."[15] The commitment of the board and management to the enforcement of sustainability principles and development of sustainability strategy and systems encourages employees to cope successfully with these challenges. The CEOs, in particular, are in a key position to motivate middle managers and other employees that achieving sustainability is an important corporate goal. A useful way of doing this is by making the internal business case for sustainability. CEOs can explain (and support by numbers) how sustainability initiatives will advance core business objectives by reducing risk, cutting costs, improving productivity, bolstering brand loyalty, improving customer satisfaction, reducing time to market, pushing up market share, helping the company recruit or retain talented people, etc. These arguments can be supported by possible risks related to sustainability, along with risks of not pursuing it.[16] This may broaden internal support for a sustainability strategy, enhance sustainability implementation, and reduce resistance from short-staffed departments.

Clear CEO commitment and straightforward communication must also be supported by leading by example and consistent support to sustainability. There may be less conflict for senior and middle managers in balancing sustainability and financial performance when these conflicts are resolved higher up in the organization. People are able to make certain trade-offs because they know they will be supported by leaders. At Nike, the CEO and other company leaders support corporate responsibility intensively. It includes leading by example, providing support, and guidance.[17]

Jeffrey Immelt, CEO of GE (General Electric), has publicly committed his company to sustainability. Immelt wants GE to be known not only for its products but also for its environmental programs. He believes that there are four things that can keep the company on top: execution, growth, great people, and virtue. Immelt emphasizes values throughout his communication with employees and the media.[18]

It is the responsibility of the CEO and board of directors to initiate, communicate, and implement sustainability values and strategy throughout the organization. To do this, they should:

- Integrate awareness of social, environmental, and economic issues into corporate decisions at all levels, and ensure such concerns have representation on the board
- Develop measures to identify, measure, report, and manage the social, environmental, and economic impacts of corporate activities
- Modify the corporate structure as needed to integrate sustainability throughout the organization
- Create incentives promoting socially, environmentally, and economically responsible behavior and integrate them into the performance evaluation system and corporate culture

Leadership and global climate change

One of the most compelling issues that currently face many corporations and society in general is climate change. Though the debate about the causes of global warming has been ongoing for many years, both governments and corporations have increasingly acknowledged the critical leadership role needed to address these issues. Through various regulatory actions and market mechanisms, incentives are increasingly in place to push companies into reducing their emissions.

Some corporate CEOs have taken leadership roles in their companies and in society on these issues; some have not. But, regardless of personal views, most now see climate change as a significant factor that must be addressed in global companies. There are several reasons why CEOs must pay attention to climate risks and address their organization's impacts on climate:

- Regulation
- Legal and regulatory liability
- Leadership obligations
- Shareholder activism
- NGO pressures[19]

The well-regarded 2006 Stern Review on the Economics of Climate Change, led by the economist Sir Nicholas Stern for the British government, examined the effect of climate change on the world economy.[20] The report provided additional support and motivation for increased public and private actions to address these issues. Earlier, in separate studies, the Intergovernmental Panel on Climate Change and the National Academy of Sciences had concluded that the Earth is warming, that humans are probably the cause, and that the threat warrants an immediate response.[21]

So what should business leaders do? As with other critical societal issues, it is important to continually scan the environment for social, environmental, and political risks. Heads of businesses must then take action—both as leaders of their corporations and as leaders in society. Investors, consumers, and governments are pressuring companies to develop strategies to address global warming and greenhouse gas emissions. Companies that are able to develop strategies to reduce their exposure to climate change risks will gain advantage over competitors. GE's Ecomagination program discussed in the Introduction is an excellent example of responding effectively to these increased concerns with profitable products to meet industry and consumer needs. Emissions trading programs have also been developed to aid executives and companies in addressing global climate change.

Ford Motor Company has seen the effect of climate change on its business. In the US, Ford's profitability has been dependent on the sport utility vehicle and light truck market. However, in 2006 Ford experienced a loss of revenue and overall market share owing to a decrease in sales of large vehicles. Ford attributes this loss to consumer concerns over fuel prices and greenhouse gas emissions. By integrating sustainability issues into business processes and products, Ford became one of the top selling brands in the US by 2013. It strengthened its presence in the electrified vehicle market with the Ford Fusion Hybrid, Ford C-MAX Hybrid and Lincoln MKZ Hybrid, the Fusion Energi and C-MAX Energi plug-in hybrids, and the Focus Electric, a pure battery electric car—along with innovative, free mobile smartphone apps to maximize efficiency and manage electrified vehicles. Since 2006, Ford improved the Corporate Average Fuel Economy (CAFE)[22] of its US vehicle fleet by 26%.[23]

CEOs can use these four steps to begin to address climate change:

1. Measure greenhouse gas emissions and track them over time. Identify and prioritize areas where emissions can be reduced
2. Assess the effect that carbon-related risks and opportunities could have on business. Consider their impact on costs and revenue
3. Develop strategies based on the knowledge gained in the first two steps. Adapt the company, as needed, to respond to the risks and opportunities it faces
4. Monitor competitor strategies and strive to do better than them[24]

Leadership and strategy come together as companies choose whether they want to play to win (PTW) or play not to lose (PNTL).[25] Aggressively seeking out opportunities can be considered a "play to win" strategy that has the explicit goal of investing in innovation to produce significant advantage that the competition will not be able to easily or quickly match. In the PTW innovation mode, a company invests in changes in technology and business models with the intent of outpacing its competition. It takes risks and manages them effectively, using innovation as a key part of business strategy. When the main organizational objective is to preserve value and address risks in order to bring them back within an acceptable range deemed by the organization, this can be considered a "play not to lose" strategy. Whether following a "play to win" or "play not to lose" strategy, the formal analysis can aid in risk management and in identifying and capitalizing on opportunities. Global climate change is one illustrative example of how companies proceed to make these choices and how significantly they are affected by leadership and how the strategy, actions, sustainability and financial results are impacted.

The role of the corporate mission and vision statements

A mission statement can be used to guide the development and implementation of a corporate sustainability strategy, and will often be included in the annual report or the sustainability report. While the CEO and other senior corporate officers must set the tone at the top, it is with a strong mission statement that awareness of corporate sustainability often begins. The mission statement represents the overall purpose the company will strive to achieve and the commitments it has made to its various constituents—employees, shareholders, customers, and others. By including sustainability principles in the mission statement, a company can declare that it considers corporate sustainability a fundamental part of its corporate strategy. The importance of a mission statement for sustainable performance can be underlined by the term "mission-driven company," which refers to a for-profit enterprise that seeks to simultaneously meet profit goals and sustainability goals that reflect the values of its owners.[26]

Some companies use principles to define sustainability and communicate vision and strategy to the organization. The Toyota Global Vision announced in March 2011, clarifies the company's value: "We want Toyota to be a company that customers choose and brings a smile to every customer who chooses it." The image of a tree has been chosen to symbolize the Toyota Global Vision—its "roots to fruits" (see Fig.2.2). The roots of the tree are the shared values expressed in the Toyota Precepts, in the Guiding Principles at Toyota, and in the Toyota Way. Toyota's business activities are based on the concept of ensuring sustainable growth by fostering the virtuous cycle "Always better cars"→"Enriching lives of communities"→"Stable base of business."[27]

Toyota's seven guiding principles are further presented in Figure 2.3. These principles reflect Toyota's commitment to providing innovative products while respecting the environment and community in which it operates.

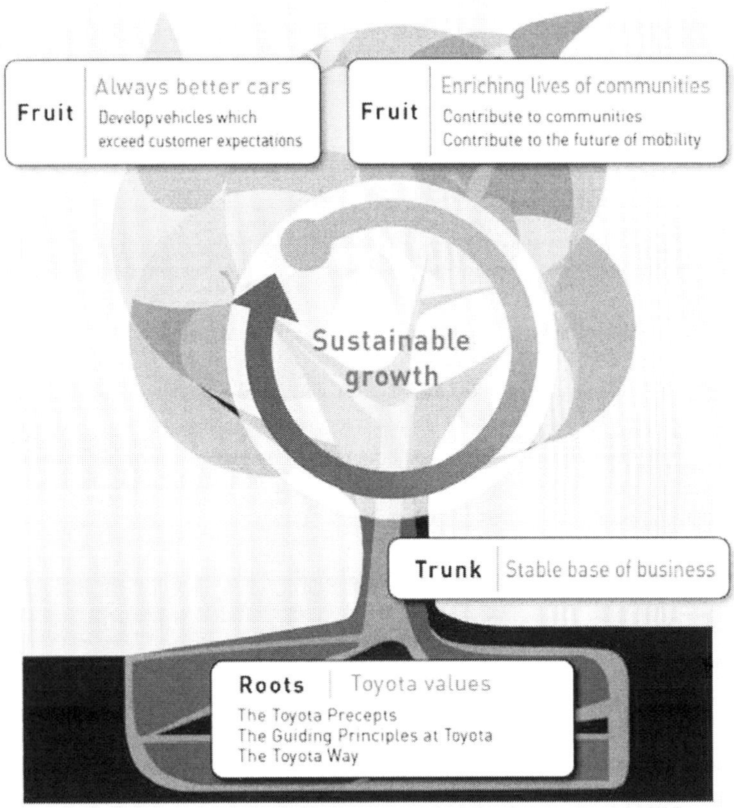

FIGURE 2.2 **Toyota Global Vision**

Source: Toyota Motor Corporation (2012) *Sustainability Report*

The role of organizational culture

A common overall organizational culture that builds on sustainability is another critical determinant of sustainability implementation. Empirical studies find that firms that integrate sustainability into their culture and business practices are better able to integrate sustainability messaging into mainstream communications.[28] Some best practices that are crucial for building the buy-in across the entire business system are: cross-functional and multigenerational interaction to create a sense of community; starting on real projects proposed and developed by the employees with short timeline (these are necessary at the early, fragile stages of the change process); and clear and consistent communication via newsletters, internal memos, articles, intranet tips of the day, blogs, etc.[29]

Guiding Principles at Toyota

(Adopted January 1992, revised April 1997)

1. Honor the language and spirit of the law of every nation and undertake open and fair corporate activities to be a good corporate citizen of the world

2. Respect the culture and customs of every nation and contribute to economic and social development through corporate activities in the communities

3. Dedicate ourselves to providing clean and safe products and to enhancing the quality of life everywhere through all our activities

4. Create and develop advanced technologies and provide outstanding products and services that fulfill the needs of customers worldwide

5. Foster a corporate culture that enhances individual creativity and teamwork value, while honoring mutual trust and respect between labor and management

6. Pursue growth in harmony with the global community through innovative management

7. Work with business partners in research and creation to achieve stable, long-term growth and mutual benefits, while keeping ourselves open to new partnerships

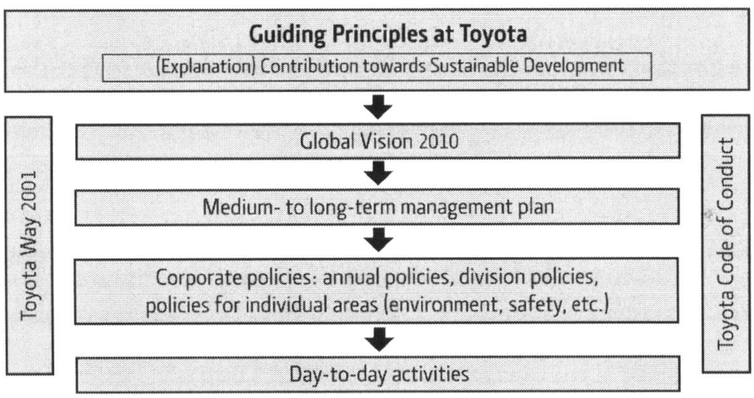

FIGURE 2.3 **Toyota's guiding principles**

Source: Toyota Motor Corporation (2012) www.toyota-global.com/company/vision_philosophy/guiding_principles.html

Teck, a Canadian-based mining company, for example, has created a cross-functional working group to develop its sustainability vision, strategy and action plan. This has resulted in an active, company-wide engagement in sustainability. Teck also conducted in-depth interviews with employees across Canada to gain even more input on how to develop a clear vision and strategy and gain buy-in.

Canadian Pacific (CP), a logistics and shipping company, on the other hand, implemented a campaign to reduce the use of bottled water and educate employees about broader sustainability issues. It was launched with presentations and walk-abouts—one could see discarded water bottles all over. The need to reduce CP's use of bottled water was thus clear. CP used many tactics to get employees engaged. It provided information about the consequences of waste from bottles through its newsletter and the intranet. Employees saw pictures of mountains of water bottles and associated dollar amounts. CP featured interviews on its intranet with employees committing to supporting the environment. Encouragement and rewards were also employed. CP gave feedback about reduced consumption and encouraged employees to see that it is doable. It introduced a yearly conference with regional awards etc. Senior managers also demonstrated their commitment to the initiative by making the switch to tap water at their meetings.[30]

Alcoa, one of the world's largest producers of aluminum headquartered in the US, organized several meetings on sustainability and innovation based on the best corporate practices of eco-friendly growth. These conversations enabled the business units within Alcoa that were further along the path to sustainable growth to serve as a reference point for units just getting started. This peer inspiration resulted in new thinking and action. PUMA, one of the leading designers and developers of footwear, apparel, and accessories, on the other hand, hired a museum curator with expertise in cultural change to drive the implementation of PUMAVision that has become central to the initiative.[31]

Employee training may be used as another tool for dissemination of new values and needed competences. To shift mindsets, Nissan, a leading global auto manufacturer, has trained 99% of its company staff in environmental issues to foster understanding and sustainability awareness. Other overall activities include monthly newsletters to raise staff awareness of the environment, workforce participation in facility inspections and lectures, and in environmental management system training.[32]

Developing a corporate sustainability strategy

The sustainability process begins with the development of a strategy that has the commitment of senior executives and the board of directors. Corporate executives decide whether the company should be sustainable, how sustainable it should be, and what resources are available to achieve sustainability. Formulating a successful sustainability strategy is, in part, about choosing which issues the company will address. Executives are responsible for prioritizing social, environmental, and economic issues and identifying those where their company can have the greatest impact. The social, environmental, and economic issues affecting a company generally fall into three categories:

1. **General sustainability issues.** Issues that are important to society, the environment, and the economy, but which the company is not able to influence
2. **Value chain sustainability impacts.** Issues that are affected by the company's activities
3. **Sustainability dimensions of competitive context.** Issues in the external environment that affect the drivers of competitiveness where the company operates[33]

Once executives have identified those aspects of business activities that have significant impact on sustainability issues (such as the industry, customer, location, and product characteristics described in the previous chapter), they should formulate a sustainability strategy that integrates the company's values, commitment, and goals. The identification of issues with significant social, environmental, and economic impacts can aid companies in minimizing risks, developing innovative strategies, and capturing opportunities to gain competitive advantage.

Sustainability strategies pass through three stages. As companies move from Stage 1 to Stage 3, the focus moves from managing compliance to full integration of social, environmental, and economic considerations into day-to-day operations.

Stage 1: Managing regulatory compliance

In this stage, organizations acknowledge the financial implications of social, environmental, and economic matters; they understand the possible risks, such as litigation and clean-up costs, associated with current practices. To offset the consequences, they develop and publish a corporate environmental policy statement and establish systems to plan for and deal with sustainability problems.

Ever-increasing numbers of regulations are forcing companies to change their practices. In the US, companies must comply with various social, environmental, and economic regulations. Other countries are pioneering approaches in such areas as packaging and the environment.

Increasingly, global sustainability standards are set by the most stringent country or region. The risks of failing to comply with these regulations have taken on new meaning. Civil penalties are common and contingent liabilities grow with increased regulatory pressures, and pressure from communities, activist groups, and the general public.

Regulatory compliance strategies avoid potential liabilities by:

- Ensuring top-management commitment and support
- Developing a corporate sustainability policy statement
- Preparing a sustainability action program
- Creating a sustainability management system
- Establishing a sustainability audit program

At this stage, companies focus more on meeting regulatory standards than on developing innovative strategies to increase competitiveness and reducing their negative sustainability impacts.

Stage 2: Achieving competitive advantage

Organizations move from a commitment to comply with legal requirements to a realization that they can gain a competitive advantage by using resources more efficiently and being socially responsible. While minimizing costs is the hallmark of Stage 1 organizations, Stage 2 companies focus on cost avoidance in approaches such as lifecycle cost management and design for environment.

Substantial competitive advantages can be achieved through improved sustainability performance. They often are reflected in improved product quality, improved production yields, and improved profitability—the result of redesigned processes and products. There is also substantial support for balancing social, economic, and environmental concerns in industry by adopting a concern for sustainability. The future of many companies depends on balancing these concerns, and the institutionalization of corporate responsibility can lead to improved operations and profitability.

In moving toward Stage 2, there are numerous ways to address an organization's sustainability needs in a more systematic and integrated manner. There may be industry-led initiatives such as the US chemical industry's Responsible Care, which gave measurable standards for pollution prevention, process safety, and emergency response. Alternatively, there may be organizational initiatives, such as the International Chamber of Commerce's Business Charter for Sustainable Development, which offers programs and frameworks to help companies build an EMS (environmental management system). There are also government-promoted frameworks such as the EU's EMAS (European Union's Eco-Management and Audit Scheme), discussed later in this chapter.

Stage 3: Completing social, environmental, and economic integration

At this stage, organizations fully integrate social, environmental, and economic components into corporate life. Social, environmental, and economic issues, large and small, become part of everyone's day-to-day decision-making. Corporate sustainability strategies are used to set corporate policies, change corporate culture, and integrate sustainability impacts in managerial decisions at all levels, in all facilities, and at all geographic locations of the organization. This strategy initiates corporate sustainability policies that can adapt to changing social, environmental, and economic regulations and changing technologies, and integrates forecasts of likely changes into management planning processes and decisions. It produces a company that is proactive rather than reactive, focusing on sustainability planning rather than on compliance. It pushes the company to change the design of products and processes to eliminate waste, reduce negative sustainability impacts, and make investments likely to improve long-term corporate profitability. Stage 3 companies create profits from antipollution efforts, "closed-loop" production, improved operational and employee efficiency and effectiveness, and improved products and services. They recognize that long-term economic growth must be socially, environmentally, and economically sustainable.[34]

To reduce any gaps between business strategy and sustainability, and to focus on the most important issues, CEMEX, a global leader in the building materials industry, regularly conducts materiality analysis. This rigorous and replicable process enables the company to prioritize sustainability issues, risks, and opportunities using stakeholder inputs and company insights. The materiality assessment is conducted according to the Global Reporting Initiative (GRI) methodology (for more on GRI, see Chapter 9) and consists of three stages: identification; prioritization; and validation. Yearly, over 1,000 interviews are completed with key stakeholders to identify material topics, and over 10,000 surveys are distributed to stakeholders across all six CEMEX regions. Surveys are used to analyze the degree of concern that stakeholders have for 20 key issues. These are then evaluated according to their importance to the CEMEX business in terms of risks and opportunities (Fig.2.4).[35]

Thinking globally

As companies have expanded around the world, becoming more multinational or global, they have often confronted additional, unanticipated organizational challenges.[36] Coca-Cola's factory in the village of Plachimada, south India, became the object of attention when it was found that the factory had been giving fertilizer containing carcinogens to local farmers. Also, the facility used much of the groundwater that the farmers needed for their crops.[37] Because of these types of incident, some activists argue that tighter regulations are necessary to govern multinational companies and those rules should apply wherever the company operates.

Though management commonly proposes that companies should "think global but act local," implementing such a system can be challenging. Companies want to think globally and develop corporate strategies that are consistent throughout the countries and business units through which they operate. But at the same time, they want to act locally and have a local presence to attract and maintain business and adapt corporate practices to country cultures and competitive conditions. This is not an easy task under any circumstances. In the case of sustainability, the rising cost associated with liabilities and the increased complexity and uncertainty of social, environmental, and economic issues pose particular difficulties. Global organizations must struggle with the balance between one worldwide corporate sustainability standard for management systems and performance, on the one hand, and widely different local government regulations and competitive pressures, on the other.

Wherever government regulations or industry standards are of particular importance, companies must:

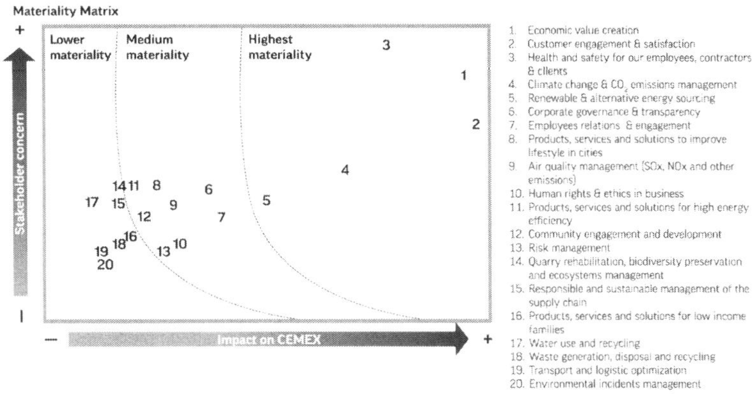

FIGURE 2.4 **CEMEX aligns its strategy through materiality assessment**

Source: CEMEX (2012) *Building the Cities of the Future, 2012 Sustainable Development Report*

- Establish policies and practices that meet local standards
- Meet international standards of various community organizations
- Meet the company's own standards or codes of conduct
- Minimize corporate costs

These issues of whether to establish worldwide corporate standards or local standards have been of significant concern in areas such as labor practices and environmental management, and exist in many industries.[38] Footwear and clothing production have been associated with low wages, child labor, and unfair working conditions. Companies with manufacturing operations in developing countries, and faced with nonWestern labor practices, find that these practices are often scrutinized by their customers or the media, both of which are sensitive to these issues.

But what are the proper standards? Should children be hired in countries where their parents encourage this employment to provide money for basic human needs? What is the proper wage rate to pay in developing countries where average wages are just a small fraction of North American and European wages? Should working hours be limited worldwide although workers in many countries need the additional money? For example, some Chinese workers of multinational corporations complain about global standards that limit their work hours preventing them from earning and saving enough money for their families.[39]

Implementing a living wage at Novartis

"We do everything we can to operate in a manner that is sustainable: economically, socially, and environmentally—in the best interest of long-term success for our enterprise." This statement is the center of the Novartis Policy on Corporate Citizenship. Novartis, a Swiss-based pharmaceutical company, is one of the first international companies to implement a living wage globally. A living wage is generally higher than minimum wage and reflects the costs of a certain group of goods that is considered to provide an adequate standard of living.

Living wages allow the workers, at minimum, the following:

• Basic food needs for employees and their immediate families

• Basic rent

• Basic health and education for employees and their immediate families

• Clothing for employees and their immediate families

• Transportation to and from work[40]

Novartis committed itself to determining the living wage in the countries where it operates and ensuring that all employees were paid accordingly. Novartis believes that paying a living wage has a positive impact on the workforce and aids in the attraction, development, and retention of employees.

To develop a methodology for determining living wages, Novartis worked in conjunction with BSR (Business for Social Responsibility). A survey of the cost of items a typical family would need was conducted in a sample of countries. Some countries conducted their own local studies to review the Novartis/BSR methodology and made recommendations based on their review.

Following these surveys and consultations, it was found that 93 employees out of more than 90,000 were not being paid a living wage. The pay of these employees was increased to the living-wage level. Wage levels will be reviewed and updated periodically.[41]

Managers must assess internal and external factors to decide whether a sustainability strategy should be based on global or local standards. How a company defines each of these factors will influence its choice of a sustainability strategy. A global integrative strategy requires the company to use similar principles everywhere it does business.[42] A locally adaptive sustainability strategy, on the other hand, does not rely on these firm principles. Instead, corporations operate differently in different cultures, based on local needs. One way to evaluate and weigh the long-term impacts of these decisions and identify and organize relevant internal and external factors that lay a foundation for the decision-making process is shown in Figure 2.5 and Table 2.1.

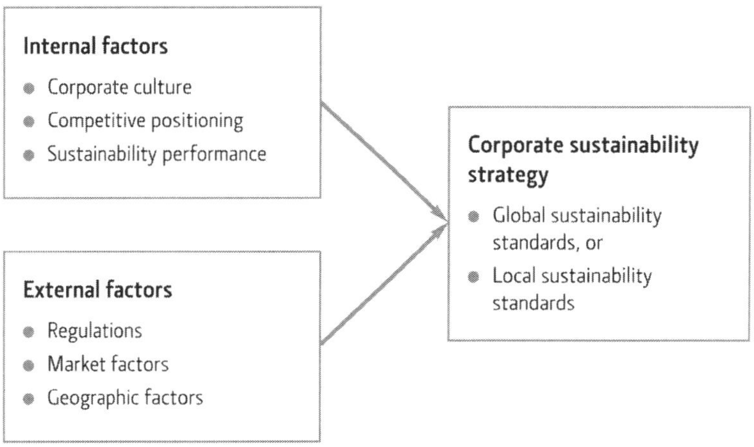

FIGURE 2.5 **Framework to evaluate alternative corporate sustainability strategies**

Source: Adapted from Epstein and Roy (1998) "Managing Corporate Environmental Performance" and Yip (1989) "Global Strategy ... in a World of Nations?"

TABLE 2.1 Determinants of corporate sustainability strategy

	Global integrative sustainability standard	Local adaptive sustainability standard
Internal factors		
Corporate culture	Low level of subsidiaries	High level of autonomy of subsidiaries
Competitive positioning	Differentiation strategy	Cost strategy
Sustainability performance	High environmental risks of operations; new facilities	Low environmental risks of operations; older facilities
External factors		
Regulations	Homogeneous regulations; anticipated standardized regulation worldwide	Fast-changing regulations; incompatible or unjustified regulations
Market factors	Standardized markets; social and environmental pressures from industry	Highly segmented markets; competitive cost pressures from the industry
Geographic factors	Homogeneous	Heterogeneous

Source: Adapted from Epstein and Roy (1998) "Managing Corporate Environmental Performance"

Internal factors

Internal factors include corporate culture, competitive positioning, and sustainability performance.

Corporate culture. The relationship between corporate headquarters and the business units is shaped by the level of autonomy given. Headquarters may choose to delegate social, environmental, and economic standards to business or geographical units to decide locally, or it may elect to centralize and maintain a low level of autonomy for the units.

Competitive positioning. The corporate sustainability strategy is affected by the company's competitive positioning as it focuses on either differentiation or cost leadership strategies. If companies choose the latter, for example, they may be less likely to invest in new technologies and tend to follow local standards since this will typically lead to lower short-term operating costs.

Sustainability performance. The last internal factor in the framework is the sustainability performance of the business units. Multinational corporations usually set strong safety and environmental technology standards in newly opened facilities, but older facilities may lag in technology. The delay in technological upgrade may motivate the company to adopt local sustainability standards. Sustainability performance is also related to the risks involved in the types of product or service provided. Operations with higher potential impacts may choose to maintain more control by adopting a global standard.

External factors

The external factors to consider when choosing either global integrative or locally adaptive strategies are regulations, market factors, and geographic considerations.

Regulations. If regulations and standards vary widely from one location to another, locally adapted sustainability standards may be suitable. However, where global standardization is anticipated or where it provides social, environmental, or economic benefit, a global standard should be implemented. For example, European standards in the car industry have already become accepted in many countries around the world. Energy Star (*ENERGY STAR* ™) is an international standard, originating in the US, for energy efficient consumer products. It has been adopted in Australia, Canada, Japan, New Zealand, Taiwan, and the EU.

Market factors. As the worldwide market has become more homogeneous in terms of customer needs and preferences, a global strategy serves to pursue global markets. However, widely different pressures from company stakeholders sometimes create difficulty in obtaining economies of scale from standardization. In that case, multinational corporations may choose to adopt local sustainability standards.

Geographic factors. Geographic and environmental conditions in a particular site may justify adopting a local sustainability standard. For example, some chemicals react differently in hot and cold climates. Soil fungicide EDB is banned for use in the US, but in hot climates the chemical becomes harmless.

Voluntary standards and codes of conduct

In developing their strategic plan, corporate executives should consider several inputs that can impact their sustainability strategies. With more and more voluntary standards, codes, and principles being developed, companies must decide which are most appropriate for their business strategies. These standards help frame sustainability issues and provide an opportunity to communicate commitment to sustainability to corporate stakeholders. Among the most prominent standards for social and environmental management systems certification are the ISO (International Organization for Standardization) 14000 series related to environmental management, the EU's EMAS, and the ISO 26000 series for corporate social responsibility. Others include Ecological Footprint, eco-labels, and the United Nations' International Council for Local Environmental Initiatives' (ICLEI) approach to triple bottom line using the ecoBudget metric. ICLEI is an international association of local governments and national and regional local government organizations that have made a commitment to sustainable development. The association was established in 1990 when more than 200 local governments from 43 countries convened at the World Congress of Local Governments for a Sustainable Future (Other guides to measurement, reporting, and verification practice, such as the GRI and AccountAbility's AA1000, are discussed in Chapter 9.)

ISO 14000 and EMAS

The ISO 14000 and EMAS series of standards help to provide companies with a structured approach to creating and implementing an EMS (environmental management system). Established in September 1996, the ISO 14000 standards are a set of process, not performance, standards. The most widely used is the ISO 14001 standard. In order to achieve ISO 14001 certification (certification that the requirements for EMSs are met), it is unnecessary for an organization to meet any prescribed levels of environmental performance; it must show only that it has completed a process of EMS development and implementation. A commitment to complying with applicable environmental regulations is required, as is a commitment to continuous improvement. The ISO 14064 series and the ISO 14065 standard provide an internationally agreed framework for measuring GHG (greenhouse gas) emissions. They support programs to reduce GHG emissions as well as emissions trading programs.

The ISO 14000 series is rapidly becoming a primary international standard for environmental certification, much as the ISO 9000 series of standards has become an international standard for quality management. As of 2011, ISO 14001 has been used by over 267,000 organizations in over 150 countries and economies.[43]

The EU introduced EMAS in 1993 as a voluntary initiative designed to improve a company's performance. Organizations registering with EMAS must be able to demonstrate that they have identified and understand the implications for the organization of all relevant environmental legislation and that their system is capable of meeting these on an ongoing basis. The scheme was revised in 2001 to add new features, such as access for organizations from all sectors, the integration of ISO 14001 as the management system of EMAS, and promoting the involvement of employees. EMAS now has over 5,000 registered organizations.

One fundamental difference between the EMAS and ISO 14000 standards is that, while the ISO 14000 series are process standards, the EMAS standards emphasize performance measurement by focusing more on significant environmental impacts or outcomes, and require an independently verified public environmental statement. Additionally, the ISO 14000 series are global standards, while the EMAS standards are EU-focused.

Companies pursue ISO 14000 and EMAS certification for a variety of reasons:

Strategic framework. ISO 14000 and EMAS provide a structured methodology for developing a comprehensive EMS. Companies with multiple facilities can use the process to systematize and standardize the company's approach to developing and implementing an EMS, as it provides a common language and a common framework for managers. The acceptance of a single international standard and external auditors and certification can also reduce the number of overlapping environmental audits conducted by customers, regulators, or registrars.

Supply chain pressure. Just as managers realized that an effective quality strategy begins with suppliers, many companies are using a similar rationale to compel their suppliers to adopt better environmental and social management practices. For example, since 2003 any company that is a first-tier supplier to General Motors has been required to certify that it has an EMS in place to meet the standards outlined in either ISO 14000 or EMAS. Sony Corporation established the Green Partner Environmental Quality Approval Program, which audits suppliers against Green Partner Standards and purchases only from suppliers who have passed the audit.

Expansion of foreign trade. Although governments have not typically issued a trade requirement for ISO 14000 certification, it is becoming a *de facto* requirement for companies conducting international business, in the same way that ISO 9000 became an international standard in the early 1990s.

Reduction of regulatory burdens. Managers at Lucent Technologies' semiconductor plant in Pennsylvania, now part of Alcatel-Lucent, decided to pursue ISO 14000 certification, in part to demonstrate to the local regulatory office of the US EPA (Environmental Protection Agency) that it had standards and procedures in place to ensure environmental compliance. Prior to certification, every time Lucent made a process change, it was faced with the cost and delay of regulatory review by the EPA and the modification of permits. With ISO 14000 certification, the company can decrease the cost and time requirement of the regulatory process, thereby helping it to get its products out of the door more quickly and at a lower price.[44]

Cost reduction. The standards stress a value-chain framework and help companies to consider and evaluate the interaction of environmental factors with operational factors from an activity, process, product, and service view. Improved corporate environmental performance has been linked with process and product cost improvements, as well as lower risk factors and lower costs of capital.

Stakeholder interests. Certification is one way to help satisfy investor and environmental group demands for corporate accountability.

Reputation. Many companies see certification as a label that they can apply to their environmental reports or to other corporate communications, signaling that the company is committed to improving environmental performance.

ISO 26000 and ISO 20121

ISO 26000:2010 or ISO SR (ISO Social Responsibility) was released in November 2010 to provide guidance rather than requirements on how businesses and organizations can operate in a socially responsible way. ISO 26000 defines social responsibility as the responsibility of an organization for the impacts of its decisions and activities on society and the environment, through transparent and ethical behavior that contributes to sustainable development. This also includes the health and the welfare of society. ISO 26000 lays out seven principles that serve as the foundation for socially responsible behavior: accountability; transparency; ethical behaviour; respect for stakeholder interests; respect for the rule of law; respect for international norms of behaviour; and respect for human rights.[45] ISO 26000 is not intended for certification purposes or regulatory or contractual use. Instead, it helps clarify what social responsibility is, helps businesses and organizations translate principles into effective actions and shares best practices relating to social responsibility, globally.[46]

ISO 20121:2012 has been developed to help ensure that events, ranging from local celebrations to "mega events" such as the Olympic and Paralympic Games, leave behind a positive legacy in terms of economic, environmental, and social benefits, with minimum material waste, energy consumption, or strain on local communities. The new management standard was created by the event industry for the event industry. It is applicable to any organization that wishes to establish, implement, maintain, and improve an event sustainability management system as a framework for identifying the potentially negative impacts of events, removing or reducing them, and capitalizing on the more positive impacts through improved planning and processes. The 2012 Olympic Games in London was the first major test of the new standard.[47]

SA8000

The SA8000 Social Accountability certification standard focuses on workplace values. The SA8000 standard was issued in 1997 and revised in 2001 to guide companies in addressing workers' rights. Social Accountability International, a human rights advocacy organization, collaborated with various trade unions, human rights organizations, retailers, manufacturers, academia, contractors, consulting, accounting, and certification firms to develop SA8000. The standard is based on international workplace norms of International Labour Organization (ILO) conventions, the Universal Declaration of Human Rights, and the UN Convention on the Rights of the Child. A brief summary of some of the SA8000 elements follows:

1. **Child labor.** No workers under the age of 15; minimum lowered to 14 for countries operating under the ILO Convention 138 developing-country exception
2. **Forced labor.** No forced labor, including prison or debt bondage labor
3. **Health and safety.** Provide a safe and healthy work environment
4. **Freedom of association and right to collective bargaining.** Respect the right to form and join trade unions and bargain collectively
5. **Discrimination.** No discrimination based on race, caste, origin, religion, disability, gender, sexual orientation, union or political affiliation, or age; no sexual harassment
6. **Discipline.** No corporal punishment, mental or physical coercion or verbal abuse
7. **Working hours.** Comply with the applicable law but, in any event, no more than 48 hours per week; voluntary overtime paid at a premium rate and not to exceed 12 hours per week on a regular basis
8. **Compensation.** Wages paid for a standard working week must meet the legal and industry standards and be sufficient to meet the basic need of workers and their families

9. **Management systems.** Facilities seeking to gain and maintain certification must go beyond simple compliance to integrate the standard into their management systems and practices[48]

As of 2012, there were over 3,000 certified facilities in 65 countries across 65 industrial sectors, employing over 1.8 million workers.[49]

United Nations Global Compact

The UN developed the Global Compact as an initiative to encourage and promote good corporate practices in the areas of human rights, labor, the environment, and anti-corruption. The Ten Principles of the Global Compact (Fig.2.6) are grounded in the values of the Universal Declaration of Human Rights, the ILO's Declaration on Fundamental Principles and Rights at Work, the Rio Declaration on Environment and Development, and the United Nations Convention against Corruption. Signatories to the Global Compact show their commitment by sending a letter to the UN Secretary General stating that they support the Compact and will apply it in their organizations. The UN does not monitor compliance with the principles but does ask that participating companies report their progress in support of the Ten Principles. Since its official launch in July 2000, the initiative has grown to more than 10,000 participants, including over 7,000 businesses in 145 countries around the world.[50]

The Global Compact has been successful in shaping human rights expectations of companies, but it has been criticized for a lack of monitoring, accountability, and enforcement.[51] Many human rights organizations such as Human Rights Watch and Human Rights First have criticized it for not evaluating companies and holding them accountable for doing what they say they are going to do. One of the primary incentives for participating in the Global Compact is to legitimize corporate sustainability strategy. However, without the effective implementation of sustainability practices and the support of NGOs, companies are not likely to achieve the desired change in corporate reputation.

Millennium Development Goals

In 2000, the UN convened world leaders to set goals and a timetable for combating poverty, hunger, disease, illiteracy, environmental degradation, and discrimination against women. The following goals were set by the committee with a deadline of 2015:

- Halve extreme poverty and hunger
- Achieve universal primary education
- Empower women and promote equality between women and men
- Reduce under-five mortality by two-thirds
- Reduce maternal mortality by three-quarters

Human Rights

- Principle 1: Businesses should support and respect the protection of internationally proclaimed human rights; and
- Principle 2: make sure that they are not complicit in human rights abuses.

Labour Standards

- Principle 3: Businesses should uphold the freedom of association and the effective recognition of the right to collective bargaining;
- Principle 4: the elimination of all forms of forced and compulsory labour;
- Principle 5: the effective abolition of child labour; and
- Principle 6: the elimination of discrimination in respect of employment and occupation.

Environment

- Principle 7: Businesses should support a precautionary approach to environmental challenges;
- Principle 8: undertake initiatives to promote greater environmental responsibility; and
- Principle 9: encourage the development and diffusion of environmentally friendly technologies

Anti-Corruption

- Principle 10: Businesses should work against all forms of corruption, including extortion and bribery.

FIGURE 2.6 **The Ten Principles of the UN Global Compact**

Source: United Nations Global Compact (undated) "The Ten Principles"

- Reverse the spread of diseases, especially HIV/Aids and malaria
- Ensure environmental sustainability
- Create a global partnership for development, with targets for aid, trade, and debt relief[52]

The combined actions of national governments, the international community, civil society, and the private sector are making the achievement of the MDGs a reality. However, while millions of people's lives have been improved by meeting targets on reducing poverty, increasing access to safe water, improving the lives of slum dwellers, and achieving gender parity in primary schools, progress toward the eight goals has been uneven, not only among regions and countries, but also between population groups within countries.[53]

Voluntary industry codes of conduct

An approach used in many industries to address stakeholder concerns is to establish codes of conduct for companies to follow. Codes can help identify critical issues in a particular industry and provide solutions for dealing with them. Incorporating codes into a strategic plan also enables companies to compare their performance with their competitors. For example, Responsible Care®, launched in Canada in 1985, is the chemical industry's global voluntary initiative under which companies, through their national associations, work together to continuously improve their health, safety, and environmental performance, and to communicate with stakeholders about their products and processes. Responsible Care encourages member organizations to share information and create a system of checklists, performance indicators, and verification procedures. It is currently run in 52 countries representing 92% of global chemical production. Voluntary codes are a proactive response that demonstrates an industry’s commitment to sustainability. They provide solutions to problems that the industry faces regularly and allow the industry to develop its own mechanisms for addressing violations. By voluntarily adhering to codes of conduct, companies can gain public trust and reduce the need for government regulation.[54]

In 2004, the international coffee community introduced the Common Code for the Coffee Community to provide a basic framework for the industry to move toward being more responsible. The components of the framework include the unacceptability of certain practices including child labor, use of banned pesticides, and forced labor. The code also includes verification of compliance and auditing by an independent third party. Semiconductor industry leaders Intel and AMD (Advanced Micro Devices), as well as others, have developed an Electronic Industry Citizenship Coalition® (EICC) Code of Conduct that established standards to ensure that working conditions in the electronics industry supply chain are safe, that workers are treated with respect and dignity, and that business operations are environmentally responsible and conducted ethically.[55] EICC members adopt the EICC Code of Conduct as operating principles for their companies and suppliers.

Three steps for developing and implementing a multinational code of conduct have been identified.[56]

First, the code should represent the concerns of the affected communities and the problems to be addressed should be defined. The standards should specify the time-frame in which improvement should take place. Second, the code must be measured and verified. The audit should enable firms to continually monitor its activities, create measurements such that two unrelated auditors would come to similar conclusions, generate data to assist management in improving performance, and enable management to report accurately and objectively on its activities. The last step is accountability and reporting. Compliance with codes of conduct should be verified by an external organization.

By following voluntary industry standards and codes of conduct, organizations can assure stakeholders that they are following all applicable laws and regulatory requirements and also adhering to minimum requirements for workplace safety and employee practices. Deciding what codes and guidelines to follow can be a complex undertaking. One method used by several companies is to develop a questionnaire for managers that assesses the current sustainability goals of the company. Based on the results, executives can identify where deficiencies lie and choose standards and guidelines that will help the company improve in these areas.[57]

Of course, any one standard or code is unlikely to address *all* of the social, environmental, and economic challenges that a company encounters. Therefore, to develop a comprehensive sustainability strategy, leaders must scan their business environment for potential risks, engage stakeholders, and implement systems to address additional concerns. Additionally, standards should be integrated only where they fit in with the existing corporate culture or are being used as part of an effort to change corporate culture or practices.

Working with government regulations

With the increasing use and success of voluntary industry standards and codes of conduct, the role of government regulations comes into question. Voluntary standards and codes of conduct offer an alternative to government regulation, reducing the costs to taxpayers. However, the question arises as to whether these alternatives might dissuade or delay government action when it is necessary.[58] For example, the clothing industry has been working with suppliers to improve the treatment of labor in facilities and this initiative has led to improvements in working conditions. However, relying on global corporations to monitor local practices could delay the development of a more effective regulatory system. Historically, much government regulation on business has been aimed at advancing social goals, particularly in the areas of the environment and labor standards.[59] These concerns will continue to dominate the regulatory agenda in many countries. It is important for government to enact and enforce laws that prevent the most unacceptable social, environmental, and economic impacts while leaving companies freedom to innovate and remain profitable.

In Australia, businesses and government have clashed in the development of a social agenda. The prime minister's vision focused on philanthropy as a means of improving societal problems such as unemployment, health, and drugs. However, corporations were faced with labor disputes and needed more guidance on integrating sustainable business operations. More recently, Australian businesses decided to take a more proactive stance and work with government to formulate more inclusive policies. This shift in public policy-making has caused Australian businesses to proactively research their social, environmental, and economic impacts and incorporate their findings not only into their sustainability strategies but also into the public policy debate.[60]

Harvard Business School professor Michael Porter and Claas van der Linde have described a useful framework for thinking about the role of government in social and environmental regulation.[61] They maintain that "government policy contributes to competitiveness if it encourages innovation ... and undermines competitiveness if it retards innovation or undermines the intensity of competition." One of their central arguments is that the entire sustainability–competitiveness debate has been framed incorrectly and the "policy makers, businesses, and environmentalists have unnecessarily driven up costs and slowed progress on environmental issues." They argue that, instead of seeing the issue as a trade-off between regulation and competitiveness and then being concerned about how the trade-off can be relaxed, the goal should be to eliminate the trade-off. Strict environmental regulations should cause companies to seek innovative solutions to minimize their cost of compliance while improving their products.

Companies can respond to or even anticipate regulations by developing innovative strategies that lower the net costs of compliance. These "innovation offsets" can easily exceed the costs of compliance if total costs and benefits are identified and measured properly. For example, avoiding the production of waste so that no money needs to be spent cleaning it up is often accomplished through a combination of capital, process, and product improvements.

Additionally, when Mexico agreed to NAFTA (North American Free Trade Agreement), the regulatory pressures from both the US and Canada positively impacted management practices and moved exporting companies to define and implement sustainability strategies. These exporting companies improved their performance compared to companies that traded only within the country. The stronger regulatory environment *pushed* the Mexican firms to improve their environmental performance to conform to the laws, or face penalties, fines, and potentially closure. NAFTA also created a *pull* effect by compelling Mexican firms to improve environmental performance in order to sell more products to customers in the US and Canada.[62]

Many managers find business–government relations challenging. However, by developing certain skills, they can get over the difficulties. The following suggestions can help develop strategies to improve governmental relations:

- **Know how your efforts support business goals.** How are programs measured and evaluated? How did the programs contribute to a public policy win or loss?
- **Integrate all communications functions.** A public relations or government affairs department can help build relationships with regulators. Tactics such as media outreach, direct lobbying, and grassroots activism can be integrated into the sustainability departments
- **Gain political influence without being too partisan.** Because governmental leadership positions can change with each election, organizations need to keep their political options open
- **Maintain a global perspective.** Companies need to understand the differences of political and social systems and gain knowledge of each country's specific issues
- **Most important, create strong personal relationships.** Reputation and credibility will be

important factors in influencing public policy. Managers and executives who are able to develop a relationship with political leaders based on mutual trust and respect have a better chance of being successful[63]

Collaborative initiatives between business and government can also be used to create a positive regulatory environment. Governments must create a collaborative environment to align public policy goals with those of business. Collaboration also allows the best optimization of resources and allows each party to apply its expertise.[64] The California Climate Action Registry is one example of how business and government can collaborate to improve environmental impacts. The Registry, established in 2000 by the state of California, encourages companies to cut GHG emissions. Members register GHG emission baselines and then measure changes against that baseline. Many companies are participating in the registry because they sense that federal regulation is coming soon and want to be prepared.[65]

Companies can take a proactive stance to influence law and be part of the public policy process. GE has focused on improving its ability to work with government. Senior GE officials and the Chinese National Development and Reform Commission have formed working groups focused on energy, aviation, rail, and water. This partnership has led to several joint seminars on wind energy and water treatment. GE has also established a Research Center in Qatar where local professionals and leaders can perform R&D activities to improve the quality of life of Qatar's citizens.[66]

To influence and respond to social, environmental, and economic regulations, leaders need to address the following questions:

- Are managers and other employees aware of major environmental legislation that affects the company?
- Is the company in compliance with the law? Is any outstanding environmental litigation under way?
- Is there a process for informing executives when the company is not compliant with the law?
- Are there processes to routinely assess the social, environmental, and economic impact of operations within the context of the regulatory process?
- How well is the company positioned to take advantage of new regulations compared to competitors?[67]

Corporate executives should monitor governmental strategies and changing regulations, and develop a corporate sustainability strategy that improves competitiveness and profitability. Innovative products and processes are needed to deal effectively with changing social, environmental, and economic conditions. Executives should focus on the likely impact of regulations on existing production and practices and develop strategies that can create a competitive advantage. Regulation on sustainability can have a positive influence on competitiveness by causing businesses to examine processes more closely, resulting in increased profitability and corporate responsibility.[68] Thus, there is an incentive for companies to be proactive in addressing social, environmental, and economic issues and to be ahead of regulatory control rather than be laggard trying to catch up.

Social investors and sustainability indices

When scanning their business environments and engaging stakeholders to develop a sustainability strategy, executives must also consider the role of social investors. More and more investors are considering social, environmental, and economic impacts when making investment decisions. Social investors include individuals, investment funds, business, nonprofit organizations, and others who want to invest in companies that achieve positive social, environmental, and economic impacts. There are two primary methods practiced by social investors:

- **Negative screening.** Eliminates companies that have practices or products that do not fit with the investor's requirements

- **Positive screening.** Invests in companies that have products or operations that fits the investor's criteria

Socially responsible investing is a booming market in both the US and Europe. Assets in socially screened portfolios climbed to US$3.07 trillion at the start of 2010, a 34% increase since 2005.[69] The most prevalent mutual fund screen is tobacco; 162 mutual funds screen out companies involved in the production, licensing, or retailing of tobacco products, including companies that manufacture products necessary to produce tobacco. After tobacco, alcohol and gambling are the most screened.[70]

To assist social investors, many socially responsible investment indices have been created listing companies that meet specific criteria on social, environmental, and economic issues. Begun in 1999, the Dow Jones Sustainability Group Indexes identify and rank companies according to their sustainability performance. The Sustainability Assessment used to rank the companies is based on five elements:

- **Innovation.** Investing in innovations to lead to a more efficient use of resources
- **Governance.** Setting high standards for governance including management responsibility and corporate culture
- **Shareholders.** Providing financial returns and economic growth for shareholders

- **Leadership.** Setting standards for best practice and maintaining excellent performance
- **Society.** Investing in local and global communities, interacting with stakeholders, and responding to their needs

For each company, information is collected from responses to a corporate sustainability questionnaire, company documentation, policies and reports, and publicly available information. Companies are then monitored through the media and stakeholder organizations for their involvement in social, environmental, and economic issues such as illegal commercial practices, human rights issues, workforce conflicts, and disasters or accidents. An external auditor reviews all the information collected. Finally, each company is assessed based on general and industry-specific sustainability criteria and can receive a score between 0 and 100.[71]

The FTSE Group creates and measures over 100,000 equity, bond, and hedge fund indices. Launched in 2001, the FTSE4Good Index Series is designed to measure the performance of companies that meet globally recognized corporate responsibility standards and facilitate investment in those companies. To be included in the indices, companies have to show that they are working toward environmental responsibility, developing positive relationships with stakeholders, and upholding and supporting universal civil rights. Using the Dow Jones Index and the FTSE4Good Index

Series is one way that stakeholders can measure companies' sustainability performance.

Numerous research firms and investors have conducted additional research. KLD Research and Analytics, an independent investment research firm, conducted an independent review of the public websites of all S&P (Standard & Poor's) 100 companies to assess their disclosure of environmental, social, and governance policies and performance to answer the following seven questions:

1. Does the company website have a separate CSR or sustainability section?
2. Does the company have an annual CSR/sustainability report?
3. Does the company reference the GRI in its report?
4. Does the company have a GRI content index?
5. Does the company report have goals and benchmarks?
6. Is the company a GRI organizational stakeholder?
7. Is the company report "in accordance" with GRI?

KLD has completed a database, which can be accessed by anyone, providing a profile of how each company responds to the seven questions.[72]

AccountAbility, a London-based sustainability think-tank, and csr network, a British consulting firm, have developed the Accountability Rating to measure the extent to which companies have integrated sustainability into their business practices. The Accountability Rating is based on scores in six categories: three external drivers (public disclosure, assurance, and stakeholder engagement) and three internal (governance, strategic intent, and performance measurement). In each of the years that have been measured, the internal drivers scored higher, but the majority of improvements are in public disclosure and stakeholder engagement.[73] These external reviews provide social investors, as well as other stakeholders, with additional information about companies' sustainability performance.

Companies may consider altering their sustainability policies in order to qualify for investments by socially responsible mutual funds or be more acceptable to social screens. In October 2000, the California Public Employees Retirement System, the largest pension fund in the US, announced that it would sell its tobacco holdings. Bank of America announced in March 2007 that it would commit US$18 billion in lending, advice, and market creation to help commercial clients with their environmental business practices and be more sensitive to sustainability issues.

Social investors can also put pressure on companies to seek changes in corporate policies. US shareholders usually file hundreds of public-interest proxy resolutions focusing primarily on labor standards, equal employment, and environmental policy. In response, companies such as Wal-Mart and The Home Depot agreed to ban employment discrimination on the basis of sexual orientation after repeated filings, and Gillette and Reebok agreed to establish a standard for greenhouse gas emissions.[74]

Investors are also concerned about companies' plans to address global warming. The GCC (Global Climate Coalition), now disbanded, was a leading lobbying group opposed to the Kyoto Protocol and government regulation of GHG emissions. Ford Motor Company was a member of GCC, but shareholder pressure led to them pulling out. Other members such as General Motors and Texaco, facing similar shareholder resolutions, also withdrew.[75]

Some companies have established their own funds to improve sustainability impacts. Danone, the French-based food conglomerate, has established Danone Ecosystem Fund that, in three years, has invested in 43 projects, some of which have already become sizeable in their own right. These projects are all conceived and managed locally by Danone subsidiaries in partnership with NGOs. They all have a direct relationship with Danone's economic ecosystem. They include, for example, the development of a high-quality milk supply chain with small producers in Ukraine, Egypt, Brazil, and China, the recycling of plastic bottles in Indonesia and Mexico, and local distribution in several countries which creates numerous jobs and helps market products to new customer bases.[76]

The rise in importance of social investors and sustainability rankings has added one more major factor in the development and implementation of corporate sustainability strategies. As the concern for these issues has become more mainstream, the need for their consideration in all day-to-day management decisions has become essential.

Summary

Sustainability performance begins with the commitment of senior company officers and the development of a mission, vision, and strategy that will be implemented. However, having the CEO and other senior corporate officers set the tone at the top is critical but not sufficient on its own. A corporate sustainability mission and vision statements should be adopted to convey the corporate commitment throughout the organization. Then corporate sustainability strategies are developed to move the company toward a full integration of sustainability. Such a move must be seen as a core corporate value, central to company operations, rather than as a reaction to current or pending governmental regulations. The implementation must continue through:

- Broad-based institutional support for the company strategy
- Development of an organizational structure to support sustainability

- Development of costing, capital investment, and risk management systems
- Performance evaluation and incentive systems
- Measurement and feedback systems
- Reporting and monitoring systems

Sustainability can improve international competitiveness and may even cause a closer examination of production processes, resulting in improved product designs, product and service quality, and production efficiency and yields, along with environmental improvements. These improvements, in turn, often result in increased employee and customer satisfaction and retention, increased social, environmental, and economic performance, and increased profitability.

In the next chapter we look at the actual structure of the company and its importance for effective sustainability strategies.

CHAPTER 3

Organizing for sustainability

Once leadership commitment is established, corporations need to implement their sustainability strategy through appropriate organizational structures, performance measurement and reward systems, culture, and people. This alignment of strategy, structure, and management systems is essential for companies in both coordinating activities and motivating employees (see our model [Fig.1.4]). In this chapter we discuss:

- The challenges for global corporations
- The integration of sustainability throughout the organization
- Information flow
- Outsourcing
- Collaboration with NGOs

The challenge for global corporations

The organizational structure around sustainability issues often entails organizing activities and resources spread throughout many locations.[1] Corporations must consider whether key resources and activities should be centralized or decentralized and decide on a level of central control versus business unit autonomy. These decisions must be appropriately aligned with corporate culture. The decision to either centralize or decentralize an organizational structure can depend on several contextual factors, including:

- Organizational size
- Product diversification
- Geographical diversification
- Social and environmental impacts

CEMEX, the global leader in the building materials industry, for example, has a centralized sustainability organizational structure. Its Sustainability Committee comprises 12 members: nine members of the management team that report to the Chairman and CEO, the Senior Vice President of Energy and Sustainability, the Senior Vice President of Global Technology, and the Corporate Director of Sustainability. This committee meets quarterly to assess and guide CEMEX in its sustainability efforts. Decisions made by the Sustainability Committee and validated by the CEO, the management team, and the board of directors are swiftly executed by the Senior Vice President of Energy and Sustainability in coordination with the different regional/country presidents and leaders of other corporate functions (Fig.3.1).[2]

The Mitsubishi Corporation (MC), Japan's largest general trading company, established its CSR & Environmental Affairs Committee as a subcommittee of the Executive Committee in 2008. Under this subcommittee sits the CSR & Environmental Affairs Review Committee, which discusses MC's basic policies concerning CSR and environmental affairs, risk management, and other issues. The same committee also reviews the company's social contribution activities. Additionally, MC has appointed a Senior Executive Vice President in charge of CSR and environmental affairs with responsibility for overseeing the aforementioned committee and generally supervising activities relating to environmental policies and philanthropic activities. MC also has a CSR and Environmental Affairs Advisory Committee with outside experts to provide advice and proposals on MC's environmental and social performance.[3]

Many companies, operating in multiple industries and multiple geographic locations, face challenging environments, which often lead to a more decentralized organizational structure. The advantages of decentralization often include greater flexibility and increased responsiveness. Specific local expertise about markets, competitors, and customers provides valuable knowledge that could translate into innovative and efficient solutions. A more decentralized decision-making process gives managers autonomy and can create an environment that is often more conducive to experimenting and developing new ideas.

Challenges facing decentralized organizations often include loss of scale economies and duplication of functions. Further, autonomy given to managers may result in inconsistencies between business units and place the pursuit of divisional profitability and short-term objectives above overall corporate performance. Decentralized organizations will also need to incorporate an information system that is able to collect data and information to disperse across business units and geography. To facilitate data collection, Canon, a global manufacturer of business machines, cameras, and other optical products, has developed a Product Environmental Information System which allows each site to enter environmental data through the company's intranet (Fig.3.2). Divisions use the system to acquire the information they need to develop environmentally sensitive products. It is also used by management as a tool for evaluating environmental results, accounting, investment, and sustainability and other management reports. Use of this system gives Canon managers access to immediate information regardless of their location.[4]

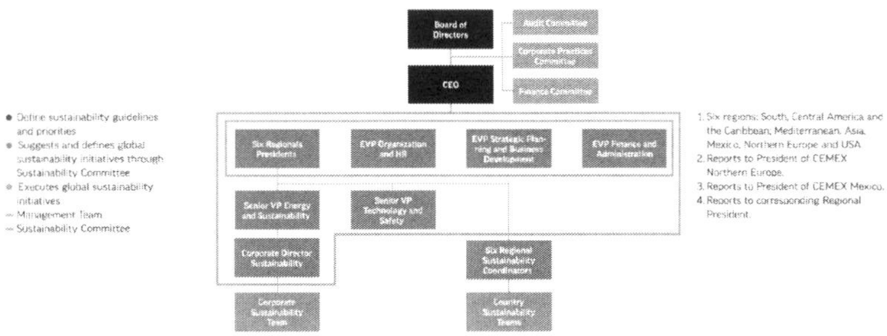

FIGURE 3.1 **CEMEX's organizational structure for improved sustainability**

Source: CEMEX (2012) *Building the Cities of the Future, 2012 Sustainable Development Report*

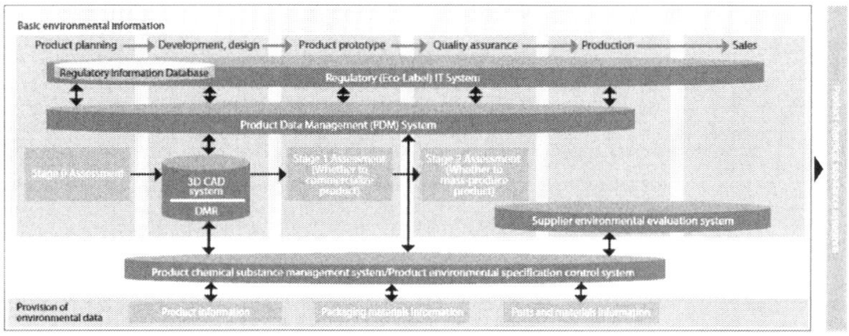

FIGURE 3.2 **Canon Product Environmental Information System**

Source: Canon (2013) *Sustainability Report*

Decisions about the best organizational structure for improved corporate sustainability performance are usually further complicated as geographical diversity increases and particular business needs, local laws, and different cultures must be confronted. A global integrative sustainability standard, as discussed in Chapter 2, implies centralization of many social and environmental functions, whereas a locally adaptive standard relates to a decentralized operation in which business units are provided with a high level of autonomy (Fig.3.3). Multinational corporations should therefore align their corporate structure and sustainability structure with their corporate sustainability strategy.

The strategy selected also determines the level of autonomy exercised by subsidiaries of the home office. In the case of global sustainability standards, the external factors simplify decision-making by allowing centralization of key management inputs related to sustainability regulations, and market and geographic factors. This is possible when regulations and external factors are homogeneous and standardized across borders.

The opposite occurs in locally adaptive standards, and local geographic and business units are empowered with high levels of autonomy regarding decision-making. Headquarters may wish to encourage and provide a high level of autonomy for its subsidiaries, respecting their understanding of local social, environmental, and economic issues and allowing them to create local standards.

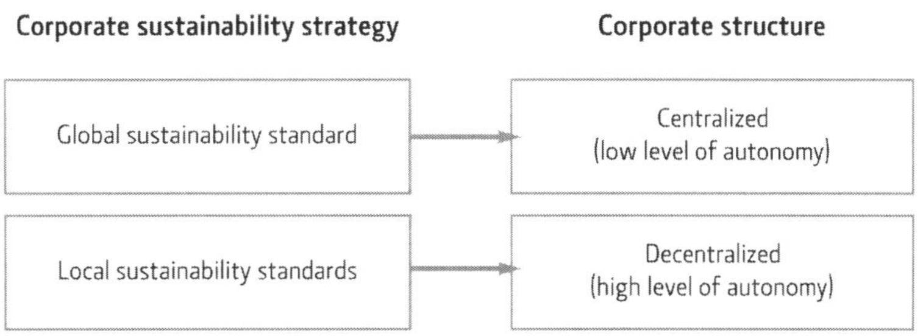

FIGURE 3.3 Aligning structure with intended strategy

Like most other global companies, McDonald's, the world's leading global food service retailer, with more than 34,000 locations in 119 countries, has a corporate structure which is decentralized, with over 5,000 of its restaurants operated by franchises.[5] McDonald's believes that its restaurant operators have a better understanding of the local customs, preferences, and business environment and, therefore, McDonald's gives them substantial discretion to decide how to contribute toward the corporation's common goals and standards.[6]

When EH&S (environmental, health, and safety) staff were initially established in many organizations, they were often part of a central corporate staff. As companies shifted their focus to sustainability concerns, it was often desirable to push primary responsibility to the business units, and many companies reduced their central staff. Now most companies have recognized that a central staff and local personnel at the facilities are both necessary. Substantial advantages can be achieved at the business unit and facility level in product and process design, operational controls, and self-audits to control and reduce waste production and other sustainability impacts. Most companies maintain environmental coordinators and community relations managers at each manufacturing site who are responsible for operating and monitoring the sites' environmental and community activities. Though coordinators for monitoring and responding to broader social issues are also typically desirable, this function remains less common and less well managed.

But a strong centralized staff is also necessary to provide overall strategic planning, guidance, and coordination for the sustainability function. Central staff is key to internal auditing and to furnish overall direction for identifying, measuring, and reporting social, environmental, and economic impacts. It is essential for developing and applying tools for costing, capital investments, and performance evaluation and for directing strategy integration throughout the organization.

In Nike, much of the CSR staff has been moved into the business units to enhance the integration of CSR principles and initiatives throughout the organization. This allows the company to rapidly communicate and operationalize initiatives and to receive information from the business units about a broad spectrum of sustainability-related challenges. Small, centralized CSR staff still creates and pushes the corporate CSR agenda, but their focus has shifted from a cost and regulatory emphasis toward a strategic approach.[7]

H&M (Hennes & Mauritz), a Swedish-based global clothing manufacturer and retailer, has a total of around 170 employees who work with sustainability as a core task. Over 20 people strong, the head office Sustainability team is responsible for developing global targets, reviewing progress, and encouraging and advising all relevant departments on the development and implementation of material Conscious Actions. With this support, all departments are then individually responsible for developing and carrying out the so-called Conscious Actions necessary to fulfill H&M's sustainability commitments. Another 100 people operate from H&M's 15 production offices around the world. They work directly with H&M's suppliers to support them in complying with the company Code of Conduct and help to make H&M's supply chain more sustainable. Around 50 so-called Conscious Coordinators work in different departments and country offices supporting the implementation of H&M's sustainability strategy in their organizations. The Head of Sustainability reports directly to the CEO and responsibility for the implementation of H&M's sustainability strategy is held by the H&M's executive management team. Major sustainability issues influencing more than one department or conflicting interests are discussed in a decision-making forum called the Green Room, with the involvement of the CEO.[8]

Some companies create regional offices to coordinate activities between different locations and divisions. Sony Corporation has managers responsible for the environment in five regions: the Americas; Europe; Japan; East Asia; and Pan Asia. The regional offices review regulations, distribute information from headquarters to the various divisions and sites, and perform audits. Thus, there may be coordination and oversight of sustainability issues from both the business unit and the geographic leadership in decentralized organizations.

Most multinational companies have established worldwide standards for social, environmental, and economic performance, creating benefits and bringing challenges. Reducing sustainability impacts and complying with various local and national regulations create challenges in organization and coordination. Thus, a strong business unit and facility staff, reporting to geographical unit managers, business unit managers, and a central management team, is important. Likewise, a strong central management team is important for planning, guidance, and coordination, in addition to setting the tone at the top and providing management commitment, as discussed earlier.

When organizing for sustainability, a corporation can choose from at least three basic forms: the same organizational structure, extended organizational structure, and new external structure. When maintaining the current structure, sustainability responsibilities are assigned to current functional areas. In this system, employees assume the additional responsibility of managing sustainability impacts, along with their current responsibilities. In the extended organizational structure, one or more levels of the organization are specifically created for the management of sustainability. Lastly, external structures, such as foundations, are formed. The external structure has a separate legal status, although it may not be completely autonomous.[9]

Organizing for sustainability at DuPont

DuPont's organizational structure has undergone changes over several years. It was one of the first companies to create an officer-level VP (vice president) of EH&S in 1989. At that time it used a decentralized EH&S structure consisting of managers for medical, safety, occupational health, and environment. These managers were supported by a small staff with specialists in each facility. In the early 1990s, DuPont created the Corporate Safety, Health, and Environment Excellence Center. The group's services were contracted by the strategic business units as needed. In 2004, DuPont was reorganized. Previously, the Excellence Center reported to the VP of engineering, under the direction of the VP of EH&S.[10] DuPont then created a VP of Safety, Health, Environment, and Engineering who was responsible for the strategic direction and administration of the group. Today, DuPont has a joint position of a VP of Safety, Health & Environment and the once separate position of a Chief Sustainability Officer. This dual responsibility reflects the way DuPont has evolved. The first is keeping DuPont operations in compliance and going beyond that to reduce its footprint. The second part is about DuPont sustainable growth, which is about finding market opportunities that are going to present themselves because of evolving societal needs. DuPont's CSO ties DuPont business strategies with some of the megatrends so that the company can identify growth opportunities.[11]

Involve the whole organization

Integrating sustainability into the organization is the process of ensuring the achievement of environmental, social, and economic goals through organization-wide efforts. To do this, organizations must assess the impacts that each of their activities has on sustainability performance.[12] It is clear that different functional areas of the organization are affected as corporations increase their sensitivity toward sustainability principles (Fig.3.4). Primary and support activities are all important in helping the organization reach its sustainability and financial goals. Each department could be used to promote sustainability in the following ways:

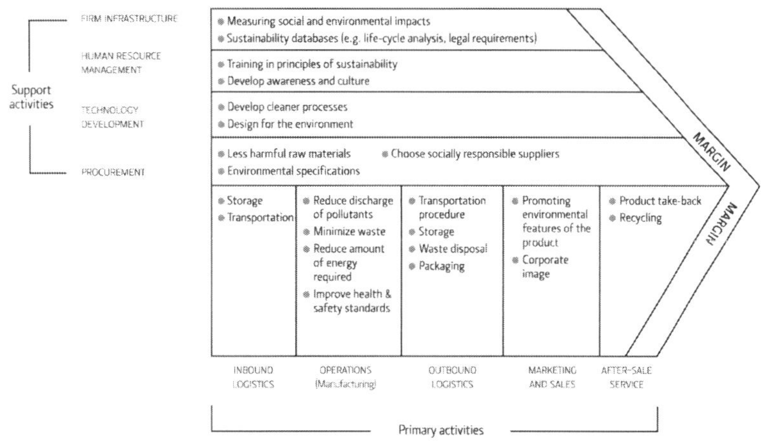

FIGURE 3.4 **The value chain and sustainability**

Source: Epstein and Roy (1998) "Managing Corporate Environmental Performance" and adapted from Porter (1990) *The Competitive Advantage of Nations*

- **Procurement** finds raw materials from sustainable sources of supply and produced with lower environmental impacts, finds ways to reduce packaging and use more recycled materials, and looks for sourcing from socially responsible factories both domestic and overseas
- **Research and development** identifies processes that use resources more efficiently by finding new uses for waste products
- **Marketing** looks at the growing consumer preference for goods that support the sustainability principles and how marketing, distribution, and selling methods can reduce adverse social, economic, and environmental impacts
- **Production** works with engineers and maintenance staff to devise processes that are more efficient and less costly in energy and resource use, as well as maintaining adequate health and safety standards
- **Legal** keeps abreast of legislation and learns how best to disseminate this information
- **Management accounting** provides managers with information so that they can make better decisions

on product costing, and pricing, product, and process design, and capital investments

- **Financial reporting and auditing** provides external disclosures related to contingent liabilities so that external users of the information can better evaluate the company's current and future prospects

In any structure, the business unit staff has a very important role in seeing that sustainability strategies have positive results. Operating personnel at the various company facilities are essential to the proper functioning of established systems. Though strategies, policies, and procedures can be developed by a central staff, it is important that business unit and facility managers and staff understand the importance to the company of excellent sustainability performance. The sustainability strategy should recognize the diversity of talents and responsibilities of different departments while at the same time creating a common identity of values for the entire corporation. For instance, human resources may have to work with manufacturing managers to develop sustainability training programs, but each will have a different contribution to the development of the program.[13]

BG Group, an integrated oil and gas company with operations in 27 countries over five continents, launched a rigorous Social Performance Standard and Management System (informed by the International Finance Corporation performance standards) with accountability structures, targets, and metrics. The business drivers for this were that BG Group has been expanding and is now operating in an increasing number of countries where it faces complex social and developmental challenges. Responding effectively to these challenges matters to the business because of rising expectations and demands for greater clarity and performance from stakeholders including governments, host communities, NGOs, and financial institutions. Following the endorsement of the new standard by the BG Group board, a training program was implemented for the CEO and senior management team and for the management at each of BG Group's key operational sites.[14]

The importance of innovation for simultaneous improvement of sustainability and financial performance has already been underlined. Companies may dedicate someone, at least partially, from the R&D or design teams to manage green innovation. This person should work with the sustainability department to develop an understanding of how environmental challenges affect the company's products and value chains. By assessing environmental impacts throughout a product's life-cycle, environmentally responsible design programs focus creative people on the areas where they can reduce footprint the most. Using this process, HP (Hewlett-Packard) developed water-based inks and a new printing technology that nearly eliminates toxicity.[15]

In the 1990s, the failure of Nike to produce an athletic shoe for low-income populations in China can be partly attributed to its organizational structure. The World Shoe was designed to attract customers who could not afford most of Nike products and the program was housed within the athletic footwear business group. Therefore, the World Shoe group was forced to use the same manufacturing, distribution, and marketing systems used for Nike's more expensive products with very different goals. In this case, it might have been advantageous for Nike to establish a separate department that would have had the freedom to design its own strategy oriented toward sustainability.[16] Having the right structure is essential for assuring positive sustainability impacts and promoting sustainability to various stakeholders.

At HP, the business unit staff plays a very important role in the company's sustainability structure. In addition to HP's Global Citizenship Council, which ensures company-wide commitment to and alignment with HP's global citizenship objectives, HP also maintains separate councils dedicated to areas such as environment, supply chain responsibility, corporate ethics, and privacy (Fig.3.5). These councils include leaders with relevant expertise from business units, regions, and functions. The business unit staff is thus directly involved in the creation and implementation of sustainability strategies.[17] This structure gets a diverse mix of employees and functions involved in implementing sustainability.

In order to achieve coherence and integration, sustainability strategies are best leveraged throughout the organization, which then needs to clearly define the relationships between the board, corporate executives, business unit managers, and functional managers.

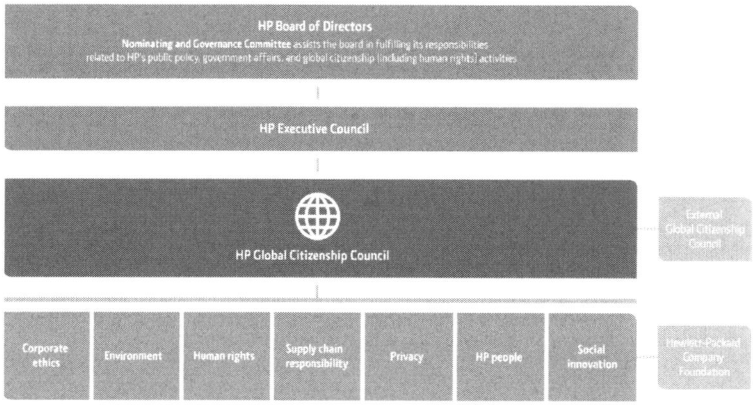

FIGURE 3.5 **Hewlett-Packard global citizenship governance**

Source: Hewlett-Packard (2012) *Global Citizenship Report*

Information flow and a seat at the table

So it is critical to facilitate communication and decision-making across the organization and to lower organizational levels to empower local managers and staff to improve sustainability performance. It is also important for senior sustainability managers to have a "seat at the table" and direct access to the CEO and the board for successful integration of sustainability into organizational decisions and processes. Information can then be provided in many ways:

- **Periodic written summary and sustainability reports.** Reports are circulated throughout the company discussing how the company is meeting established goals
- **In-person updates.** The top sustainability manager of the organization gives direct update to executives and the board. Reports should be made often enough to influence decision-making and time should be allowed for dialog
- **Use of executive committee.** The committee should consist of the CEO, COO (chief operating officer), CFO (chief financial officer), and other relevant top executives. Dell has a Global Sustainability Steering Committee made up of the chief executive, chief procurement officers, general counsel, and other managers who provide input into strategy, resources, and global policies for sustainability[18]

With any method, the lines of communication must be open from the company's managers and executives directly to the CEO and the CEO should be involved in setting social, environmental, and economic management policies and in making key impact decisions. In 2009, the VP of the Sustainable Business and Innovation Department became a member of the

Nike Strategic Leadership Team, which sets mid- and long-range corporate strategy. This change allows the Sustainable Business and Innovation Department to be integrated into decisions about new products and initiatives during the various phases of the innovation process, rather than tacked on to the back end after strategies have been drafted. In addition, the Sustainable Business and Innovation Department has the ear of the CEO and the attention of the heads of strategic business units. In the huge Nike campus, where over 7,000 people are employed, it sits in the same building and on the same floor as the CEO's suite.[19] What often matters most is not the number of people working under the top sustainability manager but the number of reporting levels above the sustainability officer.[20]

Where or with whom should the functional responsibility for sustainability lie? Some companies have a full-time sustainability leader, while others split leadership among several individuals or departments. The best location depends on the type of organization, its size, and its complexity.[21] The most important consideration is the level of authority given to the sustainability manager, who must be viewed by employees as passionate about and committed to sustainability. To be effective, the leader should be placed high enough in the organization to exercise influence and be involved in the company's strategic planning and development.[22] For example, DuPont's chief sustainability officer halted purchase of a company that was not in a sustainable business.[23] This can only happen when a sustainability manager is supported by top executives and given authority to manage sustainability across the company. As emphasized earlier, top management commitment and the "tone at the top" are critical in using the organizational structure, systems, and culture to effectively implement the sustainability strategy.

A decade ago it was common for the senior environmental officer also to be corporate general counsel because the focus was often primarily on protecting the company's assets. At that time, the only social, environmental, or economic concerns of many companies focused on outside pressures that might cause lawsuits from which the company needed protection. Many companies continue to maintain this structure with the global director of social and environmental affairs reporting to the general counsel.

However, more and more companies are creating senior executive roles and departments whose sole focus is on social, environmental, and economic responsibility and who report directly to the CEO.

Outsourcing

Some companies have chosen to outsource many of their sustainability functions to external providers. These may include strategy formulation, systems implementation, information aggregation, monitoring, reporting, verification, and other services. Some outsourcing companies guide organizations in developing metrics and applying for certifications.

A key decision for corporations is whether to seek the help of an outside consulting firm or facilitator in implementing sustainability strategies. Several questions must be answered when considering outsourcing:

- If a third party is to be involved, in what capacity: as a facilitator, a trainer, a partner, or an advisor?
- What internal resources will be committed to working with the third party?
- If the company decides not to solicit outside assistance, who will make up the internal team?

The answers to these questions will greatly influence whether the organizational structure around sustainability issues will succeed through outsourcing. As with outsourcing generally, companies must take care not to outsource critical core capabilities that should be an integral part of company strategy and success or that may be critical elements of differentiation leading to long-term competitive advantage.

Benefits of working with a third party

There are at least two main advantages to enlisting a third party's assistance in implementing a sustainability strategy and integrating it into existing management infrastructure. First, organizational inertia and ingrained business processes may make the initial transition to a focus on sustainability difficult. For example, as discussed in Chapter 4, many companies that use sophisticated financial tools for capital investment decisions throughout their organizations do not transfer these tools for use in decisions related to sustainability. A third party can step away from organizational politics and highlight the benefits of improved management systems.

Second, to identify, measure, and report external costs, the company will need to gather a significant quantity of data, which requires both time and access to appropriate information resources. In addition, the validity of conclusions depends heavily on the assumptions made and the methodologies employed. Thus, having an expert involved in the data collection can capitalize on information economies of scale, as outside parties will be gathering similar information for a number of different clients. Also, consulting companies have developed software tools and databases that can expedite the assessment of products' life-cycles and in turn provide the basis for comprehensive cost–benefit analysis.

Benefits of an internal team

Most companies rely primarily on internal staff for most of the design and implementation of their sustainability strategy. Using internal staff at the center of a sustainability strategy implementation effort has the advantage of directly involving managers in the transition to social, environmental, and economic accountability. Relying on a third party runs the risk of having social, environmental, and economic issues continue to be viewed as an activity outside the scope of business unit managers. While an expert's study may provide a large amount of data for use in decision-making, it is the managers within the company who must use that information if it is to have any impact on daily business activities. Thus, it is important to keep the entire organization aware and involved in the transition toward a more forward-thinking sustainability strategy.

The process of implementing a sustainability strategy may have positive externalities of its own. Because evaluating the entire life-cycle of a product or process requires communication across traditionally separate functional groups, the transition to sustainability may break down existing communication barriers between functions. This has broader benefits for companies that seek to improve efficiency in delivering products to market, in spreading best practices throughout the company, or in introducing other innovations.

As one considers the balance between relying on internal resources and outsourcing, the usefulness of a sustainability measurement system will vary directly with the quality of the information input into the system. This information can require substantial time and financial resources to develop. While outside experts may have access to existing databases for benchmarking, or may know where to find specialists in a given area, individual companies may find that obtaining this information is both difficult and expensive. But the identification and measurement of impacts is often most effectively completed with internal resources and can be an important element in the implementation of a sustainability strategy.

Philanthropy and collaboration with NGOs

In addition to using internal capabilities and external services as important components of sustainability strategy implementation, companies can also collaborate with NGOs to improve both sustainability and financial performance. The relationship with the NGOs can range from donations of cash or in-kind services to cause-related marketing efforts, to employee voluntarism, and significant joint projects.

As companies choose to include collaboration in their sustainability strategies, they may need to re-evaluate their use of human resources to organize for improved sustainability performance. This may include the creation of positions that focus explicitly on identifying philanthropic, community, and volunteer opportunities or positions that develop and coordinate volunteer and community programs for the organization. These efforts can involve employees from different departments, thereby integrating sustainability throughout the organization.

Integrating collaboration at Timberland

Timberland Company, a US-based clothing and footwear company, has focused much of its sustainability strategy on a partnership with City Year, a national youth corps. The partnership began in the late 1980s when Timberland donated boots to City Year corps members. After this initial donation, Timberland established its "Path of Service" community service program, offering employees up to 16 paid community service hours per year in 1992, and up to 40 paid hours in 1995, which continues today. The idea was to encourage employees to participate in volunteer activities, particularly with City Year. Over the years, Timberland's relationship with City Year has evolved. Timberland's CEO joined the City Year board of directors and City Year has established a location directly within Timberland headquarters. Timberland executives also realized that it needed to alter its organizational structure to integrate sustainability into the organization. It started the reorganization by hiring a director of social enterprise, who happened to be a City Year employee. Then the question arose as to where to house the Social Enterprise Department. Some members of senior management argued that members of the Social Enterprise Department could be located in an existing department, such as human resources. However, the decision was made for it to be its own four-person division within the marketing department. By separating the department, Timberland management could signal to the other members of the organization the importance of sustainability.[24]

To manage the collaboration efforts, it is common to designate a partner-relationship manager in both the corporation and the NGO. This person is responsible for coordinating and communicating with the partner organization. In their partnership, Visa, American multinational financial services corporation, and RIF (Reading Is Fundamental), the largest nonprofit children's literacy organization in the US, have staff dedicated to managing their relationship. Timberland and City Year, as discussed in the case study above, have account executives whose primary responsibility is to lead and manage their partnership.[25]

Corporate philanthropy is a direct contribution by a company to a charity or cause. It can be in the form of cash grants, donations, or in-kind services. These donations are critical to the operating budgets of many charities. However, they can also be beneficial to the corporations. Companies should think of philanthropy, in part, as a way of improving the business environment in which they operate, bringing social, environmental, and economic goals into alignment. Philanthropic activities have many benefits, including:

- Building reputation with respected organizations
- Creating community goodwill and national attention
- Strengthening the corporation’s industry
- Building and securing a strong brand position
- Having an impact on societal, environmental, and economic issues in local communities[26]

In deciding on philanthropic activities, it is helpful if companies think about the ways that their products and services can be used to enhance the environment, society, and the economy. Dow Chemical donated its THERMAX insulating foam to the Solar Oven Society to build 6,000 solar ovens. By using solar ovens, people in developing countries are able to cook without having to gather firewood and the hazards associated with the smoke caused from fires are minimized. Similarly, FleetBoston Financial used its expertise in financial services to donate US$725,000 to six cities where it operates. The bank provided small business financing packages to local companies and home mortgages and home-buyer assistance programs.[27]

The Boeing Humanitarian Delivery Flight program, a collaboration between Boeing, airline customers, and nonprofit organizations, delivers humanitarian supplies in the empty cargo space of new airplanes being delivered to customers. Throughout 2012, Boeing conducted humanitarian flights with airline customers EGYPTAIR, Emirates, Ethiopian Airlines, and Uzbekistan Airways, transporting more than 115,000 lbs (52,163 kg) of humanitarian supplies, including educational materials, medical goods, and winter clothing.[28]

Many companies have established corporate foundations to oversee much of their philanthropic activities. Corporate foundations usually have a special legal status and are initially established through a financial contribution from the company, and this investment is used to fund philanthropy.[29] In many cases, employees and managers of the company remain very involved with the foundation.

Boeing's Employees Community Fund (ECF) is one of the largest employee-owned and managed funds of its kind in the world. For more than 60 years, the ECF has enabled employees to make tax deductible donations to local nonprofits through recurring payroll deductions or one-time gifts. Funds are then distributed locally to nonprofit organizations across the country through employee advisory boards. All administrative costs are paid by Boeing, so every employee dollar donated goes directly to local nonprofits working to improve communities.[30]

Community volunteerism is another initiative used by corporations to get employees committed to sustainability. Participation in volunteer programs can increase corporate value. Recent studies have found that:

- Companies with volunteer programs received increased media coverage, enhanced media, and an 8% increase in sales related to community activities

- 37% of Americans cited employee volunteerism as the most impressive philanthropic activity[31]
- 91% of Americans would consider switching to a company that supports volunteer programs[32]

Volunteer programs also create value through their impact on employees. Employees who participate in company-sponsored volunteer programs have been shown to have: lower rates of absenteeism, lower turnover, higher productivity levels, higher levels of employee satisfaction, better team-building skills, and increased referrals for job openings.[33]

At THD (The Home Depot), the world's largest home improvement specialty retailer, over 300,000 employees, store managers, and associates are passionately committed to customers, colleagues, the company, and the community. Volunteerism is a critical building block of THD's corporate culture.[34]

Companies can support volunteer activities by providing paid time off, recognizing service, or organizing teams to support causes the corporation has identified. For example, Roche, a Swiss global healthcare company, encourages its employees to involve themselves in the communities in which they work and live, both as a means to give back and to engage in personally enriching experiences. Roche supports employee community service by offering flexible work schedules, brokering volunteer assignments, and coordinating group projects.[35]

Cause-related marketing is another method of working with an NGO while creating value for the organization. During cause-related marketing campaigns, corporations often donate a percentage of revenues to a cause based on product sales. Benefits from cause-related marketing include attracting new customers, increasing product sales, and building positive brand identity. This is particularly true when there is a good fit between the brand and the cause supported by the company.[36] In addition, it has the potential to raise significant funds for the cause.[37]

For example, Avon Products, a direct seller of women's beauty products, is the largest corporate supporter of the fight against breast cancer, with more than US$700 million raised and awarded globally by the end of 2010. One of the several strategies Avon uses to support breast cancer research is through its Pink Ribbon product line, which includes lipsticks, jewelry, and accessories. Funds from sales of the product line are funneled into Avon's breast cancer efforts. The company realizes the importance of transparency for this strategy and therefore it publishes the percentage of revenues that each Pink Ribbon product contributes to the cause. Through 2010, more than US$265 million in net proceeds has been raised globally through the sale of Avon fundraising Pink Ribbon products.[38]

Collaboration between companies and other organizations can also be used to encourage volunteer efforts, improve sustainability performance, and improve stakeholder reactions. Starbucks uses collaboration as a means to improve its social, environmental, and economic impacts. Coffee and Farmer Equity (C.A.F.E.) Practices, for example, Starbucks' comprehensive coffee-buying program that ensures coffee quality while promoting social, economic, and environmental standards, was developed in collaboration with Conservation International. It has created significant social and economic impacts for more than one million workers, and environmental improvements on the thousands of participating farms. Farms and mills are evaluated using a comprehensive scorecard of more than 200 indicators by third-party verification organizations, which are overseen by SCS Global Services. In 2012, 90% of Starbucks' coffee was C.A.F.E. Practices verified.[39]

McDonald's has been collaborating with Environmental Defense, a nonprofit organization dedicated to protecting environmental rights, to try to stop chicken farmers from using antibiotics to increase chicken growth, and with Eaton Corporation, a diversified industrial manufacturer, to encourage the use of hybrid vehicles.[40]

Vodafone, a British-based second-largest mobile telecommunications company in the world, partners with Pratham Education Foundation and the Vodafone India Foundation to improve the standard of education for schoolchildren in India. In 2011, the program called Learning with Vodafone was implemented in 150 schools by using innovative software and mobile internet to train teachers and help them engage students using interactive learning materials and multi-media content. The service made education more accessible to students wherever they were using mobile phones, tablets, or the Vodafone WebBox (a low-cost internet-enabled device that connects to a television). Together with Pratham, Vodafone provides all the necessary equipment and learning materials, and trains teachers to use the service. Learning with Vodafone provides access to digital educational content aligned with the prescribed curriculum. It includes multiple choice tests that can be completed via SMS text message and a notification service enabling teachers to check understanding of the content and keep parents informed of their children's progress. Teachers can also track attendance, grades, and administrative requirements using the accompanying school management system.[41]

For collaborative programs to flourish, corporations must consider the organizational structure that will best serve the programs. New structures may need to be developed to encourage collaboration. Time and resources can be wasted if the organization's structure discourages collaboration by limiting communication or making it difficult to obtain the resources necessary.[42] Organizations that are successful at collaboration have fluid boundaries to allow information to move quickly between members of the company and members outside the company.

Collaboration does present risks and challenges for the organization because of a lack of trust, a lack of communication, or different organizational cultures. Successfully managing these challenges is essential for effective collaboration. A company needs to be proactive in collaboration and to think strategically about which organizations it chooses to collaborate with, and how it will integrate collaboration into its sustainability strategy, structure, and systems.

Novartis Argentina, along with two NGOs, Cáritas Buenos Aires and Fundación Tzedaká, has instituted a drug donation program called Programa Novartis Comunidad. To determine which organizations to collaborate with, Novartis enlisted the help of the Argentine Catholic University. They conducted research and decided that the NGOs must meet two requirements: experience in handling donation of medicine to people living below the poverty line; and support for Novartis' corporate citizenship principles. This preliminary research helped identify the organizations that would best fit with the Novartis culture and strategy. Novartis attributes the success of the program to the values and experience it shares with the selected NGOs.[43]

Making collaboration work

To succeed in collaborations with NGOs, managers should:

Proactively pursue opportunities for collaboration
Managers should be proactive in building coalitions and identifying opportunities for collaboration. Engaging with corporate stakeholders such as customers and employees can help identify areas or issues where NGOs could provide assistance. It is important to identify an NGO that has significant credibility in the particular area. Routine searches for partners can help identify those that have a shared purpose or interest.

Ensure that the partnership creates value for each partner and society
Collaborators must be clear about the purpose of the partnership. A written statement of collaboration purpose can aid the organizations in staying focused on their common goals. A clear mission statement establishes the foundation for the partnership and communicates how each organization can contribute to value creation.

Recognize that the relationship requires commitment of time, talent, and resources
Each organization must be committed to the partnership. This commitment can be shown through the time, human resources, and financial resources dedicated to the partnership.

Compromise and trust are essential in establishing a good working relationship. Because trust takes time to develop, an organization should dedicate one to two years working with the NGO on the collaborative effort. Each organization should also assign an individual or a team dedicated to issues of mutual concern. By doing this, programs and communication can more easily be coordinated and the organizations can show a commitment to having a successful partnership.

Align structure, systems, and programs, as needed, to effectively manage the relationship

Just as with any other sustainability implementation, the structure, systems, and programs may need to be adjusted to effectively implement a collaborative project. The mission and strategy of the partnership should guide any changes that need to be made.

Use effective communication with each other and the community

Frequent communication is also critical. Communication should be open, honest, and consistent within and outside the organizations. Having a partnership manager, as discussed earlier, helps facilitate communication by providing one person who has the responsibility of coordinating communication between the organizations. Effective communication establishes a foundation of trust between partners and with the community.[44]

Summary

Companies should integrate social, environmental, and economic concerns into all areas of the organization. It is generally desirable for the senior sustainability officer to have direct access to both the board of directors and the CEO and not be in primarily a legal function. Organizations should provide adequate resources for the implementation and control of sustainability strategies. These include setting the appropriate structures for efficient alignment of human resources with sustainability strategies, as well as allocating technological and financial resources.

The alignment of the sustainability structure with the strategy is critical to improving sustainability and financial performance. It is important to assess the existing structure to decide the best way to integrate sustainability into the various functional and business units and whether a new department should be created. Having certain functions outsourced or using collaboration strategically are also important factors in deciding how the sustainability function should be organized. No single design is appropriate for every organization. What is critical is that the sustainability structure be aligned with the strategy and systems and encourage employees to include sustainability in their day-to-day decisions.

In the next chapter we look at how managers can use various financial management and risk assessment systems to support a sustainability strategy.

CHAPTER 4

Costing, capital investments, and the integration of sustainability risks

Once the leadership has established the corporation's sustainability strategy, it needs to implement that strategy through the effective use of various management systems. These systems are instrumental in achieving positive sustainability impacts, improving stakeholder reactions, and financial performance (Fig.1.4). The systems should take into account the organizational culture and the resources, both human and financial, available to the company. Sustainability systems should also include ways to implement strategy and measure sustainability performance. In this chapter we discuss:

- Capital investment decision systems
- Costing systems
- Risk assessment systems

In later chapters we will look at other organizational systems for sustainability, including systems for performance measurement, evaluation, incentive and reward, internal and external reporting, and verification.

The capital investment decision process

Few business decisions impact a company's long-term capabilities and operational strategies as much as capital investment decisions. Capital investment decisions influence innovation, productivity, costs, revenues, capacity availability, and quality. These decisions help to determine the company's competitive stance and long-term positioning. Most capital investment decisions require an evaluation of the cash flows associated with the costs and benefits of the decision, as well as a measure of risk. According to the AICPA (American Institute of Certified Public Accountants) poll, only 84% of companies do not formally integrate social and political risks in financial calculations and capital investment decisions.[1]

Techniques such as DCF (discounted cash flow analysis) are commonly used in corporations to evaluate general investment projects. Such techniques incorporate both the time value of money and the need to earn competitive returns on capital investments. But, in the case of sustainability projects, DCF analysis is often not used. Often, only the payback period for these investments is calculated, without consideration for the time value of money, the broad array of affected constituencies, or significant future benefits and costs associated with the proposed projects. Currently, when quantification of these risks is undertaken it is often underdeveloped and not monetized. Two principal factors contribute to this situation. First, the regulatory nature of sustainability investment projects and, second, the difficulty associated with the evaluation of social, environmental, and economic costs and benefits.

1. Regulatory nature of sustainability investment projects

Regulatory requirements are the driving force behind many capital investment decisions. Companies that are forced, because of government regulations, to invest in technologies that are more socially, environmentally, or economically responsible often do not adequately analyze the full range of social, environmental, and economic costs and benefits associated with the projects. They often evaluate a limited number of options and only check that standards and norms, such as a prescribed emission levels, are observed. The objective is often to adhere to regulations in the least expensive manner.

2. Difficulty in evaluating social, environmental, and economic costs and benefits

The analyses of risks, costs, and benefits related to social, environmental, and economic investment decisions are more complex because of the nature and timing of social, environmental, and economic costs and benefits. Future risks and benefits, such as a changing climate of sustainability awareness, changing technologies, changing costs of technology, future government regulations, long time horizons, and potential stakeholder pressures, increase the complexity of the capital investment decision-making process. Incorporating cost-management information derived using full social and environmental cost accounting or a life-cycle assessment can help managers identify and quantify impacts related to both current and future operations and current and future risks.

In some companies, large capital investment decisions are reviewed and are often subject to approval by sustainability managers before a final decision is reached. Evaluation criteria include the social, environmental, and economic impacts. Companies need to understand the costs and benefits of their activities, enabling them to make better capital investment and operational decisions. To improve decision-making, they should identify and inventory their natural resources and environmental assets including all the land and water owned by the organization, and the pollution or other environmental impacts for which it is responsible. They should determine the goods and services potentially available with these assets, since many have a significant market value, and then specify the potential value of these environmental assets.[2]

Some processes and outcomes related to the decision, such as alignment with the mission of the organization, improved community relations, or an improved regulatory climate as a result of the investment, may be difficult to quantify but should be part of the decision review process. Some companies use checklists for capital investment proposals that require managers to analyze and incorporate all sustainability impacts into the project proposal.

In Mitsubishi Corporation's strategic decision-making process, all loan and investment proposals are examined by the Executive Committee. The screening and review process takes into account both financial and legal risks as well as global environmental and social factors. All loan and investment proposals are, in addition, screened by the CSR and Environmental Affairs Office. This process is conducted using various international standards as a basis for assessment, including the Guidelines for Confirmation of Environmental and Social Considerations published by the Japan Bank for International Cooperation (JBIC) and International Finance Corporation (IFC) guidelines.[3]

Alcoa has established a capital expenditure review process for environment, health, and safety. A team evaluates the location to assess the level of risk and the ease of technology transfer. For example, when investing in a new aluminum plant in east Iceland, a sustainability team, including Alcoa executives and a number of outside experts, developed recommendations on environmental, health, safety, social, and community issues for the new facility.[4]

Sustainability issues also arise in plant closures. South Africa has instituted a policy requiring that social, economic, and environmental factors be integrated into the planning of mines, throughout the life-cycle. De Beers, a leading diamond company based in South Africa, had to decide whether to sell or close a low-producing mine. In making the decision, De Beers had to weigh the costs and benefits of each alternative. Selling the mine to a company with lower operating costs would preserve jobs and continued exploration would result in tax revenue for the government. On the other hand, closing the mine would enable De Beers to run a rehabilitation program and focus on skills development and establishing an alternative and sustainable solution. De Beers decided to sell the mine to an operating company, which operates for a probationary period during which it is required to demonstrate the value added to the community.[5] This decision was based, in part, on the ability of the buyer to deliver positive sustainability impacts, including employee training, and ensure a sustainable rehabilitation and closure process. De Beers has begun to develop a process that details guidelines and policies for closure, making future closure decisions consistent for all its operations. Planning for closure and rehabilitation should be included in the initial capital investment assessment, providing a clearer picture of the costs of capital investment.

Capital budgeting in small and medium enterprises

SMEs (small and medium enterprises) make little use of sophisticated capital budgeting techniques. The decision to adopt advanced manufacturing technologies is often not supported by any analysis tool but is driven rather by an act of faith in a new technology and the perception of an opportunity.[6] The use of a tool such as the payback method creates an important barrier to social, environmental, and economic investments since it typically ignores the time value of money. Given some of the characteristics of SMEs, this approach to capital investment decisions is not surprising. Indeed, factors such as lack of financial expertise, short-term vision, and shortage of capital partly explain their chosen approach.

As previously suggested, DCF analysis and the evaluation of social, environmental, and economic costs and benefits require some financial expertise. In the case of SMEs, there may be no staff specifically dedicated to financial analysis and that analysis may be delegated to external experts such as bankers or business advisors. Also, appropriate evaluation of social, environmental, and economic investments requires a long-term vision, while strategic management of SMEs is a process often oriented toward short-term profitability. Typical SMEs often suffer from a shortage of capital. Capital is usually supplied and ownership is held by an individual or a small group.

In addition, the owner/manager often does not have the time nor the resources to think strategically or manage with a long-term view. In larger corporations, organizational structures distribute responsibilities among different managers, allowing top executives more time to pay attention to long-term issues and strategic planning. Larger corporations have long used strategic planning to integrate the objectives and activities of their diverse business units. SMEs do not have the planning staff and personnel that larger corporations have, and managers of SMEs are often so busy fighting fires that they do not have time to think about strategic long-term issues such as social and environmental management.[7]

Costs in the decision-making process

One of the first steps in the approval process for making capital decisions is to evaluate the costs and benefits of the decision.[8] Companies are investing large sums of money in sustainability programs. Even more significant are the general capital investments and the fact that most companies do not have an adequate system for the identification and measurement of social, environmental, and economic impacts of new products, projects, processes, and facilities. In some instances, companies do not separately track or accumulate the social, environmental, and economic costs, and thus do not know the total amount or the causes of those costs.

Within a cost-management and decision-making framework, companies must distinguish and account for three categories of social, environmental, and economic costs:

- Costs (both current and future) related to past operations
- Current costs related to current operations
- Future costs related to current operations

Costs related to past operations

A substantial amount of social, environmental, and economic expenditure relates to management and production decisions that were carried out years or even decades ago. For example, companies have often been held liable for cleaning up pollution that was previously generated by the organization, or for product liability claims related to previous design and production choices. Also, as companies are increasingly required to manage post-consumer product use, they often incur additional costs that were caused when the product was produced, sold, and used. Costs related to past operations also often include liabilities for closed facilities with newly discovered environmental impacts and employment claims from prior employment.

Many companies did not anticipate these future liabilities, and therefore did not account for the costs at the time of production. As a company incurs these costs, it often includes them in current production costs, either as a direct charge to activities, processes, products, or services or through a manufacturing or administrative overhead allocation. But doing so distorts current product, facility, or division profitability and negatively affects performance evaluations and compensation. It also distorts decision-making, as managers are relying on cost information that mixes the impacts of current production decisions with the impacts of past production decisions.

Just as past product costs were understated because relevant future costs were not accounted for at the time of production, current product costs are overstated because they are now bearing these costs related to past production. Many believe that current managers should not be held responsible for costs that they do not control, such as costs related to past operations. Many sustainability and EH&S managers now recognize that including costs related to past production in current product costs is inappropriate to effectively measure the performance of products, facilities, and divisions.

One solution is to capitalize these costs and amortize them over a period of years, indicating that the expenditure provides future value and may relate to a "license to do business." This treatment also reduces current division and product profitability but has a smaller annual impact. Another solution is to charge these costs directly against shareholders' equity or to a corporate overhead account that is not allocated to divisions or products. Current product costs would not be distorted, and performance evaluation of the division would be based on costs related to the current operation. However, this treatment could lead to the business units showing a profit while the corporation shows a loss. It also does not highlight to managers the extensive life-cycle impacts of production decisions, and the importance of planning for future social, environmental, and economic impacts.

Current costs related to current operations

How to account for current social, environmental, and economic costs related to current operations is less controversial; these costs should be reflected in operating activities, processes, and products. However, the difficulty for many organizations has been to separately identify and account for those costs as social, environmental, and economic costs. In some companies, social, environmental, and economic costs related to production are accounted for as manufacturing overhead costs and are arbitrarily allocated to activities, processes, and products using a cost driver that does not reflect the relationship between the cost incurred and the activity, process, or product. Still other social, environmental, and economic costs are accounted for as administrative overhead costs, and are never allocated to activities, processes, or products. This makes it difficult to understand the social, environmental, and economic cost impacts of operational decisions, which again impedes effective decision-making. Tools such as lifecycle costing, activity-based costing, and full social and environmental cost accounting can help managers to better capture and assign these costs.

Future costs related to current operations

It can be difficult to accurately predict the future social, environmental, and economic benefits, costs, and liabilities related to past or even to current production. Estimating future impacts depends on many factors that may be unclear today, including changing social and legal structures. It is unlikely that Philip Morris understood, 40 years ago, that changes in the social and legal climate in the US would result in extensive product liability costs for cigarettes. Recognizing potential future liabilities may cause a company to modify its strategy, product or production processes, or its accounting and management decisions.

The difficulty of predicting changes that may occur in the social and legal climate, along with the inability to reasonably estimate and measure the economic impact of those changes, is one reason why many future costs are not accounted for in the formal accounting system. However, there are some future costs that can be reasonably understood and should feature in the decision-making process, such as post-consumer use and recycling costs, disposal costs, facility decommissioning costs, natural resource restoration costs, and risk and legal liability costs. Other costs that are less predictable, such as those related to changing social and legal structures or reputational costs and the changing costs of technology, also need to be factored into the decision-making process.

Many managers find that practices such as life-cycle analysis and full social and environmental cost accounting are useful in helping them to identify and evaluate the longer-term impacts of current decisions. Other approaches identified in this book provide ways to measure and integrate social, environmental, and economic costs and benefits into operational and capital investment decisions.

Costing systems

Identifying the full range of corporate sustainability impacts is an important step toward better management decision-making. Once identified, the impact of these costs on the company's activities, processes, products, and services can be analyzed using available tools. A number of companies have begun the transition to improved social and environmental cost accounting in two ways: by clarifying their understanding of internal social and environmental costs through ABC (activity-based costing), and by placing a value on significant external costs, through LCC (life-cycle costing) or other approaches. Other companies have chosen to use FCA (full cost accounting) to include a broader set of external costs along with future costs into management decision-making.

Activity-based costing

Two often-stated reasons for unreliable accounting data are the tendency to allocate social, environmental, and economic costs to overhead and the tendency to combine social, environmental, and economic costs in cost pools with nonenvironmental costs. This hampers management's ability to assess social, environmental, and economic costs and make informed decisions. For example, AMP Ltd., an Australian-based global financial services organization, analyzed its environmental accounting and identified areas where costs were being inaccurately aggregated. Costs for waste collection and disposal and wastewater were included in the rent expense paid for buildings. The aggregation of these services made it difficult to identify opportunities to reduce waste and its associated costs. The company conducted a waste audit of one of its offices and identified that general and kitchen waste could be reduced by 65–80% through recycling.[9]

Increasingly, companies have seen the benefit of methods such as ABC to identify, measure, and track social, environmental, and economic costs and to assign them to activities, processes, products, services, customers, and channels. While traditional cost accounting assumes that producing products and services causes costs, ABC assumes that activities performed for products, services, and customers cause the costs. ABC first assigns costs to the activities performed by the organization (direct labor, employee training, regulatory compliance), and then attributes these costs to products, customers, and services based on a cause-and-effect relationship.

Better cost management requires the accumulation of social, environmental, and economic costs and tracing those costs to the activities that cause them. Carefully identifying all social, environmental, and economic costs has often produced totals that are four to five times the estimated amounts. These costs often hidden in manufacturing overhead include: permits, penalties and fines, water and air treatment costs, energy costs, waste treatment and disposal, training, inspections, and protective equipment. Also frequently overlooked are social, environmental, and economic costs that are buried in administrative overhead, such as record-keeping costs, community relations costs, site studies, legal costs, and audits. By attributing social, environmental, and economic costs to the activities that generate them, managers and employees can be motivated to find alternatives that lower those costs and increase profitability.

An ABC methodology provides detailed activity-cost and related information, and is especially useful for an organization that has many social, environmental, and economic costs embedded in its manufacturing and administrative overhead cost structures, and that also has some degree of either process or product variation. An ABC analysis provides a better understanding of a company's costs, links social, environmental, and economic costs to management objectives and activities, improves decision-making, and supports full cost accounting as well as LCC.[10]

Life-cycle costing

LCA (life-cycle assessment) is a design discipline used to minimize the environmental impacts of products, technologies, materials, processes, industrial systems, activities, or services. LCC, an extension of the basic LCA, attempts to identify all the costs—internal and external—associated with a product, process, or activity throughout all stages of its life. Life-cycle cost has been defined as the amortized annual cost of a product, including capital costs, and disposal costs discounted over the lifetime of a product.[11] With regard to social and environmental costs, LCC consists of monetizing social and environmental impacts throughout a product's life-cycle. It requires the measurement of present and future costs and benefits of a company's products, services, and activities and can be an important part of the implementation of a sustainability strategy.

Canon assesses the CO_2 emissions of its products over their entire life-cycle (Fig.4.1) and implements concrete plans based on findings. Entire life-cycle CO_2 emissions in 2012 were approximately 4,890,000 tons, an approximately 14% decrease over 2011.[12]

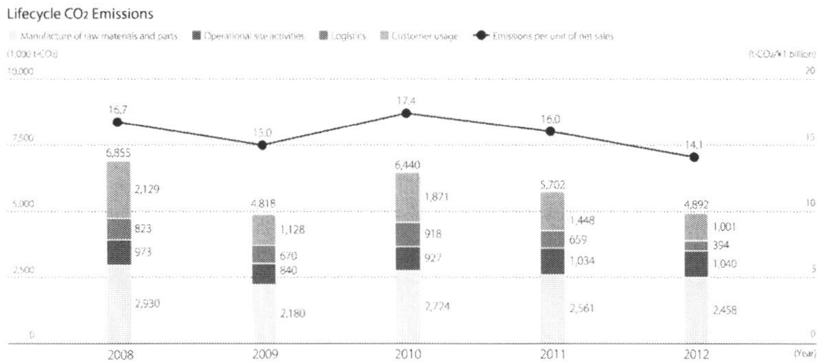

FIGURE 4.1 **Canon evaluates life-cycle CO_2emissions of its products**

Source: Canon (2013) *Sustainability Report*

Full cost accounting

Some companies use FCA to include a broader set of external costs along with future costs into management decision-making. FCA allocates all direct and indirect costs to a product or product line for inventory valuation, profitability analysis, and pricing decisions. In other words, LCC translates social and environmental performance into financial currency, and FCA integrates these values into the framework of accounting. For example, Baxter International calculates and reports its positive and negative sustainability impacts as subsets of traditional accounts, allowing sustainability items to be easily identified.[13] The combination enables managers to integrate sustainability impacts into decisions such as product costing, product pricing, capital investments, product design, and performance evaluations.

An FCA framework allows for consideration of external or societal costs and benefits (e.g. costs to human health and the natural environment) along with internal or private costs and benefits in the decision-making process. This requires a company to integrate present and future social, environmental, and economic impacts into its process and product costing system, including costs related to contingent liabilities and image and relationship costs and benefits.

FCA adapts existing management decision support systems to accommodate the new information generated through LCC. An important element of FCA is the consideration of future social, environmental, and economic costs and allocation of these costs to products. Then, present and future environmental costs should be integrated into the product costing system.

Full cost accounting versus full cost pricing

A common misconception is that FCA implies the expression of full costs in prices as well. It is important to separate the decision to adopt FCA methods (or any of the other methods discussed here) from the decision to incorporate these costs into product pricing. Corporations should adopt FCA so that they will better understand both the present and future costs of current production and can use that information to guide decisions throughout the value chain. Whether to make this new information transparent in product pricing is another issue. Prices may continue to be determined by the market, but an assessment of the company's profitability must use more complete information about present and future social, environmental, and economic costs.

Summary of costing systems

Companies are increasingly trying to improve their costing of social, environmental, and economic impacts. In 2003, Canon introduced a program in which each department bears the financial burden of its own waste processing. Prior to this program, the general affairs division handled all the costs of waste disposal. In the new program, waste, including papers and plastics, generated by each workplace is collected at a recycling center where the department, type of waste, and amount are recorded. Each department is then assessed a waste-processing fee for the waste produced.[14] Using a full environmental costing system is beneficial because:

- Many environmental costs can be eliminated by simple changes
- Some environmental costs add no value to the process or product and usually constitute cost savings
- Understanding the environmental costs can lead to better pricing and creation of value of goods and services[15]

Part of the reason that more companies have not adopted FCA is the difficulty in valuing social, environmental, and economic impacts.[16] However, an estimation of these impacts (discussed in Chapters 6 and 7) can help companies internalize external costs. As companies improve the costing of social, environmental, and economic impacts, they gain a clearer understanding of the complete costs of products, services, processes, and other activities. This should lead to a better understanding and improved management of both sustainability and financial performance.

Risk assessment

Today, risks are both larger and more varied than previously thought and have been seen in companies and countries that thought they were shielded.[17] With globalization increasing rapidly, a common challenge is how to integrate social, environmental, and political risks such as political instability, political corruption, business corruption, child labor practices, anti-corporate sentiment, terrorism, and environmental pollution into management decisions.

Some businesses are prone to social, environmental, and political risks because of the location of their facilities, their product and customer characteristics, the nature of their employment relationships, or industry characteristics. Well-known examples include companies such as Nike, Walmart, and Shell, and the notorious social, environmental, or political risks associated with industries such as mining, footwear, apparel, toys, and chemicals. Also, varying social, environmental, and political risks, and degrees of risk, affect companies located in specific countries or regions of the world. More globally, devastating terrorism attacks such as that on September 11, 2001 have dramatically increased risk, resulting not only in a terrible impact on individuals and governments but also in an overwhelming impact on businesses.

Understanding what the critical components of ongoing business operations are, and planning for disruptions in these processes, increases organizational resilience. In addition, innovation is a critical component of mitigating risk and creating value. Creating an innovation strategy and the management control systems within which to develop this innovation is part of the process that balances defensive risk mitigation and offensive opportunity capture.

Corporations hoping to properly manage risk require more analysis, evaluation, preparation, mitigation, and response planning.

To weigh up the costs and benefits of a capital investment project, corporations need to identify and measure the social, environmental, and political risks of the decision. The development and implementation of an appropriate model for decision-making and measurement of social, environmental, and political risks are critical for improving the ability of organizations to more effectively anticipate, prepare for, mitigate, evaluate, and manage alternatives.

Effective risk management includes identifying the corporate environment that might impact the risk, identifying risks, evaluating potential effects, measuring these impacts, identifying and analyzing possible solutions, adopting the most appropriate solutions for managing risks, communicating results, and monitoring risks as they continue to evolve. Figure 4.2 shows a general risk management process model.

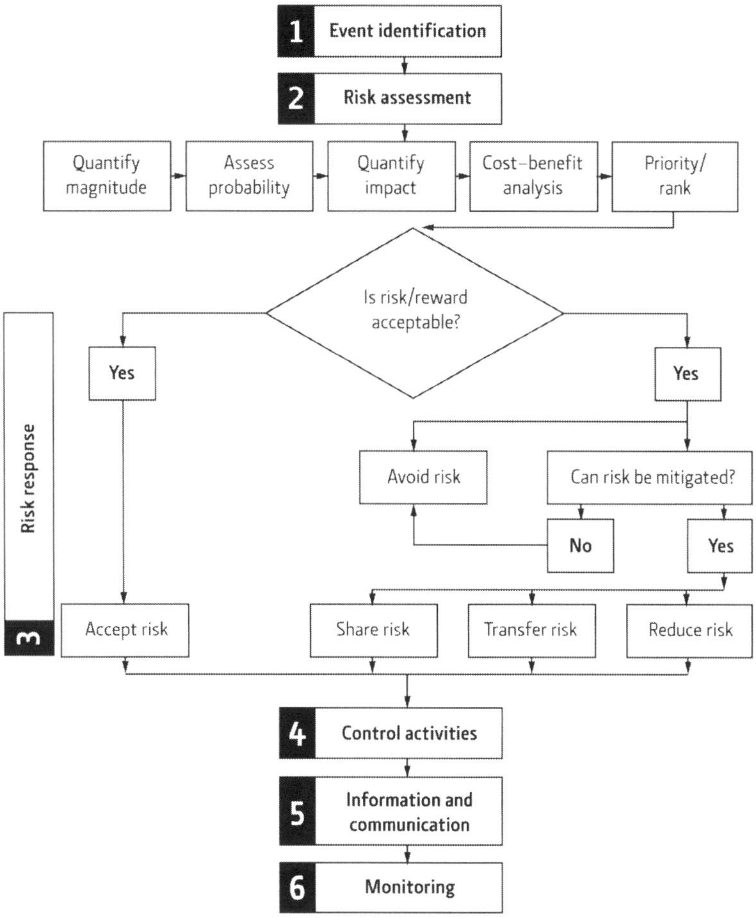

FIGURE 4.2 **Risk management process**

Source: Epstein and Rejc (2005) *Identifying, Measuring, and Managing Organizational Risks for Improved Performance*

Complex social, environmental, and political issues often affect company operations. Identifying risks that can affect company value is an important first step in managing social, environmental, and political risks.

Sustainability risks

There are many social, environmental, and economic issues that can impact a company doing business in the international context, particularly in developing countries. Some industries are more prone to experiencing these risks than others. For example, businesses with large installations such as factories, mines, and refineries can be the target of unrest in a local population when:

- There is a perception that local expectations are not being met
- The surrounding area is being polluted
- Business is undertaken in a region of general political unrest, where the military is protecting a site and using its presence to harass the local population for reasons unrelated to the business. The local population may associate the company with these practices, and target it as a proxy for the government or the military

If any of these are issues in ongoing operations or may potentially emerge when a new plant, road, or mine is constructed, they will constitute a red flag and be listed on the company's or project's risk catalog for consideration, measurement, and mitigation.

Rio Tinto, a leading international mining group headquartered in the UK, developed a screening process to understanding the social, environmental, and economic implications of its new investments so that it can optimize benefits and reduce negative impacts for local communities and for regional and national economies. Identifying potential risks and opportunities as well as evaluating social impact assessment is part of this screening process.[18]

Sometimes, the arrival of a large company in a relatively isolated or underdeveloped area creates unintended consequences. These can include unmet community expectations or a sudden influx of people, often unskilled, looking for work. When no jobs are available, some turn to violent behavior. For example, unemployed Nigerian youths have been known to attack oil pipelines and, in one case, seize an offshore oil rig, demanding that they be given jobs. Another potential unintended long-term consequence of new operations is that the local economy focuses exclusively on one industry over a long period of time and becomes dependent on one company, operation, or industry sector. When the company leaves or the industry is eclipsed, the surrounding area is often economically devastated. This has happened frequently in company towns in the US and abroad in industries ranging from steel milling to coal mining. For example, Detroit, Michigan, saw a tremendous dip in its standard of living when the automotive industry became less competitive. Although this is less of a problem for some industries, companies that plan to be in one location for the long term have adopted a risk-mitigation convention of planning an exit strategy covering social, environmental, and economic issues to diminish negative consequences.

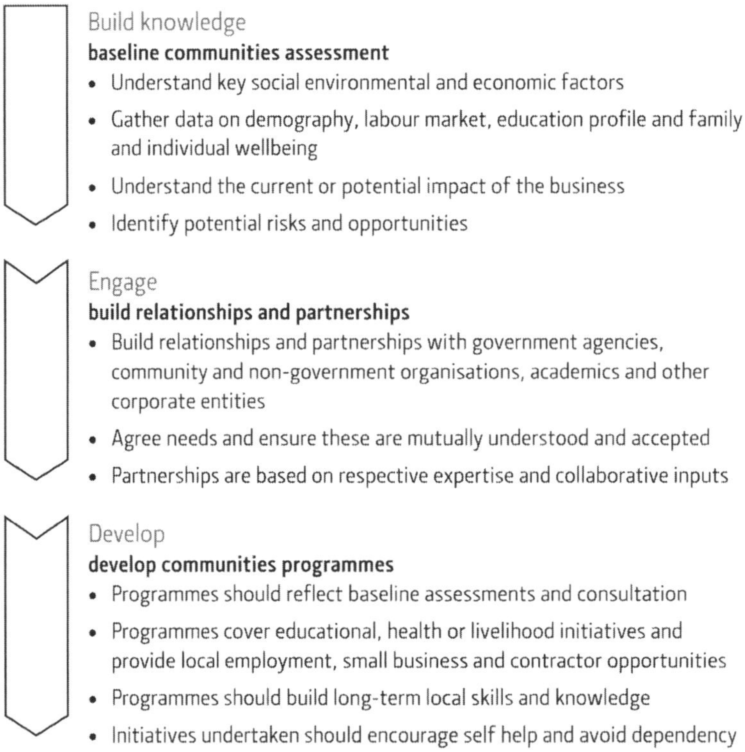

FIGURE 4.3 **Rio Tinto's approach to identifying risks and opportunities of new investments in local communities**

Source: Rio Tinto (2012) *Sustainable Development Report*

Political risk

Political risk, generally, can be understood as when political power is exercised in such a way that it threatens a company's value. Mass antigovernment protests, then, may not be considered a political risk to a company if they have no effect on current or future operations or value. However, changes in the legal framework governing contracts could have a significant negative impact on the company. There are two types of political risk that are relevant to corporations doing business internationally: company-specific and country-specific.

Company-specific political risk is directed at one organization, such as the government nationalizing an oil company or terrorists targeting a plant. Country-specific political risk does not affect just one company but rather is spread widely across a country. It can include a civil war, drastic changes in foreign currency rules, or sweeping changes to the tax code. These two types of risk can be generated directly from the government of the host country or emerge from an unstable social situation within the country. Regardless of the source, understanding political risk as it affects a company means recognizing the difference between political issues that can impact on corporate performance and situations that appear dramatic but have no financial impact on the company.

Developing a risk profile

The first step in risk management is to identify risks facing the company and integrate them into a larger risk management framework. There are three steps to this process:

1. Identify background risk sources

2. Identify real versus perceived risk
3. Identify company-or project-relevant social, environmental, and political risks

1. Identifying background risk sources

When developing an integrated risk profile that includes social, environmental, and political issues, it is critical to acknowledge the differences depending on a company's sector, industry characteristics, product, customers, geographic location, and employment. For those companies that must or choose to operate in risky environments, identifying these risks is the first step to accounting for and managing them effectively. Based on studies and campaigns, De Beers, the world's leading diamond company, estimated in 2006 that 10.2% of its employees in South Africa could be HIV-infected. An impact analysis revealed a cost to the company of between 1% and 2% of gross payroll over the next 10–14 years when the full impact of the disease would be realized. This included direct employment-related costs such as absenteeism, lost productivity, medical costs, training, and replacement costs, medical incapacity costs, and the costs of HIV/Aids management intervention, including treatment. De Beers calculated that the cost of inaction would ultimately far exceed the cost of its Anti-Retroviral Treatment Program and treatment costs. While De Beers' strategic approach to HIV/Aids was borne out of care and concern for its employees and the communities in which the company operates, it was also based on sound commercial considerations.[19] De Beers' striving for a holistic strategy for employee wellbeing and community development started in 2005 with an industry-leading HIV/Aids strategy and its ongoing support of HIV/Aids prevention and treatment.[20]

Developing a risk profile is an important step in better management of sustainability and political risks. Beyond identifying threats, the risk profile can also identify sources of opportunity for innovation. The identification of sources of risk (Fig.4.4) helps to hone the list of issues that could impact the company and enables managers to manage the issues at their source. When these risks and opportunities are identified and prioritized, managers can decide how the company should respond, innovate to beat the odds, or change the market entirely.

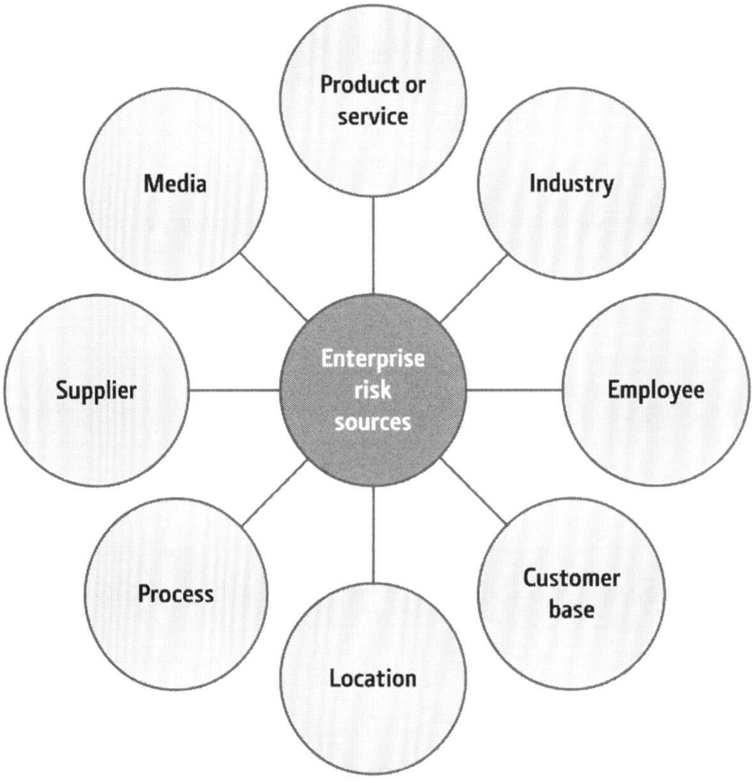

FIGURE 4.4 **Sustainability and political risk sources**

Source: Bekefi and Epstein (2006) *Integrating Social and Political Risk into Management Decision-Making*

In order to identify sustainability and political risks that may impact a company or product, managers must understand the setting in which they are doing business, and how that might generate risks. This process does not necessarily have to be costly and time-consuming, though investment in risk identification would probably be correlated to the size and importance of the project. The sustainability and political risks generated by product, customer, geographic location, employee base, and industry characteristics, as well as examples of industries that have been affected by these issues, are seen in Table 4.1. Risk can be divided into: (1) risks to society (and the environment) that could create dissatisfaction; and (2) other issues that could negatively affect and so pose a risk to the company. Analyzing the characteristics of these two kinds of risk helps companies to understand their potential impacts on the company or project, a critical first step in developing a risk profile and estimating the effect on profitability.

2. Identifying real versus perceived risk

Social, environmental, economic, and political risks can be grouped into two categories: real and perceived. It is important to identify real and perceived risks, as well as their sources, in order to manage them effectively. Though both real and perceived risks may carry financial costs to the company, the preparation and response to these risks differ.

- **Real risk** includes all social, environmental, and political issues that occur either to the corporation or because of corporate operations and impact the business
- **Perceived risk** includes all issues that stakeholders, including consumers, employees, and communities, consider a company responsible for, whether or not evidence bears out the perception

For example, the local population in Cajamarca, Peru, considers mining at Yanacocha, owned by Newmont Mining and Compañía de Minas Buenaventura, to be responsible for the contamination of drinking water and the depletion of water supplies. This is despite two independent environmental audits of the region that show this is not the case. Local distrust is at such a high level that, when Newmont began exploration of a nearby mountain as a potential for further mining activities, rioting began and the company abandoned its expansion.

There are three alternatives of real and perceived risk that can impact companies.

Real and perceived risks. Some issues are real and perceived. When a Union Carbide plant in Bhopal, India, leaked methyl isocyanate in 1984, approximately 3,800 people in the surrounding community died and several thousand others were permanently disabled. What followed was nearly eight years of court battles, both in the US and India, establishment of the Bhopal Trust, creation of a local hospital and ongoing anti-Union Carbide rallies where victims of the disaster continued to fight for compensation and medical care. Immediately following the leak Union Carbide's stock fell 12 points[21] destroying 27%, or almost US$1 billion, of its market value.[22] Also, opinion polls of Americans aged 18 to 29, its potential recruitment base for hiring, were uniformly negative.[23] By the time it was acquired by Dow Chemicals in 1999 for US$11.7 billion, Union Carbide's employee base had eroded from over 100,000 in 1984 to 11,600.[24] Union Carbide's experience illustrates how manifestations of risks, even those that take place far away, can destroy a company.

TABLE 4.1 Examples of risk sources and correlated potential risks

Source	Examples	Risk: Risk to society	Risk to companies	Potential company accountability
Product	Diamonds	Diamond trade and revenues being siphoned off by corrupt governments and rebel groups, thereby fueling civil wars in Africa, e.g. Angola	Reputation: profits from legal diamond mining fueling civil wars in Africa	Accusations of profiting from trade of "conflict diamonds" may impact sales and product reputation. Consumer boycotts/protests and pressure from employees (both existing and potential)
	Petroleum products	Negative environmental impact	Reputation: fossil fuel emissions correlated to climate change	Imposition of legislation to manage emissions creating a cost to the company

Source	Examples	Risk: Risk to society	Risk to companies	Potential company accountability
	Shoes, clothing, toys	Potentially poor working conditions, including long hours and little pay	Reputation: accusations by consumers of sweatshop conditions leading to boycotts of products	No legal liability but cost of negative impact on public opinion once the issue becomes public. Consumer boycotts/protests and pressure from employees (both existing and potential)
	Chocolate	Slave labor, child labor, and people trafficking in West Africa	Reputation: boycotts of products and bad publicity connected to use of slave and child labor, as well as human trafficking	Lawsuits under the Torture Victims Protection Act and the Alien Tort Claims Act (U.S. court). Consumer boycotts/protests and pressure from employees (both existing and potential)
	Chemical	Negative environmental impact	Fines by government, lawsuits, remediation	Imposition of fines, legal demands for remediation

Source	Examples	Risk: Risk to society	Risk to companies	Potential company accountability
Customer	Advocacy consumers (particular correlation with products produced in developing markets)		Reputation issues	Potential reactions to perceived "bad behavior:" product boycotts and/or lawsuits
Geographic location	Stable developing country		Potential for corruption, which creates difficult situations when trying to uphold home-country law, legal framework where contracts cannot be enforced	Overstepping the Foreign Corrupt Practices Act (U.S.) or similar home-country anti-corruption measures, tainted in-country courts where law is not applied equally, which creates uneven playing field

Source	Examples	Risk: Risk to society	Risk to companies	Potential company accountability
	Unstable developing country	Government is supportive of company but local population could be dissatisfied and generate social and political risks	All of the above and targeting by predatory government and/or by insurgents, etc.	Potential for nationlization of assets (Bolivia LNG, Venezuela industry more generally), targeting of infrastructure by insurgents if seen as colluding with government (e.g. Colombia)
Employee base	Children	Working at young ages (though this may not be as much of an issue for local population where the alternative to child labor is child prostitution or homelessness)	Reputation: anger in consumer markets about use of child labor	

Source	Examples	Risk: Risk to society	Risk to companies	Potential company accountability
	Women	Exposure to hazardous materials that cause birth defects, social issues with women working	Loss of skilled employees and potential for reputation impact: health problems among employees and potential that consumers will react negatively to these conditions	
	Men	Working, particularly in extractive and transport industries, away from their families and communities risk exposure to STDs and increased drug and alcohol use. The STD is often transmitted to others, including women along the trucking route or wives	Increased absenteeism due to illness, higher turnover of workers due to HIV-related deaths, more industrial accidents because of intoxication by drugs or alcohol	

Source	Examples	Risk: Risk to society	Risk to companies	Potential company accountability
	Nondiverse workforce in North America, Europe		Lawsuits field under laws such as the Racial Discrimination Act, the Sex Discrimination Act. Reputation loss that impacts ability to hire talented workers	Legal fees, inability to recruit, potential obstruction in capturing growing minority markets
Industry characteristics	U.S./European-based firms		Threat of terrorism because businesses are a target	Loss of infrastructure, interruptions in production or getting product to market, loss of employee life

Source: Bekefi and Epstein (2006) Integrating Social and Political Risk into Management Decision-Making

Real risks but not perceived. Some issues are real but not perceived by either the company or by society at large for some time. In the short term the risk is not real, but once stakeholder perceptions change, either because they are more informed or because sensibilities have shifted, the risk becomes manifest and can become a longterm issue for the company. Sometimes stakeholders identify the risk before the company understands what is at stake and the company is taken by surprise, with very negative consequences.

Coping with political and environmental risks at Coca-Cola

Coca-Cola worked in Kerala, India, for years and its water use was not recognized as an issue by the company, by its stakeholders, or by the public at large. Then, in the mid-1990s, residents of 50 villages surrounding Coca-Cola's bottling plant claimed that the company was siphoning off drinking water and depositing waste with high cadmium and nickel content in the surrounding areas. Soon international activists joined in. Though the final outcome of the legal battle may be unclear, information about the issue has spread quickly and efficiently to both the US and Europe and has sparked anti-Coca-Cola protests. As a result of the ongoing lawsuits, trouble with the local community, and worldwide protests, Coca-Cola did a cost–benefit analysis weighing the revenues produced by doing business in Kerala against the cost to its reputation worldwide and decided the price it was paying was too high. As a result it decided to leave Kerala.[25]

By 2011, Coca-Cola's system in India had achieved full balance between the groundwater used in beverage production and the groundwater replenished to nature and communities—ahead of its global target date. Bottlers throughout India have improved water use efficiency by 25% since 2005. Coca-Cola's efforts to replenish groundwater in India are focused on harvesting rainwater, constructing check dams, restoring ponds and other natural water bodies, as well as supporting agricultural water efficiency improvements. For example, at the end of 2011, Coca-Cola had installed more than 600 rainwater-harvesting structures across 22 states in India to capture monsoonal rains for aquifer storage. NGOs and local communities help Coca-Cola's bottlers identify priority areas, implement projects, and mobilize community members to ensure local input in project planning and assessment.[26]

Perceived risks that are not real. Some issues are perceived but not real and Shell's Brent Spar experience in 1995 is a case in point. The company decided to dispose of its decommissioned oil platform by sinking it in the North Atlantic. In response, Greenpeace activists carried out intensive campaigning in northern Europe claiming that Shell was being environmentally irresponsible and that sinking the Brent Spar would dump 5,500 tons of oil in the sea, wreaking havoc with the environment. In addition, 25 activists occupied the platform and Greenpeace encouraged boycotts of Shell stations that resulted in some violent attacks and threats to Shell workers. In the face of falling sales and a drop in share price, the company commissioned a third party to investigate Greenpeace's allegations, which turned out to be inaccurate and led to an apology by the activists to the company. The Brent Spar incident cost Shell US$100–170 million, when short-term loss of sales was considered.[27] This calculation does not include damage to the company's reputation, time by management dedicated to managing the Brent Spar incident, the internal company resources applied to the Brent Spar, or the total cost of the diversion to ongoing operations.

3. Identifying company-or project-relevant sustainability and political risks

Company- or project-relevant sustainability and political risks can vary, depending on specifics such as location within a country, and can be more nuanced than the more general risks discussed above. The discussion of risk management often focuses on financial issues, to the exclusion of other equally important matters. Financial risk is usually managed in a specific and established manner, but the intensification of social, environmental, economic, and political risks often requires an integrated risk management process across a firm to adequately identify emerging issues. Although the CEO and the board are the ultimate risk managers in a company, many different employees can integrate risk management into their jobs. Personnel who may become aware of risk at an early stage include a line manager at a plant in a developing country. He or she may be aware of negative community reactions to the corporation through discussions among workers, or from personnel in public affairs who learn of negative government attitudes to the firm while lobbying. These first signals can herald much larger issues if left ignored.

De Beers uses a sustainability risk matrix (Fig.4.4) to evaluate the short-, medium-, and long-term risks that are most material to its stakeholders and to De Beers' business. The matrix is developed using outputs from the company risk identification process and from engagement with company stakeholders. Business risk (y axis) includes both the likelihood of a risk manifesting itself and the potential impact that it would have on the company's business (i.e., direct risk). Stakeholder materiality (x axis), on the other hand, is based on the degree to which stakeholders consider an issue relevant and material, and the ability of those stakeholders to have an impact on the company's business (i.e., indirect risk). In 2012, this process identified a total of 29 sustainability risks. The upper right quadrant represents high-level risks that are prioritized for reporting.[28]

To anticipate social, environmental, economic, and political risks, personnel must be aware of what constitutes risk to the firm, and understand how to identify these risks. Identification is a two-step scanning process:

1. Generate a **risk profile** for the corporation. A variety of risks can materialize from factors such as sector, industry characteristics, product, customers, geographic location, and employment.

A risk profile is simply a list of risks generated from these contextual issues

2. Generate a **risk catalog** for the corporation. Risks can be specific to a particular project or location. For example, a new shoe-manufacturing plant in Bolivia will face different risks than an existing refining facility in Oman. A risk catalog is simply a more specific list of "red flag" issues, developed from the general risk profile, that are connected to a certain project or location

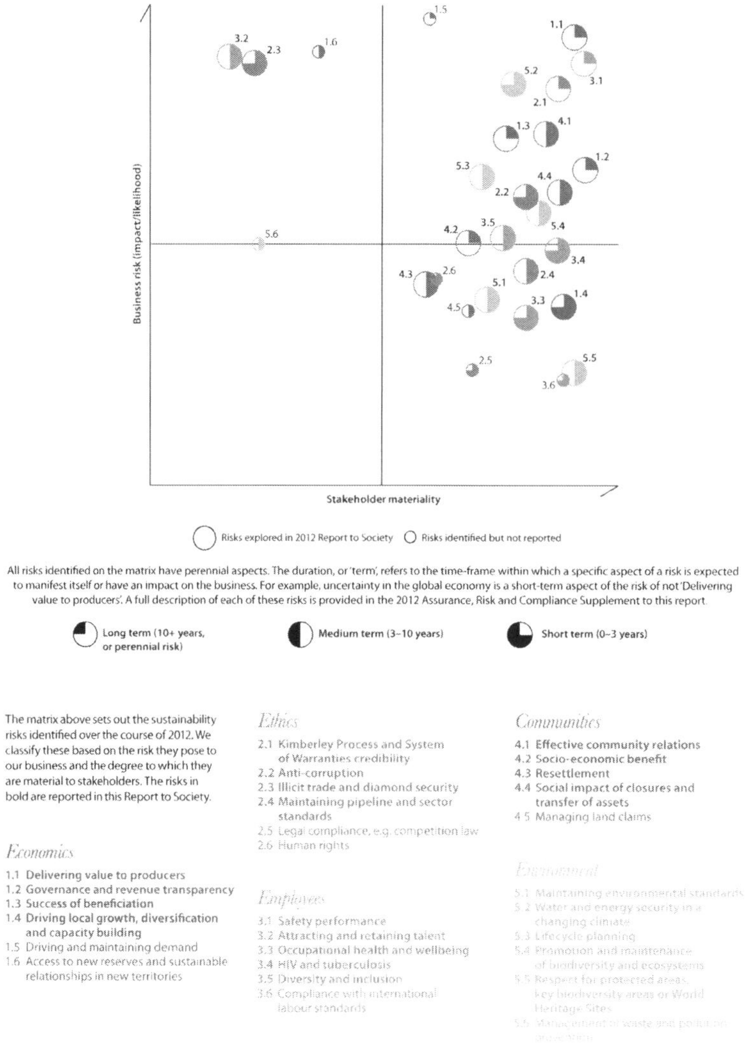

FIGURE 4.5 **De Beers' sustainability risk matrix for 2012**

Source: De Beers (2012) *Report to Society*

The Equator Principles

Developed in 2003 by ABN AMRO, Barclays, Citigroup, and West LB with the IFC (International Finance

Corporation), the World Bank's private-sector lending arm Group IFC, the EPs (Equator Principles) emerged in response to growing awareness of the risks of socially and environmentally irresponsible investing. EP is a risk management framework adopted by financial institutions for determining, assessing and managing environmental and social risk in projects, and is primarily intended to provide a minimum standard for due diligence to support responsible risk decision-making. Since the launch of the EP framework in June 2003, and a subsequent revision in 2006, there has been significant growth in the number of EP adopters from the original 10 to 75 financial institutions from 32 countries across the globe.[29] A newly revised and updated set of EPs was released in June 2013.

The principles apply to four financial products when supporting a new project: project finance advisory services where total project capital costs are US$10 million or more; project finance with total project capital costs of US$10 million or more; project-related corporate loans; and bridge loans. The EPs apply globally and to all industry sectors. Projects are categorized by their degree of environmental and social impact. Projects can fall into one of three categories, A, B, or C, with category A indicating the highest degree of environmental or social risk. EPFIs (Equator Principle Financial Institutions) require their borrowers to demonstrate in their social and environmental assessments, and in their action plans, the extent to which they have met the applicable World Bank and IFC sector-specific EHS Guidelines and IFC Performance Standards, or to justify deviations from them. EPFIs will insert into the loan documentation for high-and medium-risk projects covenants for borrowers to comply with the action plan. When a borrower is not in compliance with its social and environmental covenants, EPFIs will work with the borrower to bring it back into compliance to the extent feasible, and, if the borrower fails to re-establish compliance within an agreed grace period, EPFIs can determine the solution they consider appropriate.[30] Institutions that adopt the EPs should be able to better assess, mitigate, document, and monitor the credit risk and reputation risk associated with financing development projects.

The Mikheyevsky copper mine project was a greenfield development by Ruskaya Mednaya Kompaniya (RMK), a Russian mining company. HSBC, one of the world's largest banking and financial services organizations, lent US$100 million to Gazprombank, one of the top three banks in Russia, which used the funds to support RMK to buy mining equipment for the project. At HSBC, the policy was always to treat corporate loans for projects in the same way as project finance loans as the impacts are the same. By following the EPs, HSBC found that the project had potentially significant impacts on the environment and on the mine workers, including the possibility that dust from the mine could be harmful to them. To achieve safe operations, it was necessary for RMK to devise technological measures to suppress dust and minimize the potential danger to employees. Applying the EPs to this project meant that potential environmental and social impacts were properly addressed.[31]

Summary

Techniques are currently available to incorporate social, environmental, and economic costs, benefits, and risks into operating and capital investment decisions. Project and product decisions can be improved by:

- Identifying and measuring a broad set of social, environmental, and economic benefits and costs and considering current and future impacts on both the company and society
- Integrating all current and future social, environmental, and economic costs and benefits into decisions
- Integrating the assessment of social, economic, political, and environmental risks into the evaluation of product, process, and project decisions

Costs and benefits should be identified and measured before investment decisions are made and strategy implemented. This should include costs and benefits related to both current and future operations but should not include current costs related to past operations. These present and future risks, costs, and benefits can be more accurately measured for more effective costing and investment decisions. Measurement approaches and extensive sample measures are described in detail in Chapters 6 and 7. Through these models and measures, and the systems to implement them, managers can make more effective decisions to improve both sustainability and financial performance.

In the next chapter we look at ways to measure and reward sustainability performance.

CHAPTER 5

Performance measurement, evaluation, and reward systems

In developing strategic responses, it is important for senior executives to understand the causal relationships between sustainability performance, and financial performance, to understand the payoffs from social, environmental, and economic improvements, and to create a culture where employees understand and work toward corporate social, environmental, and economic goals. Corporate incentive and reward systems can be a critical tool to implement sustainability and align the interests of the corporation, senior managers, and all employees. These systems are usually a part of a broader set of systems to evaluate the performance of the organization, its various units, and individuals. They will probably measure success in numerous areas, including both sustainability and financial performance. Systems that measure performance and encourage employees to pursue sustainability are necessary to improve social, environmental, and economic impacts, to communicate the value of sustainability to the organization, and to hold employees accountable for their contribution to the sustainability strategy. In this chapter we discuss some of the systems that encourage performance and aid in performance measurement:

- Corporate, strategic business unit, functional, facility, and individual measurement and evaluations
- Compensation, incentive, and reward systems
- Internal waste taxes
- Emissions trading
- Strategic management systems (such as the balanced scorecard, shareholder value analysis, or other dashboards and performance measurement systems)

Performance measurement and evaluation systems

One important tool for linking corporate objectives with results is the company's performance measurement and evaluation system.[1] Measurement is critically important because it links performance to the principles of sustainability and facilitates continuous improvement. Managers may use indicators to define goals and targets when they implement new programs to improve their sustainability performance; they can then compare these indicators to actual performance, along with various benchmarks, and measure success. Managers need to use feedback constantly to challenge their assumptions about the viability of various decisions and their long-term implications for both the company and society. Appropriate measurement systems provide the proper tools for feedback and corrective actions. For example, in 2012, Colgate, a leading consumer products company with over 37,000 employees, began evaluating new products using a product sustainability scorecard to drive improvement across the product life-cycle. The scorecard rates products with 35 parameters across eight focus areas: responsible sourcing; materials; energy and greenhouse gases; waste; water; ingredient profile; packaging; and social impact.[2]

For an organization intent on changing its corporate culture and achieving sustainability, performance measurement is extremely important. Best-practice companies achieve superior sustainability performance by sending a clear message that these issues are critical to company success. The challenge in performance measurement is that many systems in place are missing relevant and comprehensive measures of performance. Systems that extend beyond the financials to nonfinancials deliver maximum value to shareholders, customers, and other stakeholders.

A measure for individual or business unit performance can be determined primarily by two factors: the corporation's strategy and the action taken by a person or business unit that contributes to the success of the strategy. This can be centralized or decentralized. In a decentralized method, the corporation prescribes the performance measure for the individual or the business unit, and then they decide what the performance drivers are and how to manage them. In a centralized method, the corporation sets the performance measure by giving the individual or business unit the performance drivers and the weight each driver has in the determination of the performance measure.[3] This becomes an important issue in both the formulation of strategy and the organizational design of decision-making discussed earlier. Corporate decisions on whether sustainability performance, strategy, and goals will be determined centrally or be delegated to the discretion of business unit or geographical unit managers will have an impact on performance and performance evaluations and on the incentive systems to be used.

Senior managers can use organizational performance indicators to evaluate whether the sustainability strategy is achieving stated objectives and contributing to overall corporate performance. A weak performance on the organizational metrics signals a need to examine the inputs and processes and determine whether they have been poorly specified or just poorly executed. It can also provide an opportunity to identify potential benefits to organizational effectiveness and profitability from sustainability that may have been overlooked. This is an opportunity to examine how well sustainability programs are contributing to corporate value and should unveil specific opportunities, directions for improvements, and standards of performance.

The social, environmental, and economic performance of the entire corporation, individuals, facilities, and business units is an integral part of performance measurement and evaluation systems. If sustainability performance is truly important to corporate leaders, evaluations should highlight that component. When performance measurement and evaluation systems are aligned with sustainability strategy, executives gain a key source of information. That translates into increased performance and payoffs from sustainability investments. Thus, the sustainability performance of corporations, business units, facilities, teams, managers, and all other employees should be measured and be part of the way they are evaluated for success.

The corporate-level measurement system sets the organization up for brainstorming complementary sets of measures down through the organization. Managers should cascade measures down through the hierarchy. By taking a cue from the family of measures developed by corporate executives, every unit of the organization should address sustainability measurement in a coordinated way. Business units, functional groups, facilities, teams, and even individuals obtain guidance from measures that dovetail with corporate strategy. When people's efforts to execute strategy are aligned in this way, a company can expect to join leading organizations in enjoying the benefits of increased sustainability performance.

A prime challenge is to create a "performance logic" among all measures. From the bottom of the organization up, managers must ask: How does each variable measured contribute to a higher-level variable and, in turn, contribute to organizational results? From the top down: What variables drive the economic profit figure and, in turn, what variables drive those variables? The critical step is to configure the measurement system so that measures at corporate, functional, and team levels connect.

Devising the right performance measures

It can be difficult to devise measures that send the right signals and prompt the right actions. Measures should have the following six objectives:

1. Make strategic objectives clear
2. Focus on core cross-functional processes
3. Focus on critical success variables
4. Act like early warning signals for problems ahead
5. Identify critical factors going awry
6. Link to rewards[4]

Managers should also include a mix of input, process, output, and outcome measures. Input includes the money and people used to implement a sustainability initiative; process includes any systems used to deliver an output; output includes intermediate results achieved; and outcomes are the final results that may include both sustainability and financial performance. Each element of the Corporate Sustainability Model from Chapter 1 (Fig.1.4) should be converted into a performance indicator and measured.

Workable measures need to serve not just management but the people who actually execute the strategy, no matter what level of the organization they work in. Setting the top-level measures is only the beginning. Top managers must challenge business unit managers to create measures of their own, aligned with the top-level set.

Every team and operating unit needs a family of measures to motivate workers to act in concert with the strategy developed for the whole company. The idea is to cascade the measures down through the organization so they logically connect one to the next. Each group of employees should customize their own measures. Sometimes the measures are the same as corporate measures, sometimes entirely different, but in any case they both come from a global strategy and serve local needs.

Performance measurement and evaluation systems fulfill at least three vital roles. The first role is to capture the logic behind a sustainability strategy and facilitate agreement about what is important, how day-to-day activities add value, and how each person contributes to the mission. Making the strategy explicit through a measurement system has at least three advantages:

- It allows discussion about the underlying assumptions, and provides agreement in the organization about the strategy
- It encourages communication of the strategy and its execution throughout the organization. Communication clarifies expectations and it becomes clear to staff why certain actions add value and others do not
- It tracks the evolution of the organization and the strategy. Sustainability efforts frequently span long periods of time; a performance evaluation and measurement system identifies if the organization is on the right track to achieve its sustainability objectives and whether the strategy is working

Volvo Car Group, a Swedish car manufacturer, has developed a sustainability scorecard to communicate strategy and measure performance. It also provides an overview of trends that are relevant to Volvo’s contribution to sustainable development (Table 5.1).[5]

The second use, and probably the most commonly thought-about function of measurement systems, is monitoring progress (in Volvo’s case exemplified with trend monitoring). Organizations often develop performance measures to help them gauge the sustainability performance of strategic business units and company facilities. Regulatory requirements and external stakeholders prompt some of these measures, such as toxic substance data, energy consumption, number of industrial accidents, employee safety, and workforce diversity. However, companies also recognize that it is important to develop measures that are aligned with the internal mission and objectives of the organization. Integrating measures of sustainability performance will ensure that statements of social responsibility articulated by the CEO and in corporate mission statements are implemented properly.

For example, Kingfisher, a leading home improvement retailer in Europe and Asia, has developed a performance evaluation system to measure its so-called Operating Companies' progress against 50 targets. These cover Kingfisher's Net Positive priorities (timber, energy, innovation, communities) and three other issues (employees, suppliers & partners, environment). Kingfisher has developed a questionnaire to monitor Operating Companies progress against these targets. The nine Operating Companies complete the questionnaire twice a year and average scores for each are published every six months on Kingfisher's website (for January 2013 results, see Fig.5.1). There are multiple targets for each Net Positive priority and the three other issues. The score for each issue is an average score, calculated from the individual scores for each relevant target. Since Kingfisher just started its ambitious Net Positive journey, results for January 2013 reflect relatively low scores in several areas.[6]

KEY SUSTAINABILITY DATA	2012	2011	2010	2009	2008	2007	Trend[1]
Creating value							
Total sales (retail deliveries)	421,951	449,255	373,525	334,808	374,297	458,323	(–)
ASSUMING SOCIAL RESPONSIBILITY							
Product responsibility							
Safety test results Share of independent tests where Volvo Cars received the highest rank (%)	95	89	88	80	70	69	(+)
Occupational health and safety							
Health Sick leave per available hours (%)	4.4	4.4	4.5	4.7	5.0	5.5	(=)
Occupational injuries Number of injuries resulting in at least one day of sick leave per 200,000 worked hours	0.55	0.7	0.6	0.5	0.9	1.5	(+)
Diversity and equal opportunity							
Gender balance. Share of women in leading positions (%)	21.3	21	19.6	18.7	18.5	18.0	(+)
Gender balance. Ratio of basic salary of women to men (white collar, average for eight salary grade levels 'SGD' – 'SGK')	0.97	0.97	0.790	0.994	0.974	n/a	(=)
Ratio of basic salary of women to men (bluecollar, average for five salary grade levels "41-42" – "49-50")	0.99	0.99	1.010	1.034	1.027	n/a	(=)
Employment							
Total workforce	22,715	21,512	19,494	19,650	22,732	24,384	(+)
Rate of employee turnover	2.8	2.3	3.3	12.8	9.2	9.1	(=)
PROMOTING ECOLOGICAL SUSTAINABILITY							
Emissions from product							
Fuel efficiency Fleet average CO_2 in EU (g/km)	143	151	157	173	182	190	(+)
Energy use in car production							
Total energy consumption in car production (MWh)	798,487	815,301[2]	837,785[2]	713,079	816,581	916,669	(+)
Emissions from production							
Total carbon dioxide emissions (tonnes)	61,670	62,922[2]	67,585	58,980	68,367	126,735	(+)
NOx emissions (tonnes)	72	80	85	71	90	101	(+)
SOx emissions (tonnes)	<1	<1	<1	<1	<1	1	(=)
VOC emissions (tonnes)	796	828	738	527	712	740	(+)
Hazardous waste (tonnes)	10,837	11,439	9,087	5,594	9,320	11,395	(+)

1 Trend indicates our progress in relation to Volvo Cars' goals and vision. A plus sign (+) indicates that the company is moving in the right direction toward our goals, a minus sign (–) indicates that actions need to be taken for the company to develop towards our desired direction.
2 Restated.

TABLE 5.1 **Volvo sustainability scorecard**

Source: Volvo Car Group (2012) *Sustainability Scorecard*

A third role of measurement systems is to facilitate the ongoing discussion within an organization that will lead to better performance. Niagara Mohawk Power, a New York-based power company now part of National Grid, developed a comprehensive self-assessment program to focus the organization's efforts on performance areas that would create value for the company's stakeholders and that would help to sustain longterm improvements. Its three primary objectives were:

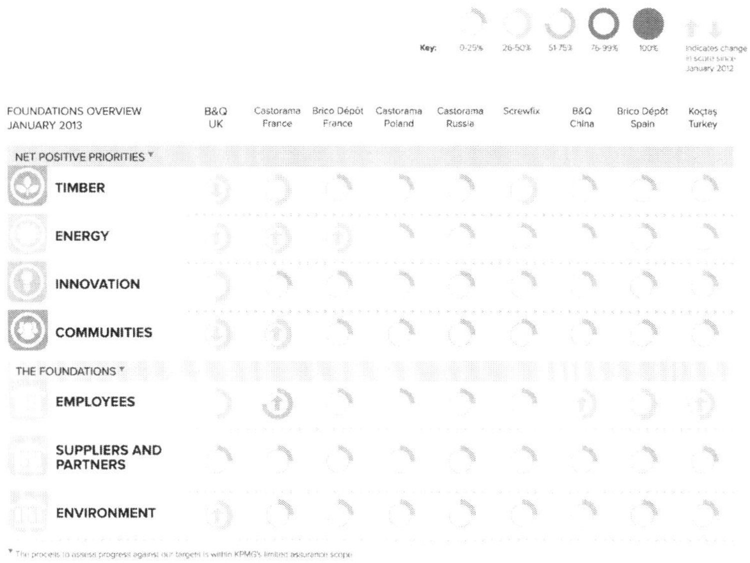

FIGURE 5.1 **Kingfisher's monitoring of Operating Company performance**

Source: Kingfisher (2013) *Net Positive Report*

- Responsiveness to customer needs
- Efficiency through cost management, improved operations, employee empowerment, and safety
- Aggressive, responsible leadership in addressing environmental issues

The company also developed an environmental performance index that established targets and measurable improvements based on a baseline of performance. Improvements made toward meeting the three objectives determine how large a financial reward is available to company employees. Establishing solid benchmarks against which environmental performance can be measured encourages management and staff to improve compliance with environmental regulations. It also leads to a decrease in costly noncompliance issues and corrective actions. At Niagara Mohawk, three categories of performance were measured: emissions/waste, compliance, and environmental enhancements.

In specific industries, industry leaders are exploring ways to evaluate, communicate, and share tools for measuring social, environmental, and economic performance of their products. Nike, for example, has been cooperating with the Sustainable Apparel Coalition (SAC), an industry-wide group of leading apparel and footwear brands, and with retailers, manufacturers, NGOs, academic experts and the US Environmental Protection Agency, to reduce the environmental and social impacts of apparel and footwear products around the world. Its core initiative is developing the Sustainable Apparel Index, a common, industry-wide tool for measuring the environmental and social performance of apparel products and the supply chains that produce them. Apparel retailers and brands can compare the performance of products and upstream supply chain partners, and those partners will have a single standard for measuring and reporting performance to their downstream customers. Eventually, this approach can provide a foundation for reporting to consumers on the sustainability of the products they purchase.[7]

Subjective measures

Objective measures are the bread and butter of most performance measurement systems. However, subjective measures of performance should be used to complement objective measures. There are numerous advantages to subjective evaluation. Managers can:

- Include information not foreseen before the project started
- Observe the actions and decisions of the person evaluated
- Evaluate tasks that are hard to quantify and judge whether they are beneficial to the company
- Discount the effect of uncontrollable events
- Adjust the importance of different measures and observations with changing priorities for the sustainability project
- Use what they know about the person evaluated to better assess performance, because people interact in various issues and over time

However, subjective measures do have their limitations. They rely on the availability of information and the ability, knowledge, and effort of the person doing the evaluation. Subjective evaluation also relies on the evaluator having the right incentives to provide a fair evaluation and on his or her reputation, fairness, and ability to judge. A person without credibility will hardly lead to satisfactory evaluation. Subjective evaluation can be the best performance measure when the person evaluating is competent, trustworthy, and committed—and the worst performance measure if any of these conditions are not met. A mix of objective and subjective measures for evaluation is the best approach. Over-reliance on either one distorts incentives and behavior.

Evaluating the CEO and senior executives

One of the primary functions of the board, as discussed in Chapter 2, is oversight and evaluation of the CEO. Given recent concerns over excessive executive compensation, conducting a rigorous performance evaluation and explicitly linking it to compensation can provide improved governance and accountability on the part of both CEOs and boards. The identification of the performance objectives should be a joint effort between the board and the CEO, and should identify objectives and goals that reflect the CEO's roles and responsibilities.

Dean Foods, one of leading food and beverage companies in the US, expressly charges board members with governing sustainability. The Audit Committee oversees corporate social responsibility policies, including those covering sustainability, ethics, compliance, and reputation, while the Compensation Committee evaluates executive officers on the basis of these factors, weighted 40%. The CEO is measured on the basis of instilling a culture of ethical behavior and social responsibility. The Chief Supply Chain Officer is measured on the basis of saving water, improving energy efficiency, and reducing waste output in the supply chain. As a result, Dean Foods has cut water use by 5.6% in 2010, and has reduced greenhouse gas emissions by 6% over the period 2008–2010.[8]

At National Australia Bank, one of the four largest financial institutions in Australia, individual senior executive performance is assessed against a number of key measures supporting the bank's strategy and business objectives. Measures and targets are tailored to the individual's role.[9] Table 5.2 details some of the key measures used in 2012 to assess individual performance outcomes.

A particular challenge when constructing a performance evaluation system for the CEO is to develop a system that will adequately capture the inherent distinctions between corporate performance and the CEO's performance. Performance evaluations and rewards should not result in rewarding poor performance or in overlooking superior performance. For example, CEO performance can be marginal, even though the stock price is rising. Or the reverse may be true. It is the board's responsibility, through the development of effective measurement and evaluation systems, to distinguish between the performance of the CEO and the performance of the corporation.[10]

Why performance measurement and evaluation is important

Performance measurement and evaluation is an important tool in the implementation of a sustainability strategy and aids in the alignment of strategy, structure, and other systems to achieve success. It is critical to set objectives and targets and measure success against them. It is also critical to measure success of not only the results (outcomes) but also the inputs, processes, and outputs that lead to those outcomes. Explicitly identifying corporate goals and setting specific targets improves corporate sustainability performance and focuses attention on areas of concern and priority. These are some of the benefits a company can gain by including social, environmental, and economic indicators in its performance measurement evaluation at all levels and in all areas:

- Comparison of performance over time
- Highlighting of optimization potential
- Derivation and pursuit of social, environmental, and economic targets
- Evaluation of sustainability performance between firms (benchmarking)
- Communication tool for corporate reports

Key Business Driver	Measure[1]	2012 Achievements
Financial and risk management	Group cash earning ROE ROTAE Tier 1 ratio Risk appetite	• Group cash earning[2] of $5,433 million for 2012 year decreased by $27 million or 0.5% against 2011. Excluding the $250 million ($175 million post tax) uplift to the economic cycle adjustment, cash earnings increased by $148 million or 2.7%. This was largely driven by higher earnings in Wholesale Banking, Personal Banking and NZ Banking, partially offset by significantly lower earnings in UK Banking as a result of higher charges for bad and doubtful debts. • Group cash ROE decreased by 100 basis points to 14.2% due to lower earnings, coupled with higher levels of capital being held as the Group continues its transition to Basel III. • ROTAE was below budget. • The Tier 1 capital ratio has increased by 57 bps to 10.27%, consistent with the Group objective of maintaining a strong capital position. • Maintained sound capital, funding and liquidity positions. • Adherence to risk appetite across the Group and strong risk culture.
Strategic projects and process quality	Technology and process transformation Cross sell	• Several key milestones achieved in the multi-year project, including launch of nabtrade. • Cross sell targets met.
Employees and culture	Employee engagement Diversity targets	• Strong gains in employee engagement, placing the Group above the financial services average.[3] • On track to meet diversity targets (for more details see the Diversity section).
Customer and community	Customer satisfaction Corporate responsibility objectives	• Achieved the highest bank customer satisfaction score of the four major Australian banks since the customer satisfaction survey began.[4] • Maintained or improved customer satisfaction across all Australian business segments.[5] • Improved outcomes on majority of corporate responsibility measures (for more details see the Company's 2012 Annual Review available at www.nabgroup.com)

(1) Refer to the Table of key terms for definitions of cash earnings, ROE and ROTAE and to the Glossary for a definition of Tier 1 ratio.
(2) Refer to the Financial report for statutory net profit attributable to owners of the Company, and to Note 2 of the Financial report for a reconciliation between cash earnings and statutory net profit attributable to owners of the company.
(3) May 2012, measured through an annual employee survey conducted by external consultants (Hay Group).
(4) August 2012, Roy Morgan Research Customer Satisfaction Report. NAB compared with ANZ, CBA and WBC.
(5) DBM Business Financial Services Monitor April 2009 to September 2012, six month rolling averages

TABLE 5.2 **National Australia Bank links remuneration and performance**

Source: National Australia Bank (2012) *Report of the Directors*

- Feedback instrument for information and motivation of the workforce
- Technical support for certification programs
- Most importantly, providing the information to change managerial actions to improve performance

Despite their importance, many performance measurement systems are inadequate at most companies. They tend to rely on historical information and lack predictive power, failing to give managers the information they need to make decisions. And measures should not be an end in themselves.[11] Instead, they should feed back into management systems to facilitate change and improvement in individuals, business units, and the organization as a whole. The challenge is to look past financial performance toward a more thorough integration of sustainability performance. A balanced family of measures can evolve into a powerful system for executing strategy. The measures help define the strategy, communicate it to the organization, and direct its implementation, from the corporate level to the individual. They also help keep everyone's efforts aligned, because they link strategy to budgets, resource allocation systems, and to pay programs. In the best cases, they route high-quality feedback through the organization so executives can make critical, mid-course adjustments in strategy.

Measures should communicate to employees the values of the company and how performance will be judged. A lack of communication and understanding of what is important to the organization, along with too much emphasis on short-term results, can lead to low levels of commitment and reduced performance.[12] Shared understanding of what is important is critical to improving sustainability and financial performance. Relating the measures to individual compensation might also be desirable, as an explicit system that directly affects individual pay provides strong incentives for employee performance. Examples of sustainability performance measures are provided in Chapter 7.

Incentives and rewards

The traditional accounting system often provides a disincentive to report potential hazards or violations of environmental laws, corporate goals, and corporate practices. Employees sometimes believe they will be penalized if they notify a manager of a potential hazard because eliminating the hazard might cause the business unit to suffer a short-term financial loss. This expenditure typically is viewed as an expense rather than an asset, investment, or value creator and often has a negative impact on a manager's overall rewards.

To confront this disincentive, many companies have programs that provide awards to employees for exemplary sustainability performance. In some cases awards are given to teams rather than individuals. They vary from cash gifts and various methods of acknowledging the achievement to banquets, plaques, and so on. Seiko, Japanese-based manufacturer of watches and precision and optical products, for example, has established an environment prize in recognition of employees' environmental contributions. It is a positive incentive for employees to go beyond their job responsibility and to become eligible for a cash award of US$500 to US$5,000. Awards can be useful but only in connection with a more comprehensive program of performance evaluation that includes other motivations for improved sustainability performance among divisions, their managers, and their support staff.

Some companies have tied individual performance reviews and compensation explicitly to social, environmental, and economic performance including performance on climate change. They have established sustainability performance as a critical variable for compensation in incentive systems. For example, Alcoa, a leading producer of aluminum, champions pay for performance to achieve specific sustainability objectives to ensure the integration of sustainability into core business strategies. During 2012, 20% of the variable compensation was tied to achieving significant aspects of Alcoa sustainability targets. Across the entire workforce, the targets focused on safety and CO_2 emission reductions through process improvements and improved energy efficiency. Management-level employees had an additional target to improve the diversity of Alcoa workforce. Specifically, these sustainability targets and their share of the variable compensation plan at the corporate level were: CO_2 5%; safety 5%; and diversity 10%. The 2012 achievement payouts at the corporate level were 9.1%, 2.5%, and 16.3% for CO_2, safety, and diversity targets, respectively.[13]

At Henkel, individual target agreements with employees also include sustainability criteria, such as reducing energy and water consumption or accident rates, if these fall within the sphere of influence of the employee concerned and have a clear bearing on business performance.[14]

Many other corporations across the globe from Intel, XcelEnergy, a utility supplier of electric power and natural gas service, National Grid, an international electricity and gas company and one of the largest investor-owned energy companies in the world, Suncor Energy, a Canadian integrated energy company, to ING, a Dutch insurance conglomerate, are making executive compensation decisions based on how well the company's business units perform in relation to its sustainability goals.

The weight given to social, environmental, and economic targets also helps to signal the importance of their impact to the employee and to the organization. At UK-based Anglo American, one of the world's largest diversified mining and natural resource groups, 10% of performance-related compensation at the executive level is tied to safety, while at the operational level it increases to 25%.[15] This communicates that safety is highly valued at the company and that it benefits both the employees and the company to perform well in this area.

XcelEnergy has a significant stake in environmental preservation and marks its commitment to sustainability at the highest level by applying specific quantitative metrics to incentive rewards. One-third of the CEO's annual bonus is tied to environmental performance, as measured by renewable energy, emission reduction, energy efficiency, and clean technology.[16]

Another way to involve individual employees in improving social, environmental, and economic performance is to give them a stake in the performance of the company. One such company is UPS where approximately 40% of employees own shares in the company and own about 40% of the outstanding shares.[17] Additionally, stock is rewarded to the management team based on the company's performance on key goals, including sustainability. UPS believes that a stock-and profit-sharing plan is one way to align employee interests with company goals. However, it is important that the employees understand how and why positive performance in sustainability increases shareholder value. Otherwise, this strategy may not achieve its intended goals.

Performance goals and incentives can also be used for subcontractors. Nike's subcontractors must comply with employment standards set by Nike. These standards are enforced by Nike inspectors and subcontractors who continually violate Nike's requirements risk the penalty of losing their contracts, even if they are in compliance with local laws and practices.[18] So the subcontractors have a financial incentive to follow Nike policies. By instituting this policy and regularly monitoring performance, Nike reinforces sustainability performance as a core component of its strategy and encourages subcontractors to also value sustainability performance.

The problem with many incentive systems is that they reward the wrong behavior and provide disincentives for the right behavior. Incentive systems can also fail because they are overused. Putting too much emphasis on pay-for-performance without considering the risks involved and pressures created may lead managers to shy away from taking risks. An additional challenge with incentive systems is their potentially negative effect on intrinsic motivation—people's internal drive to do something because they love doing it. Sometimes the most important reward for performance is the act of doing the job itself, and improving sustainability performance can provide significant personal rewards for many employees. Lastly, the level of risk taking that a company encourages is an important issue to consider in addition to measuring and rewarding. Risk-taking behavior is necessary for successful sustainability strategies but can be dysfunctionally reduced if failure is punished economically.

Using an environmental multiplier to drive performance

BFI (Browning-Ferris Industries), a North American waste management company, part of Allied Waste Industries, decided in the 1990s that it needed to make a fundamental change to its corporate culture in order to meet its environmental objectives. The core corporate and district-level objectives related to both business and community needs. The company developed a set of AC (awareness compliance) tools for each of its three major lines of business: landfill operations; solid waste; and medical waste. The AC tools included a detailed training manual that described the objectives, explained the problems, and outlined the role of all employees in achieving corporate environmental compliance and responsibility, and training videos and other tools to help all employees understand and meet the performance goals.

In addition to providing the AC tools, the company changed its incentive program to tie environmental performance directly to employee compensation. Under this system, one-third of compensation became at-risk pay, whereby the incentive pay earned would be based on the employee's score in meeting environmental goals. The table below illustrates the multiplier scale used by BFI to convert environmental performance to incentive pay. An employee who scored 95 points would receive 100% of the incentive pay, and an employee who only scored 75% would only receive 50% of the incentive compensation. Employees who scored lower than 70 points would not receive any incentive compensation. This incentive pay system applied to employees at the level of district manager; however, district managers themselves used incentives to motivate their subordinates to achieve district-level environmental goals.

Points earned	District environmental multiplier
95–100	1.00
90–94	0.90
85–89	0.80
80–84	0.75
75–79	0.50
70–74	0.25
Below 70	0.00

BFI believed this emphasis on environmental compliance boosted the company's image and, ultimately, its financial performance. This system worked partly because all employees understood that environmental compliance was nonnegotiable and a critical success variable for both their own and the company's performance.

Source: Epstein (1996) *Measuring Corporate Environmental Performance*

Internal waste taxes

An internal waste tax is a practical application of activity-based costing at organizational level. It introduces more direct accountability by making each business unit responsible for the waste it produces. This could also be developed and applied to other social, environmental, and economic costs. With an internal waste tax, waste treatment costs and fines are charged back to product lines creating the waste. This reduces the internal subsidies created when environmentally efficient divisions are allocated similar monetary amounts for environmental costs as divisions that cause more waste-related costs. In fact, internal subsidies need not exist, as all business units are responsible and accountable for their own costs.

An example of the link between full cost accounting and performance evaluation is Dow Chemical's waste tax. In the 1990s, Dow Chemical built a waste landfill at its Michigan division that was then expected to last until 2007. After development, the company began charging each plant a fee according to the actual waste it brought to the landfill. Plants discovered that it was more economical to introduce process improvements to reduce the quantity of waste. This Dow internal waste tax has reduced solid waste significantly. The Michigan landfill is now estimated to last until 2034.

Some companies have argued that a waste tax works better in highly centralized organizations than in less centralized ones. There is concern that in decentralized organizations a central tax imposed on business units may not fit in with the corporate culture and will meet with resistance. Decentralized organizations often allow managers to make their own trade-offs of business and environmental improvements where necessary rather than dictating local actions. But, even in decentralized organizations, corporate managers often do provide incentives such as penalties or additional resources to motivate excellence in sustainability performance.

Another innovative example of managing environmental performance through a waste-tax mechanism is the company-wide emissions trading program launched by BP in 2000. The company set as a corporate objective the reduction of its GHG (greenhouse gas) emissions such as CO_2 and methane, contributors to global warming. BP-Amoco operating units were given internal targets for allowable emissions, and units reducing emissions to levels below the targets could sell the emissions credits to other BP-Amoco units that had not made deep enough cuts, improving both its environmental and its financial performance.[19]

Internal taxes force the business units that cause negative social, environmental, and economic impacts to be financially accountable for the waste they generate. This accountability motivates managers to evaluate their processes and products for opportunities to minimize the social, environmental, and economic impacts that are creating the additional costs.

Emissions trading

A very different incentive for performance is developing rapidly with emissions trading programs. They provide powerful inducements for corporations to reduce emissions, and it seems likely that these incentives and the financial impact will continue to increase. As emissions trading practices become more established, they are likely to have a more important effect on the evaluation of both corporate financial and corporate sustainability performance. The effect is also likely to cascade through organizations, influencing the evaluations and the rewards of CEOs, senior corporate managers, and managers throughout the organization as pressure to reduce the financial cost of emissions intensifies. This may be an important emerging development in using performance evaluation and rewards to drive improved corporate financial and sustainability performance simultaneously.

The Kyoto Protocol requires nations to cut their GHG emissions and countries divide the burden among their industries. Companies who do not meet the standard can buy credits from companies who cut their emissions by more than what is required. The EU set up its carbon trading system in 2005, and trade in GHG permits doubled to more than US$26 billion.[20]

There are several options for companies to consider when developing strategies on how to use their emissions credits:

- **Emission offsets.** Companies may increase the level of a pollutant if they also do something that is good for the environment, such as planting trees
- **Bubble policy.** Companies may increase pollution at one source as long as they reduce pollution at another source
- **Banking.** Companies store emissions allowances for later use or lease them to another firm[21]

Launching its trading platform in 2003, CCX (Chicago Climate Exchange) was the world's first, and North America's largest, legally binding rules-based GHG emissions allowance trading system, as well as the world's only global system for emissions trading based on all six GHGs. CCX members made a voluntary but legally binding commitment to meet annual reduction targets for GHG emissions. Those who reduced below the targets had surplus allowances to sell or bank; those who emitted above the targets complied by purchasing CCX CFI (Carbon Financial Instrument) contracts.[22]

Strategic performance measurement systems

Numerous approaches can be used to organize, identify, measure, and report sustainability performance for improved managerial decision-making. The balanced scorecard and shareholder value analysis are two approaches currently used by many managers to help implement corporate strategy. Both of these systems can also be used to help managers implement social and environmental strategies, drive organizational change, and evaluate and improve performance.

The balanced scorecard

The balanced scorecard is a strategic management system that links performance measurement to strategy using a multidimensional set of financial and nonfinancial performance metrics. The term "balanced scorecard" refers to the framework first described by Kaplan and Norton in 1992 and further expanded upon in numerous other books and articles.[23]

The traditional model contains four dimensions or perspectives that relate to the strategy and core values of the company. These dimensions are financial, customer, internal business processes, and organizational learning and growth. In practice, many managers use the term "balanced scorecard" to refer to any set of financial and nonfinancial measures that link performance indicators to corporate objectives. The four perspectives in the balanced scorecard represent four key components of creating and sustaining corporate value:

- The **financial perspective** focuses on the shareholders' interests and shows the link between strategic objectives and financial impacts
- The **customer perspective** focuses on measures that reflect how the company is creating customer value through its strategy and actions
- The **internal business processes perspective** contains measures that indicate how well a company performs on key internal dimensions
- The **learning and growth perspective** stresses measures of how well the company is preparing to meet the challenges of the future through leveraging its organizational and human assets

Many companies include sustainability key success factors and key performance indicators in each of the four dimensions of the balanced scorecard, choosing perhaps one or two key measures in each dimension. The choice of where to include sustainability indicators on the balanced scorecard depends on the challenges facing the organization. Figure 5.2 shows an example of a sustainability-focused scorecard. This example broadens the customer dimension to include other stakeholders of the organization, better reflecting a sustainability focus.

Financial dimension • Percent of sales revenues from "green" products • Recycling revenues • Energy costs • Fines and penalties for pollution	**Stakeholder dimension** • Sustainability awards • Funds donated for community support • Number of community complaints • Employee satisfaction
Internal business process dimension • Percent of suppliers certified • Volume of hazardous waste • Packaging volume • Number of community complaints • Cost of minority business purchases • Number of product recalls	**Learning and growth dimension** • Diversity of workforce and management • Number of volunteer hours • Cost of employee benefits • Percent of employees trained re sustainability

FIGURE 5.2 **Balanced scorecard for sustainability**

Source: Epstein and Wisner (2006) "Actions and Measures to Improve Sustainability"

Companies that have identified sustainability as a key corporate value or strategy may choose to expand the balanced scorecard by creating a fifth perspective. This dimension would include social, environmental, and economic performance indicators that link with the other four perspectives, and would serve to highlight the importance of social, environmental, and economic responsibility as a corporate objective.

The weight given to this fifth perspective would depend on the relative priorities of the organization, and the measures included would depend on the drivers of performance that managers of the company have identified. These are some of the reasons why companies establish a separate balanced scorecard perspective for sustainability:

- Social, environmental, and economic responsibility is seen as core to the strategy of the organization, creating competitive advantage (through factors such as corporate image, reputation, and product differentiation), as opposed to being seen as a means to improve operational efficiency
- The fifth perspective becomes a tool to focus managers' attention on social, environmental, and economic responsibility as a core corporate value. It communicates management's strong concern about these issues and objectives
- When a company has high-profile or high-impact sustainability issues, a fifth perspective helps to highlight the importance of these issues. Companies in industries that have had problems (chemicals, oil, and apparel, for example) may be more likely to focus internal attention on sustainability resources and company strategy

- When the resource allocation to social, environmental, and economic responsibilities is relatively large, companies may want to highlight the link between the use of those resources and company strategy[24]

Sustainability strategies reflected in the corporate balanced scorecard should be cascaded down to the SBUs (strategic business units) of the organization and ultimately to the support functions, including EH&S. While the corporate-level scorecard clarifies corporate values and beliefs and identifies actions that create corporate synergies, the scorecards for SBUs can be customized to reflect the market and operational challenges faced by each SBU.

For example, Unilever, the multinational consumer products company, has a corporate goal of minimizing its "environmental imprint." Each Unilever SBU links to this corporate goal but in ways that are relevant to the SBU. For example, some Unilever SBUs are challenged by the availability and quality of water, and focus their measures on reducing water use and effluents. Reducing packaging waste is a priority for other SBUs, especially those operating in northern Europe and North America. These SBUs focus on bottle weight reductions, developing concentrated product formulations that require less space, and developing a line of refillable products.

A cascaded set of balanced scorecard measures that reflects the strategy to reduce packaging waste could be expressed as follows:

Corporate	• Percentage decrease in environmental impact
Geographic unit	• Percentage reduction in packaging waste
Business unit	• Number of product reformulations (concentrated) • Percentage of refillable products
Manufacturing unit	• Percentage decrease in packaging • Container weight reduction
EH&S department	• Percentage of life-cycle analyses on product lines • Number of product designs with integrated environmental concerns • Tons of waste

Balanced scorecards for support functions, including sustainability, community affairs, and EH&S, should align with the strategies and objectives of the corporation and the SBUs, thereby reinforcing performance alignment. Many companies are now extending their sustainability oversight activities to their suppliers as issues such as child labor practices and environmental responsibility pass through the supply chain. A number of support functions could link their scorecard measures to this objective; for example:

- EH&S

–Number of supplier audits

–Percentage of suppliers with environmental certifications

- Purchasing

  –Percentage of materials purchased from ISO 14000-certified suppliers

- Human resources

  –Number of audits of contract labor firms

  –Number of suppliers complying with corporate codes of conduct

A complete balanced scorecard for sustainability, EH&S, and other departments would probably contain performance measures in each of the four scorecard perspectives, reflecting each department's role in supporting corporate objectives and strategy. Thus, successfully cascaded balanced scorecards provide clear linkages between the strategies and performance metrics at the various levels in the organization and provide guidance to employees throughout the organization as to how they can contribute to overall corporate financial and sustainability performance.[25]

Implementing a balanced scorecard causes managers to integrate financial measures with other key performance indicators around customer, internal business processes, organizational learning and growth, and perhaps sustainability perspectives. It increases social, environmental, and economic accountability by explicitly including performance metrics related to sustainability goals, and by recognizing their interconnection with a multidimensional set of corporate objectives. Companies using the balanced scorecard can position themselves to generate the profitability, and demonstrate the accountability, demanded by customers, shareholders, employees, and the communities around them.

Shareholder value analysis

Increasing shareholder value is a key objective of most companies, and managers have begun to recognize that shareholder value is improved by creating value for employees, customers, suppliers, the community, and other stakeholders. Many companies have expanded their method of measuring shareholder value creation by using measures that reflect economic value created by an organization.

Perhaps the best-known metric of shareholder value analysis is economic value added. This financial metric of economic profit takes into account the cost of the capital and assets involved in creating profits. The traditional measurement of net profit does not take into account the cost of capital provided by shareholders and is also distorted by applying GAAP (generally accepted accounting principles) that govern corporate financial reporting. Shareholder value calculations include the costs of equity capital and also adjust for GAAP-related distortions.

As illustrated in the simplified equation in Figure 5.3, shareholder value is created by sustainability initiatives that generate profits, minus the capital charge for the utilization of assets. Profit is generated from growth initiatives that increase revenues, such as product innovation and market development, and/or efficiency achievements that reduce costs: for example, waste reduction. The capital cost of assets is a function of the amount of resources used and the risk involved when using such resources.[26]

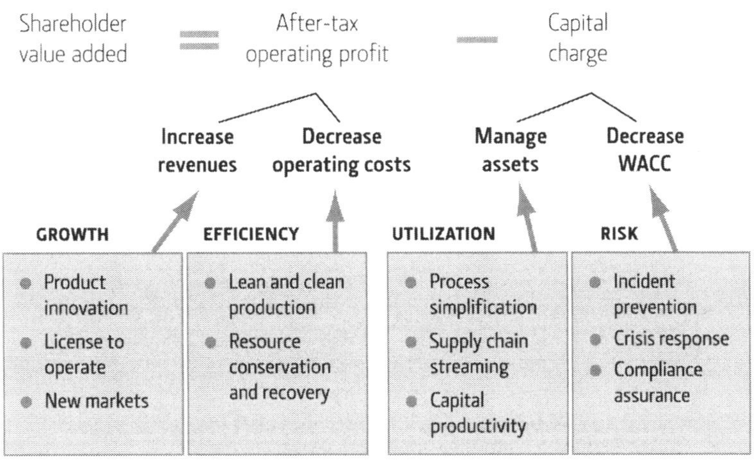

FIGURE 5.3 **Shareholder value creation**

Source: Fiksel (2003) "Revealing the Value of Sustainable Development"

Like the balanced scorecard, shareholder value analysis is a system that can be implemented throughout the organization, not just at senior levels, with the expectation that all employees will be directed toward creating shareholder value. It can also be used to measure and report performance in the capital markets, for capital investment projects, and in the evaluation and compensation of performance.

DuPont uses a metric called "shareholder value added per pound of production" or SVA/lb. SVA is defined as the shareholder value created above the cost of capital. A company increases SVA by adding material, knowledge, or both. SVA/lb emphasizes the addition of knowledge, rather than material. DuPont has used this metric to evaluate its business units and set goals to increase its SVA/lb based on those evaluations.[27]

Using shareholder value analysis

Georgia-Pacific, the large forest products company, used shareholder value analysis to align the company's goals of creating shareholder value and environmental responsibility. The EH&S department at Georgia-Pacific, as well as individual environmental projects, has been evaluated using shareholder value analysis. Included in each environmental project evaluation is an assessment of the project's impact on revenues, operating costs, such as consulting fees, fines and administrative costs, and capital costs. Using shareholder value analysis, Georgia-Pacific has been able to identify environmental investments that create financial and shareholder value for the company. For example:

• A project to use boiler fly ash generated at the plants reduced landfill and transportation costs, generating a shareholder value analysis of US$800,000

• An aerator optimization project reduced energy usage, generating an shareholder value analysis of US$102,000

• By re-engineering the process for complex environmental permitting, the cycle time for new permits and construction projects was reduced, as well as external consulting fees, generating increased shareholder value of US$2.1 million[28]

Shareholder value analysis provides an incentive for sustainability managers to pursue investment opportunities to create shareholder value. It also helps to communicate the potential value of sustainability initiatives to managers who must justify the allocation of scarce resources. By better identifying and including broader and longer-term social, environmental, and economic impacts that affect corporate profitability into a single performance measure such as shareholder value analysis, executives can improve the likelihood that an organization's sustainability objectives will be pursued.

Summary

Performance measurement systems communicate management priorities by signaling throughout an organization the expected outcomes that management has determined to be important. "What gets measured gets managed" is an adage that represents the signaling capability of performance measures. Also, the actual performance outcomes provide feedback to management about the efficacy of the strategy.

The performance of all employees, teams, facilities, and business units should include a sustainability performance component where appropriate. By defining specific social, environmental, and economic work goals for the individual and measuring progress toward these targets an organization is signaling that sustainability performance is an important driver of corporate value. Incentives are often necessary to motivate employees to integrate social, environmental, and economic impacts into their decisions. Sustainability performance can often be improved if it is integrated into the performance evaluation system for all employees, teams, and business units. Empowering and rewarding managers and production workers can improve social, environmental, and economic planning and compliance activities. Better alignment of corporate and sustainability strategies with company-wide performance measures and rewards can improve sustainability and financial and operational processes and performance.

Taken together, the impact of management commitment and leadership, organizational structure and rules, systems, communication, performance measurement, and the incentive structure all are key factors in establishing the culture of the organization toward sustainability initiatives. It is through effectively establishing and managing these strategic management systems that an organization establishes a culture of sustainability and can most effectively move toward its strategic sustainability goals.

The next chapter gives an overview of the approaches that can be used to effectively measure social, environmental, and economic impacts of products, services, processes, and other corporate activities.

CHAPTER 6

The foundations for measuring social, environmental, and economic impacts

Measuring the payoffs of sustainability initiatives is challenging even without specifically identifying the appropriate inputs, processes, outputs, and outcomes. However, to know if sustainability strategies are succeeding, measurement of these elements is critical. In addition, surveys indicate a growing market in impact investments that are made into companies, nonprofit organizations, and funds with the intention to generate measurable social, environmental, and economic impact alongside a financial return.[1] Although it is difficult to precisely measure sustainability performance, social science, economic, and financial analysis techniques that provide reasonable estimates for social, environmental, and economic performance do exist. These measures provide substantial and valuable information that enables managers to more accurately evaluate the trade-offs made in day-to-day management decisions.

In this chapter we look at the conceptual foundations for measuring social, environmental, and economic impacts and risks before discussing the practical applications of these approaches in Chapter 7.

The costs and benefits of a sustainability strategy are cross-dimensional throughout an organization, not firmly lodged in any one functional area. Furthermore, many economic benefits of sustainability initiatives are often seen as intangible and therefore difficult to measure. Measuring hazardous waste generated is relatively straightforward, measuring employee satisfaction is harder, and measuring the impact of a company on society is even more difficult. And converting these impacts into monetary terms provides additional challenges. However, for each of these, we know the number is not zero and they each represent an output that relates to the success of a sustainability strategy. Sustainability benefits are also often longer-term in nature, making them more challenging to relate to current organizational performance.

Organizations also have to consider the differing and multiple objectives of stakeholders. Some of these objectives relate to the social, environmental, and economic impacts of organizational actions. Where once managers might have made a routine capital-investment decision based on estimated cash flows including such traditional items as capital outlay, cost of capital, and reduced expenses or additional sales, managers now must consider the social, environmental, and economic impacts of the decision as well. While research suggests that organizations need to evaluate multiple, diverse stakeholder interests, be aware of social, environmental, and economic impacts, and integrate this into decision-making, there is little guidance on the underlying process. But, although sometimes managers think that sustainability is more difficult to measure and integrate into investment decision-making models, there is a solid academic foundation for measurement.

Collecting these data differs from obtaining traditional financial measures from a cost accounting system since organizations must first identify multiple stakeholders and understand their objectives. Furthermore, relevant measures might rely on methods more typically used in sociology, social psychology, and economics, and which are only now being applied to management decision-making. Companies should first identify the potential impacts to their stakeholders. Examples of these not-so-obvious impacts are often defined as externalities and include changes in landscape due to a construction project or the effects on biodiversity. Externalities need to be incorporated into management decision-making. Although usually viewed as negative impacts, externalities can provide benefits to stakeholders. For example, a tree forest planted by a lumber company is a scenic landscape until it is harvested.

DB (Deutsche Bank), a leading German bank and one of the world's "greenest" banks, recently introduced its Global Impact Tracking to assess all global and regional flagship projects (with minimum investment of €25,000). This enables DB to evaluate whether its investments as a corporate citizen are efficiently and effectively aligned with its strategic goal of building social capital in its key areas of activity: education; social investments; art; and music. Figure 6.1 shows DB's three-step impact measurement process (input–output–impact). In Step 3, DB measures the impact of its various corporate citizen projects on beneficiaries, project partners, society, the company, and employees.

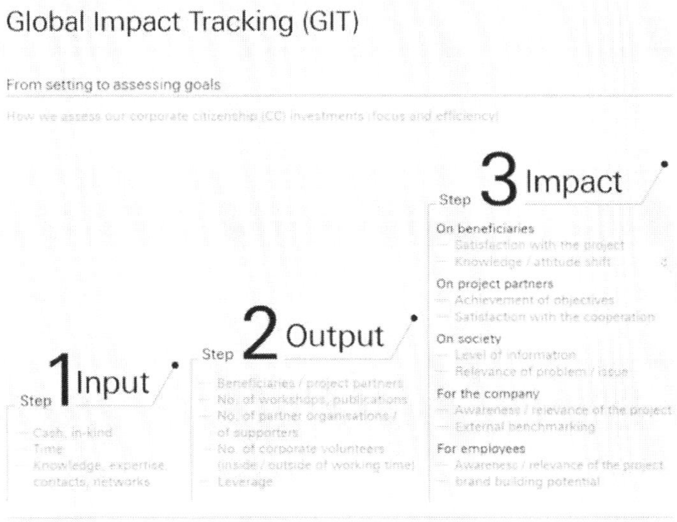

FIGURE 6.1 **Deutsche Bank Global Impact Tracking**

Source: Deutsche Bank (2012) *Corporate Responsibility Report*

A separate exhibit (Fig.6.2) presents an example of a matrix with measurements of awareness and relevance of DB's support of corporate citizenship (CC) projects as perceived by cultural professionals in Germany, UK, and the US.[2]

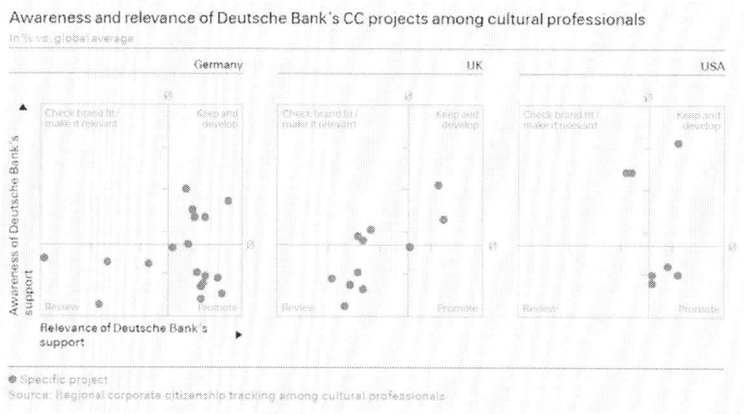

FIGURE 6.2 **Deutsche Bank is assessing corporate citizenship projects**

Source: Deutsche Bank (2012) *Corporate Responsibility Report*

Measuring environmental impacts at PUMA—PUMA's E P&L

PUMA, one of the world's leading sport lifestyle companies, has developed an important approach to value and report the environmental externalities caused by the corporation and its entire supply chain. Called E P&L (Environmental Profit and Loss), it measures and values both reductions in ecosystem capacity and increases in environmental impacts which occur as a result of PUMA's operational and supply chain activities. PUMA published the first E

P&L in November 2011, assessing the cost of its environmental impacts at **€**145 million. The E P&L covers water use, GHG (greenhouse gas) emissions, air pollution, land use, and waste.

Figure 6.3 sets out in monetary terms the changes in human welfare which result from PUMA's environmental impacts. The impact on climate change, for example, was measured by tons of GHG emissions and these were monetized through an estimate of the so-called SCC (social cost of carbon) to PUMA's 2010 operational and supply chain emissions. Estimates of the SCC look to value the damage resulting from current and future climate change (e.g., reduced crop yields, damage to infrastructure, or increased incidents of extreme weather) attributable to each ton of CO_2 equivalent (CO_2e), released in a given year. PUMA's global estimate of the 2010 SCC is **€**66 per ton of CO_2e.

The Environmental Profit and Loss

EUR million	Water use	GHGs	Land use	Other air pollution	Waste	TOTAL	% of total
	33%	33%	25%	7%	2%	100%	
TOTAL	**47**	**47**	**37**	**11**	**3**	**145**	100%
PUMA operations	<1	7	<1	1	<1	8	6%
Tier 1	1	9	<1	1	2	13	9%
Tier 2	4	7	<1	2	1	14	9%
Tier 3	17	7	<1	3	<1	27	19%
Tier 4	25	17	37	4	<1	83	57%
Regional analysis							
EMEA	4	8	1	1	<1	14	10%
Americas	2	10	20	3	<1	35	24%
Asia / Pacific	41	29	16	7	3	96	66%
Segments							
Footwear	25	28	34	7	2	96	66%
Apparel	18	14	3	3	1	39	27%
Accessories	4	5	<1	1	<1	10	7%
Intensity	Environmental impact (EUR) per EUR 100 of sales						
Footwear	1.8	2.0	2.4	0.5	0.1	6.7	
Apparel	1.9	1.5	0.3	0.3	0.1	4.1	
Accessories	1.2	1.5	0.00	0.3	0.00	2.9	

FIGURE 6.3 **PUMA's E P&L**

Source: PUMA (2011) *PUMA's Environmental Profit and Loss Account for the Year Ended 31 December 2010*

- PUMA operations include offices, shops, warehouses, business travel, logistics, and IT
- Tier 1 Suppliers include shoe, apparel, and accessory manufacturing
- Tier 2 Suppliers include outsole and insole production, textile embroidery and cutting, and adhesive and paint production

• Tier 3 Suppliers include leather tanning, petroleum refining, and cotton weaving and dyeing

• Tier 4 Suppliers include cattle rearing, rubber plantations, cotton farming, petroleum production, and other material production

PUMA has identified that most of its environmental impact occurs far down its supply chain, mainly in Asia, where the impacts were assessed at €96 million. The greatest environmental cost arises from production and use of raw materials, for example the methane produced during cattle farming, and the water consumed during the subsequent transformation of cattle hides into leather. PUMA is considering including social impacts in sustainability arising from decent/fair wages, security and stability, standard of living, empowerment, and community cohesion in future reports (PUMA's Environmental, Social, and Economic P&L).

Various approaches have been used to identify and measure the sustainability impacts of a company's products, services, and activities. These approaches provide an important conceptual foundation for measuring sustainability. They include methods such as cost of control and shadow pricing, damage costing, market price and appraisal, hedonic pricing, travel costing, and contingent valuation. Managers need guidance in applying these techniques to identify metrics that facilitate the implementation of sustainability and an informed decision-making process. Measuring these impacts, monetizing them, and including them in management decisions permits improved analyses of benefits and costs and better decisions for both the social benefit of stakeholders and the long-term profitability of the firm.

The concept of value

The benefits related to social, environmental, and economic impacts are often categorized as either **market** or **nonmarket impacts.** Market benefits include:

- Increased sales quantities due to increased market demand
- Increased prices due to quality and reputation
- Reductions in costs due to increased efficiencies
- Increased productivity
- Reduced future costs related to environmental clean-up, internal control and ethics breaches, and employee and customer problems related to lack of social sensitivity

Examples of nonmarket benefits include:

- Increased recreational benefits from cleaning up waterways (boating, swimming, and fishing)
- Enjoyment of greater species diversity

- Increased life-span and quality of life

To measure these impacts, we need to understand how stakeholders place value on social, environmental, and economic assets. The concept of value is based on the preferences that people have for the services and products they use. Preferences are in theory substitutable—one service or product can be exchanged for another if individuals perceive that they are no better or worse off than before. The trade-offs made by individuals indicate the value placed on social, environmental, and economic goods and services.[3]

The value given to goods and services can include:

- Use values

  –Consumptive value: for food or recreation

  –Nonconsumptive value: observing, photography

- Nonuse values

  –Option value: personal opportunity to use the resource in the future

  –Existence value: importance of the resource to others in the present and in the future

Use value is defined as the economic value associated with human use of a resource. Use value may be further categorized as having either consumptive value (logging of forests or use of water for drinking or farming) or nonconsumptive value (recreational use such as bird watching or photographing which leaves the resource unchanged). If a resource such as a river is used more often and more effectively because it is clean, then a nonmarket use value has been created. When a company takes actions that improve the environment and create a water resource that is more suitable for swimming, drinking, boating, or washing, and is of no cost to users of the resource, it provides a value in use that can be measured and included in resource decisions.

Nonuse value refers to any values not directly associated with human uses of natural resources and includes two types: **option value** and **existence value.**

Option value. If the future benefits that a resource might yield are uncertain and the depletion of the resource would be irreversible, one might value preserving the option to use the resource in the future. For example, the pharmaceutical industry relies on plants and animals for potentially curing diseases. As the industry gains more information about a particular species, it may begin to place value on having the option to use the species in the future. The magnitude of the uncertainty and the extent to which people are risk-averse determine the magnitude of the option value.[4]

Existence value. Also called conservation or intrinsic value, existence value is independent of people's present use of the resource. These values arise from a sense of environmental stewardship related to a responsibility to preserve natural resources for future generations. Even if a resource does not have any clear value in use in the present (use value) or in the future (option value), people may wish to preserve the resource because they believe it has a right to exist and should be protected.[5]

Therefore, the total value of a resource is the sum of the three components:

Total value=use value+option value+existence value

In many cases, the distinction between these values is unclear because individuals can sometimes derive both use and nonuse values from a resource. For example, a person's interest in preserving a wilderness area may be motivated by the anticipation of hunting *and* the pleasure of conserving it for future generations. Additionally, placing a value on the continued maintenance of a species could be considered an existence value or a use value because the user obtains some value from knowing the resource exists and there may be a clear and present benefit to the user.

So, how can managers measure these values and incorporate them into organizational decisions to reduce social, environmental, and economic impacts and also to improve long-term profitability? How can they include these measurements in reports to various stakeholders to improve their ability to monitor and evaluate the performance of the firm on various dimensions including past, current, and future financial, environmental, economic, and social performance? How can managers include these measurements in decision-making and evaluate the importance of various social, environmental, and economic impacts (both costs and benefits) in various operational and capital investment decisions?

One step is to express the use and nonuse values in terms of individuals' WTP (willingness to pay) for the resource or WTA (willingness to accept) compensation in exchange for the resource. Then managers can use the economists' approach of consumer surplus to estimate what constituents are gaining from the resources available:

- **Willingness to pay.** One way to measure consumer benefit from social, environmental, and economic improvements is to compare what they are willing to pay for them with actual price for these services. Thus, if a social, environmental, or economic benefit is provided at no charge, the stakeholder benefit can be measured by the amount that they would be willing to pay for it. Aggregated, this provides an estimate of the total benefits provided

- **Willingness to accept.** An alternative approach is to examine the amount of money stakeholders would be willing to accept that would make them indifferent to degradation in the environment, the society, or in ethical values or practices

- **Consumer surplus.** Consumer surplus is the basic approach that economists often use to measure consumer benefits. It is the difference between what one is willing to pay and what one actually must pay to acquire a service or product. Thus, when stakeholders are provided with a benefit (for example, a reduction of pollution in the environment) at no cost or at a cost that is less than they would be willing to pay, they receive a consumer surplus

Though both WTP and WTA have been found to be good approximations of social, environmental, or economic impacts, studies have determined that the results are somewhat higher in WTA analyses as stakeholders state how much they need to be compensated for the damage from social, environmental, or economic declines from the status quo. By using the status quo as their reference point, stakeholders require higher compensation to allow social, environmental, or economic degradation than they are willing to pay for making improvements. However, WTP questions can often result in higher estimates of values on commodities (for example, a quality improvement on a TV) and could capture an attitude that is not appropriate for economic translation to a stated preference in terms of a WTP; that is, the theory that underlies WTP, economic consumer theory, may not always be appropriate.[6] But, although the method may not be entirely precise, it does provide an effective approach to measuring stakeholder reactions and a relevant quantifiable measure of corporate social, environmental, and economic impacts and performance.

Both WTA and WTP measures are based on the assumption of substitutability of goods and services, but WTP is constrained by the individual's income and tends to lower its value of social and environmental goods and services. WTA has no upper limit on what a person might ask as compensation for giving up the right to use public services, and so the goods or services tend to be overvalued.[7] It is important to assess this ambiguity when designing surveys to measure social, environmental, and economic values. Because WTP is constrained by the realistic limitations of an individual's income level, but WTA is not subject to any such constraint, the US Department of the Interior and the US EPA (Environmental Protection Agency) have both endorsed the use of WTP, instead of WTA, to achieve conservative estimates in costing studies of environmental damage. A recent study again proved that consumers are in fact willing to pay more for ethically produced goods and will demand a substantial discount from companies that produce goods in an unethical manner.[8]

In Chapter 4 we discussed the costing systems (ABC, LCC, and FCA) used to analyze and integrate social, environmental, and economic costs that would otherwise go unaccounted for. Understanding the magnitude of internal costs through application of these types of costing system is important. But to fully evaluate the impact of social, environmental, and economic costs on its operations, a company must also evaluate external costs, especially for potentially large liabilities or for impacts that are likely candidates for future regulation. External costs present another level of complexity in that, unlike conventional internal costs, they are not transaction-based and often cannot be directly observed in the marketplace. Once monetized, the cost estimates can be integrated into the costing systems for improved decision-making. Using concepts such as WTP, WTA, or consumer surplus can aid managers in placing value on how their products and services affect company stakeholders. By internalizing these external costs (externalities) managers have a better understanding of the long-term sustainability and financial impacts of their actions.

Methodologies for measuring social, environmental, and economic impacts

A variety of techniques have been developed to collect data on WTP or WTA. The first type is categorized as **revealed preference methodology.** Revealed preference methods use estimation of actual behavior to determine the value people place on social, environmental, and economic products and services. When expenditures vary with the level of sustainability impacts posed by the product, then the value of the impact can be estimated. The value can be estimated using various methods including the travel cost method and hedonic pricing (discussed later in this chapter). The second type is known as **stated preference methodology.** In this approach, people respond to hypothetical questions rather than observations of real-world alternatives. It is used to evaluate potential social, environmental, and economic policies or when nonuse values are involved. The primary approach used for stated preference is contingent valuation. Table 6.1 summarizes several of the approaches.

TABLE 6.1 Methodologies for measuring sustainability impacts

	Description	Advantages	Disadvantages
Cost of control and shadow pricing	• Cost avoiding damage before it occurs	• Avoid difficult-to-determine actual costs	• Shadow pricing assumes legislators accurately value costs of damage
		• Simplicity of calculations	
Damage costing	• Actual costs of damage	• Recognizes external damages	• Difficult to assess monetary effects
Market price and appraisal	• Resources are traded in existing markets	• Uses LCA	• Requires existence of a competitive market
Contingent valuation	• Hypothetical questionnaire	• Assesses passive use values	• Lacks precision
		• Helps identify impacts	

	Description	Advantages	Disadvantages
Hedonic pricing	• Property value of wages as proxy of costs	• Values an entire range of impacts simultaneously	• Precision is often challenged
Travel cost	• Cost of travel to recreation sites	• Data are available	• Difficult to measure hypothetical alternatives

While none of the methods described represents a perfect proxy for the cost of social, environmental, and economic damage, each gives a sense, at the very minimum, of the general magnitude of the cost. Each approach to monetizing external costs provides information about market and nonmarket values that is important for effective decision-making.

Cost of control and damage costing

Many companies have considered two major approaches to monetizing social, environmental, and economic externalities: the **cost-of-control approach** and the **damage-costing approach.** The cost-of-control approach is defined as the cost of reducing or avoiding damage before it occurs. Damage costing focuses on attempting to assess actual cost incurred from social, environmental, and economic damage.

Cost of control

The cost-of-control approach is a measure of the cost of reducing or avoiding social, environmental, and economic damage before it occurs to place value on the damage itself. For example, a facility could assign a value to soil contamination from a leaking underground storage tank by estimating the cost of the equipment needed to prevent the leak, or the cost of implementing a process redesign that would eliminate the need for such tanks. The cost-of-control approach avoids difficult-to-determine actual costs of environmental damage by replacing them with more easily estimated costs of installing, operating, and maintaining environmental control technologies.

Advocates of cost of control contend that control costs are an acceptable substitute for damage costs. In certain situations, cost of control can represent the most realistic estimate of that dollar value which will eventually be internalized by the organization. The use of child labor in some parts of the world has led to product boycotts, ruined company reputation, and multi-million-dollar lawsuits. When considering the use of child labor by itself, contractors, or licensees, a company might assess the costs of using part-time workers, overtime, or exclusively adult workers to control or avoid the cost of any negative social impacts. Cost of control can also be seen as the cost to mitigate risk. The mitigation could be accomplished through insurance or various actions to control or avoid the cost.[9]

A variation of cost of control, **shadow pricing,** deduces the cost of avoidance from existing regulations. Shadow pricing implies society's willingness to pay for sustainability performance from the cost of specific measures that have been required under regulations. In other words, the basis for valuation under this approach is the cost of complying with regulations. Like the cost-of-control approach, shadow pricing uses the cost of controlling sustainability impacts to monetize social, environmental, and economic damage. This approach regards existing and proposed social, environmental, and economic regulations as estimates of the value that society implicitly places on specific sustainability impacts and extrapolates the cost of future impacts of the same type from this implied willingness to pay for avoidance.

The most significant advantage of cost of control and shadow pricing lies in the simplicity of the calculations. Cost information about control technologies is readily available, and a given control strategy can usually be linked to a quantifiable reduction in the sustainability impact being controlled. However, the most significant weakness of this approach is that the resulting value may bear little relation to the true cost to society of the impacts being avoided. Ideally, the cost of the social, environmental, or economic damage itself should be quantified, and the cost of the control technology should be used to evaluate the cost-effectiveness of investment in preventing that damage.

In addition, these approaches do not account for national or regional differences or site-specific characteristics associated with various options. This information can be critical in determining the extent of damage. For example, the cost of control for two similar power generation stations would be the same even if one were located close to an urban center with high population density and the other situated in a rural area.

Damage costing

In contrast to cost of control, which uses the remediation cost of a sustainability impact as a basis for decision-making, damage costing attempts to assess the actual economic cost of the social, environmental, or economic damage. The loss of value attributable to the damage is estimated by the public's willingness to pay to avoid the damage. This willingness to pay can be extrapolated from market-based data on the commodity or impact in question (as in the market price, hedonic pricing, and travel cost methods discussed below) or can be observed through a survey that replicates the commodity in the form of a valuation scenario (contingent valuation).

Because damage costing focuses on site-specific impacts and provides a realistic estimate of external damages, it is able to assign an economic value to nonmarket commodities. The downside of damage-costing approaches, and the reason that so few companies use them today, is that they are complicated and require substantial data. Collecting adequate data to perform the analyses could require significant time and financial resources. Because proper design and execution of a contingent valuation or other damage-costing exercise is critical to its usefulness as a valuation tool, organizations commonly turn to third parties for research and advice. This can translate into a substantial expense. A company attempting to estimate a potentially large social, environmental, or economic liability would need the expertise of professional survey designers and trained interviewers to produce accurate results. Sometimes, though, approximations can be developed simply and quickly and aid in better understanding the scope of the impacts and improve decisions. And, in many instances, large or complex projects do warrant the resource expenditure to do a complete analysis.

The basic principles of damage costing have broader applicability for an organization. The estimation of passive-use values or other external costs through solicitation of consumers' willingness to pay can be applied informally in the preliminary phases of internal decision-making. Damage costing can be helpful even in the absence of a formal study. Conducting focus groups or brainstorming with employees or community constituencies to estimate order of magnitude, if not a specific value, can clarify a company's vision of its environmental priorities.

The approach has often been applied specifically for cost–benefit analysis of health, safety, and environmental policy and for assessing damages in civil cases. One of its main applications, however, has been in health valuation, defined as the cost of illness approach. Valuing the cost of illness requires the identification of direct and indirect costs associated with illness, injury, or death. Direct costs include the resources used to diagnose, treat, rehabilitate, or support ill or injured persons affected by adverse social, environmental, or economic conditions. Valuing these costs is done by identifying the relevant categories of healthcare costs such as hospital care, physician services, nursing, and home healthcare, estimating utilization rates by persons with the same condition, and multiplying by cost estimates of each category.

Indirect costs, often calculated separately, relate to forgone earnings from morbidity and mortality. Morbidity may affect earnings through increased absenteeism, reduction in the amount of time a person works, or impairment of a person's ability to perform specific activities. Mortality losses are estimated from the value of statistical life derived from the willingness to pay to reduce fatal risks.[10] For example, a firm that has operations in South Africa could be faced with a growing number of employees who are becoming symptomatic and dying of Aids. The company realizes that it will bear a large overhead cost due to absenteeism, high turnover, and the need to consistently train new skilled personnel for the jobs that are being left open because of Aids-related deaths. Using a damage-costing approach, the company could estimate the earnings lost by doing nothing to mitigate the situation or it can estimate the costs of undertaking a variety of programs such as workplace education, condom distribution at work sites, HIV testing at facilities, or medical treatment for workers and families (see the De Beers example in Chapter 4).[11]

In the 1990s, Ontario Hydro, at that time Ontario's hydroelectric power company, used the damage-costing approach, rather than the cost-of-control-approach, to identify, quantify, and, where possible, monetize the external impacts of its activities. Its reasoning for using damage costing was that the approach considers specific environmental and health data, uses modeling techniques that take into account how emissions and effluents are transported, dispersed, or chemically transformed in the environment; and then considers who or what (for example, people or fish) are affected by these emissions. The company could then apply economic valuation techniques to translate physical impacts into monetary terms.

Although the company realized that there was a degree of uncertainty associated with the quantification and monetization of externalities, it concluded that uncertainty was pervasive in many areas of business and the measurement of environmental externalities must be placed in that context. Using the damage-costing approach, Ontario Hydro attempted to assess the actual costs from environmental damage. Although calculation may be difficult, this approach can provide a realistic estimate of external damages, including human health problems, animal herd losses, and crop damage from toxic air and water emissions.

Market price and appraisal

Adverse social, environmental, and economic impacts identified during a life-cycle assessment can often be linked with the damage, depletion, or loss of resources that do have market values. Air pollution from an electric power plant may cause acid rain, which leads to crop losses in the region. The market-pricing approach directly measures the market value of resources damaged or lost as a result of social, environmental, and economic impacts. For example, the utility could assess the cost of air pollution using the market value of the resulting crop losses. In evaluating human health impacts, the cost of medical treatment can serve as a useful proxy.

The market-pricing approach is useful because it expresses social, environmental, and economic damages in terms of concrete, tangible losses of economic value. Clearly, virtually all corporate activities and decisions translate into a variety of associated ultimate impacts, and ultimate impacts can usually be traced to a range of sources. However, developing and implementing a model such as the one discussed in this book can be helpful in understanding the impacts of corporate activities and the subsequent effects on stakeholders and company profits. Such an analysis can be used to estimate both acute and chronic social, environmental, and economic risks. The market-pricing approach can broaden the evaluation of long-run losses, making these values present in short-term decision-making as well.

The market-price method requires resources or services to be actually traded in a reasonably competitive market through voluntary exchanges between buyers and sellers. The value of the service is directly revealed through the market process. Where such exchanges do not exist, a professional appraiser's knowledge of markets may be used instead of directly observed values. This method also needs a market to exist to provide the appraiser with knowledge of market outcomes.

Hedonic pricing

Most social, environmental, and economic impacts cannot be expressed strictly in terms of damage to private goods. Therefore, valuation of more general indicators of social, environmental, and economic quality must supplement market price assessments. Hedonic pricing is one method of valuing consumers' willingness to pay for superior social, environmental, and economic quality. This technique applies information derived from surrogate markets for private goods, traded in a competitive market, which may bear some relationship to a public social, environmental, or economic good.

The most commonly used surrogate markets for social, environmental, and economic quality are real estate and labor markets. Hedonic pricing has been used to estimate the impact of environmental deterioration by examining the decline in real-estate values after a contaminant has been discovered. The method can also be applied to decreases in real-estate values based on an airport expansion that changes a flight pattern or changes in government policy affecting the desirability of living in a particular neighborhood. A company facility and its impact on the community will probably also affect real-estate values as residents desire to be close to or distant from the facility depending on its level of pollution or its appearance. Additionally, a company can also affect real-estate values through its investment in other areas in the community such as schools, community centers, and various community programs. The hedonic-pricing method assumes that consumption of housing depends on the characteristics of the house, neighborhood characteristics such as parks and crime rates, and location-specific social and environmental impacts.[12]

Hedonic pricing also uses labor markets to determine salary scales and premiums for riskier jobs. Labor costs can reflect environmental differences: holding all else constant, workers in positions requiring exposure to environmental hazard would demand a risk premium. This is, in effect, a proxy for willingness to accept environmental risk. Thus, companies can use information from these markets to place a monetary value on environmental quality.

The main advantage of hedonic pricing is that it allows the entire range of impacts associated with an activity, as reflected in overall environmental quality observed by homeowners or wage-earners, to be valued simultaneously. Hedonic pricing applies statistical modeling techniques to identify differences in property values or wage rates specifically attributable to environmental quality, separating out other attributes that influence the decision to purchase property or accept a job in a given region.

Travel-cost method

The use of recreational sites can also lend insight into the value the public places on social, environmental, and economic quality. TCM (travel-cost method) uses observed expenditures and behavior to develop an indirect measure of the economic value of nonmarket goods. In particular, travel to recreational areas can indicate the value of maintaining those areas. TCM most commonly serves as a tool to evaluate alternative management plans for recreational areas. The difference in travel costs between two management alternatives illuminates the potential gain or loss in economic value associated with changing from one practice to another. Travel costs typically include both direct travel expenses and some measure of the opportunity cost of scarce time, although a variety of methods have been applied to approximate opportunity costs. Using the cost per visit and the number of visits in a given time period, a demand function is estimated.[13]

In addition to its direct use by recreational planners, TCM can be a valuable tool for business decision-makers faced with potential impacts on nearby recreational sites in two ways. First, facility managers can use TCM and hypothetical TCM to monetize their direct impacts on recreational sites in the region. Deterioration of surface water or air quality as a result of industrial pollution may manifest itself as a decrease in the use of nearby recreational facilities, and the value of this loss use can be used as a proxy for the damage done.

Second, the money spent on travel to recreational sites can be an indicator of poor social, environmental, or economic quality in the region from which people are traveling. The opportunity cost of this travel—money that would have been spent within the region if social, environmental, or economic conditions did not motivate people to leave the area—represents a loss of economic value to the region. For example, a company might use TCM, in conjunction with other valuation methodologies, to estimate the economic impact of a spike in crime rates or pollution from fossil-fueled power stations, evidenced as increased travel away from the affected region.

While TCM permits valuation of an existing scenario, using current management practices and travel frequencies, decision-makers frequently need to compare current practices with a hypothetical alternative. Traditional TCM alone does not provide a structure for evaluating how individuals will value a recreational site with a decline or improvement in environmental quality, or under management options that have never been implemented. Others propose use of a hybrid methodology that blends traditional TCM techniques with contingent valuation. This approach, called the hypothetical travel-cost method, uses traditional TCM to estimate current demand for recreation opportunities delivered by a given site. In this approach, respondents are asked about actual trips taken to the site and to substitute sites, actual expenditures per trip, and hypothetical questions about what the respondent would have done if they had not taken the trip in question. These questions about decisions under actual conditions are asked prior to questions about hypothetical behavior, reminding respondents about their actual behavior. Respondents are then given hypothetical scenarios and asked how many trips they would make under those circumstances. In contrast to typical contingent-valuation methods, price and payment vehicle are not explicitly stated.

TCM has several limitations. First, it provides an estimate of WTP for the entire site, but people may value only specific features within the site. If the analysis were completed to enhance or add a particular feature, TCM would not serve the purpose unless the survey specifically addressed preferences within the site. Another limitation is that the method is limited to evaluating sites to which people from different zones have significantly different costs. It could not be used to evaluate a site where people attending have easy access because there would probably be little variation. TCM also relates the total cost of traveling as WTP to attend the site, but the persons traveling could have multiple destinations and in such case only a fraction of the travel costs should be attributed to the site. The estimation of opportunity costs is also a difficult task but needs to be factored in if visitors face radically different costs for their time.

Contingent valuation

The cost-of-control method, the market-price method, hedonic pricing, and travel costing all rely on existing market values. However, in determining the full range of costs associated with social, environmental, and economic impacts or improvements, some costs evade extrapolation from market transactions. At the same time, limiting the analysis to market-driven values has the potential to severely underestimate some environmental, economic, and social costs. While the value of national parks and wilderness areas might in part be inferred from annual collections in visitors' fees and other expenses related to visiting these sites, the existence of such areas has inherent value even to those who do not visit them. Likewise, pressure to protect spawning grounds for threatened fish species may derive largely from populations other than commercial or recreational fishermen, the direct users of the resource. For some commodities, "passive use" value or "existence" value—the benefit perceived by people who do not directly use the commodity—is large and can have a significant impact on decision-making.

CV (contingent valuation) is a method that has been used since the 1960s to estimate passive use values and can be used in conjunction with the other methods described above.[14] CV assesses WTP for a defined benefit or WTA payment for a defined loss, by presenting consumers with a hypothetical market in which they have the opportunity to buy or sell the goods or services in question. Though there are questions about the precision of CV estimates, the method has been used extensively in valuing social, environmental, and economic impacts and can be used effectively to improve management decision-making.[15]

CV studies take the form of a questionnaire describing a hypothetical "valuation scenario" and moving through a series of WTP or WTA questions. CV studies are often presented as a referendum upon which participants in the study will vote on what they are willing to pay or accept. CV obtains the estimate of the benefits of a public good, which can then be used in a cost–benefit analysis.

Despite its usefulness in assessing passive use values, there are some potential challenges and limitations with the CV process:

- **Inconsistency with rational choice.** Some CV studies have found that WTP does not increase with the magnitude of the benefits
- **Uncertainty of responses.** Because a CV study typically ask participants to consider only one valuation scenario at a time, critics have claimed that individuals give WTP responses that are unrealistically large, considering the multitude of environmental commodities from which a consumer must typically choose at any given time
- **Absence of a meaningful budget constraint.** Individuals may respond without considering how much disposable income they have available to allocate
- **Information provision and acceptance.** Some CV studies have not provided respondents with adequate information to make informed valuation choices. Even if substantial information is provided, its usefulness is limited by respondents' ability to process and use the information in formulating their responses[16]

Nonetheless, CV studies, when properly designed, tested, and executed, constitute an important tool for gaining insight into passive use values. Managers have found that, even though CV usually lacks precision, it is very useful for determining the impacts of a product, service, or activity on the community and other stakeholders. Furthermore, the first step of the identification of the impacts provides useful information and the CV will at least provide a sense of both the direction of the number of the impacts (are they positive or negative and do the costs exceed the benefits?) and the scope of the numbers. This usually provides sufficiently precise information for the decision and usually more precision than most other methods. In fact, in most cases, companies have not been including any of these measurements in their management decisions. Though the measures are imprecise, the analysis is critical for improved managerial decisions. Currently, companies tend to ignore these significant impacts in their capital investment decision-making (and thus implicitly consider that these impacts have no value). Decisions are improved with more rigorous measurements and reporting of the risks including the impacts on both sustainability and financial performance even when included as range of estimates instead of point estimates.

Evaluating impacts of natural gas drilling

Sublette County, Wyoming, contains one of the largest natural gas reserves in the US. Six main energy companies operate in Sublette County: Anschutz Exploration, EnCana Corporation, Questar, Shell, Stone Petroleum, and Ultra Petroleum. The rapid expansion of gas production and development has triggered tremendous environmental concerns. Development, due in part to natural gas drilling, threatens the migration corridor of pronghorn antelope, elk, and mule deer. The population of the sage grouse, which makes use of this habitat during summer months, has declined by 90% over the past century owing to the loss and degradation of sagebrush habitats.

Residents are concerned about protecting the environment for future generations but also realize the value in having gas production in the area. To partially alleviate concerns, there is a winter moratorium on drilling natural gas wells. During this annual moratorium (November 15 to April 30), energy companies operating on public lands cannot drill any new wells. In addition to the moratorium, there are restrictions on the spacing of wells in an attempt to preserve the habitat for big game animals and sage grouse.

So how can the energy companies balance development with the need to be environmentally responsible? It is important that these companies analyze and evaluate the effect of their strategies and systems on both sustainability and financial performance. The sustainability performance impacts financial performance most prominently through stakeholder reactions. It is therefore critical to identify the likely stakeholder reactions and acknowledge how the stakeholders make necessary trade-offs. An estimation of willingness to pay to offset the development of the environment is an important part of understanding stakeholder reactions. To elicit data on stakeholder reactions, a contingent valuation study was conducted with a broad range of stakeholders including homeowners, ranchers, hunters, conservationists, environmentalists, local businesses, and government. Details of the survey used, survey methods, and measurements will be discussed further in Chapter 7.[17]

Methodologies for measuring sustainability and political risks

Social, environmental, economic, and political issues pose risks to companies that must be measured and managed. Bob Dudley, the CEO of BP, wrote in his letter to stakeholders: "Following the Deepwater Horizon incident, our employees have worked systematically to enhance safety and risk management. And we have turned the insights gained into new oil spill response plans and technologies, which we are adopting within BP and sharing with others. As someone who has worked in the oil and gas industry for more than 30 years, I know that risk can never be entirely eliminated. But it can be managed effectively, and in increasingly sophisticated ways."[18]

In fact, risks should be monetized for inclusion in ROI calculations, and to improve resource allocation and investment decisions. Product take-back and producer responsibility (requirements that companies accept responsibility for final disposal of their products such as computer goods, cartridges, and appliances) is increasingly common throughout the world. Similarly, site clean-up has become mandated in many locations, and companies are now recognizing that they did not consider these sustainability and political risks when making costing decisions. This has led to underestimating total product cost. Better forecasting of potential changes in the social, natural, economic, and political environment can lead to improved decision-making on process, product, and capital investment.

Not only do managers need to know the impact that their products, processes, and services have, they also need methods to measure the risks they undertake when making decisions. Contingent liabilities can constitute a substantial risk, even if the associated probability is very small, and it should be emphasized that scenarios of very low probability should not be ignored. Social, environmental, economic, and political risks are typically low-probability, high-cost events—often with long time horizons. This makes the analysis of these risks quite critical. In the case of a nuclear power plant, the risk of meltdown, however improbable, is so potentially disastrous that it merits considerable precaution. Social, environmental, economic, and political assessments that omit measurements and discussions of risk can create future legitimacy and credibility problems for the company.[19]

Volkswagen quantitatively and qualitatively evaluates all risks against eight criteria, and assigns ratings. The results of the risk evaluation process are reported to the board of management, and the process itself is reviewed annually. This wide-ranging evaluation is at the heart of Volkswagen's climate protection strategy which takes into account several factors: firstly, regulatory aspects, in particular EU penalties for failing to meet fleet-average emission targets; secondly, market-related requirements, resulting in particular from increased public awareness of climate issues; and, thirdly, physical aspects, such as potential supply-chain or production disruption. Water shortages pose a significant risk to Volkswagen's operations, particularly in light of the company's plans for new production facilities in Asia, Africa, and Central and South America.[20]

An organized methodology for quantifying uncertainty in risk assessment and NPVs (net present values) should comprise technical evaluations, programmatic interpretations, and mathematical computations, with the joint goal of measuring the degree of confidence with which the estimate is held. Every estimate of risk is in actuality the sum of estimates of risk levels of a large number of contributing factors, each of which is itself uncertain to some extent. In fact, the simple act of conducting such an analysis often calls attention to possible risks that previously had been unnoticed.

To measure risk, management must first identify the potential liabilities. The scheme in Figure 6.4 classifies risk into four broad categories—strategic, operational, reporting, and compliance:

- **Strategic risks** relate to an organization's choice of strategies to achieve its objectives
- **Operational risks** relate to (1) threats from ineffective or inefficient business processes for acquiring, financing, transforming, and marketing goods and services, and (2) threats of loss of company assets, including its reputation
- **Reporting risks** relate to reliability, accuracy, and timeliness of information systems, and to reliability or completeness of information for either internal or external decision-making
- **Compliance risks** address the inadequate communication of (1) laws and regulations, (2) internal behavior codes and contract requirements, and (3) information about failure of management, employees, or trading partners to comply with applicable laws, regulations, contracts, and expected behaviors[21]

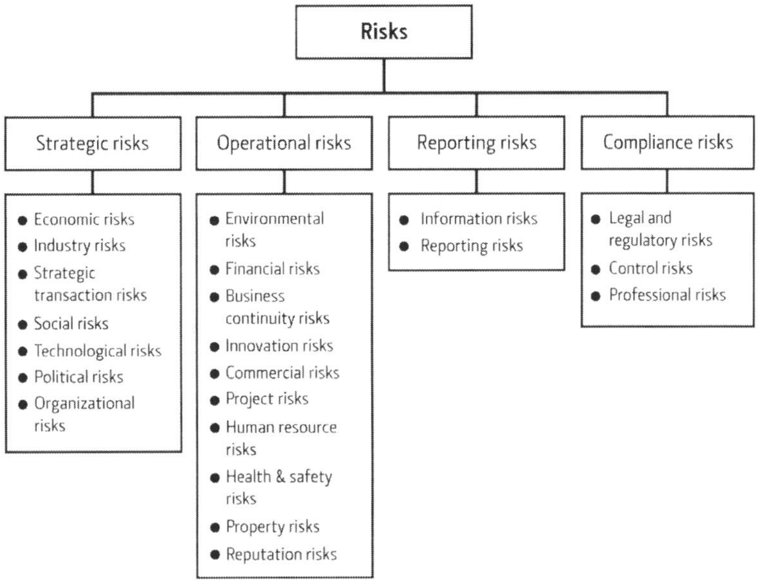

FIGURE 6.4 **Risk classification**

Source: Epstein and Rejc (2005) *Identifying, Measuring, and Managing Organizational Risks for Improved Performance*

This classification is used in Chapter 7 as a guide to specifically identify, measure, and manage social, environmental, economic, and political risks. These contingent or probabilistic costs can then be evaluated in terms of their expected value—the cost of the impact, weighted by the probability of its occurrence. Using a decision tree to structure the various potential outcomes of different management options, managers can use expected-value calculations to provide a realistic monetary value for use in decision-making. Several methods of valuing contingent liabilities are available.

Scenario-based methods

A tool used by several companies to identify social, environmental, economic, and political issues and opportunities is scenario analysis. The approach is based on anticipating stakeholders' reactions to and concerns about sustainability in order to determine the underlying issues. Those issues that could have an impact on the business are then grouped, and different scenarios are developed and forecasted. In companies with high levels of uncertainty, where change is imminent and diversity of opinion exists, scenario forecasting can be useful to clearly identify the various choices for decision-makers. Some have suggested that scenario forecasting aids in assessing and managing risk, broadens corporate thinking, and makes managers focus on the long-term impacts of their decisions.

Royal Dutch Shell has a history of using scenarios to forecast alternative future events and identify potential challenges associated with current decisions. Shell's global scenarios are prepared every three years by the Global Business Environment Unit. The scenarios are carried out in four phases:

1. **Research.** This phase addresses knowledge gaps and attempts to reframe thinking in order to identify new challenges for the company

2. **Scenario building.** Interviews are conducted with leaders and workshops are held to build scenarios

to address the challenges identified during the first phase

3. **Application.** Participants take ideas back to their respective business units. Once finalized, scenarios are presented at workshops and used to test strategies

4. **Dissemination.** Copies are distributed to staff and presentations are made to staff in locations around the world

Shell has used this process to write scenarios on topics including de-integration of the international oil business, entry into inaccessible countries, and the inclusion of renewable energy sources in Shell's portfolio.[22]

National Australia Bank uses tools such as stress testing, scenario planning and economic modeling to help the group, business lines and support functions to understand their resilience in the event of a significant event or shock, and to help monitor and prepare for future opportunities and threats.[23]

Risk mapping is also commonly used and plots the expected frequency, severity, and degree of exposure of various risks on a graph, with probable frequency on the horizontal axis and expected severity on the vertical axis.[24] Calculations are made according to the following formula:

Exposure=(event)þ(hypothetical likelihood)þ(hypothetical consequence)

The benefit of such modeling is that it permits measurement of various types of risk, and enables managers to visualize where to allocate resources for risk management. In addition, mapping is a valuable communication tool, providing a comprehensible visual review of exposures, though they are often not expressed in monetary terms. For this reason, mapping as currently practiced does not provide a link to the financial statement, or to the ROI calculation that is critical for comparisons between possible project options. With some modification, including assignment of monetary values to the hypothetical consequences, however, axis points on such a risk map could correlate to financial data and be integrated into ROI calculations.[25] The integration of these risks into ROI calculations is discussed in Chapter 7.

Some companies may consider not only the scenarios of impacts produced by the corporation's products and processes but also the effects that third parties have on sustainability. For example, when a company's staff travels on business, the airline generates emissions of carbon and nitrogen oxides. These gases can be calculated from total air miles traveled by company executives and could then be monetized. From this information, companies could not only improve their environmental performance (gas emissions) but could also cut costs by avoiding travels when possible and using other sources of communication (tele- and video-conference, for example).[26] For example, British company Vodafone, the world's second largest mobile telecommunications company, reduced its carbon emissions from business travel by 24% in 2012/13. Business travel was limited to only the most essential trips as part of the group-wide cost reduction efforts, bringing accompanying carbon savings. At the same time, Vodafone has increased the use of its remote collaboration technologies by almost 30%, with employees using video conferencing for more than 75,000 hours a month. Vodafone's flexible working programs in several markets have also helped to reduce energy use and emissions from offices and employee commuting.[27]

Companies should also think about scenarios involving contractors in the supply chain. As contractors engage in activities such as poor manufacturing practices or working conditions, other companies in the supply chain are often affected. Use of a scenario-based model would monetize the risk to the company in continuing a partnership with this supplier.

Fuzzy logic

Fuzzy set theory is a branch of mathematics dealing with sets of information that do not have precise boundaries. To account for uncertainty, a "best" estimate is provided to establish the "most likely" dollar value that will be required to cover the foreseeable consequences and the most probable to occur of the uncertain consequences. Next, the most optimistic (best case) and pessimistic (worst case) monetary value limits are estimated.

To use fuzzy logic, identified possible magnitudes of future social, environmental, and economic liability independently are assigned a DOB (degree of belief), between 0 and 1. These future liabilities might include various externalities such as water and air pollution of a residential area near a plant or internal disposal costs for chemicals used directly in the manufacturing process. DOBs are also assigned to possible interest-rate levels for each period. All possible combinations of circumstances that define the range of possible realizations of future financial liability are considered. NPV is calculated for each such realization, and a DOB is derived by combining the DOBs attached to each circumstance and associated with that NPV level. The fuzzy logic analysis results in a set of possible NPV levels, along with a DOB for each. One way to use the results of a fuzzy logic analysis is to rank the possible NPV levels according to DOB magnitudes. Though fuzzy logic has many limitations, it does provide an alternative approach to identifying and measuring future sustainability impacts.

Monte Carlo simulation

For complex decision trees, Monte Carlo simulation can be used to calculate the probability distributions of outcomes. First, the user expresses a given social, environmental, or economic risk in terms of a probability distribution. That risk can increase or decrease depending on changes to social, environmental, or economic regulation or improved information. Once probability distributions are established for all inputs required for an NPV analysis, a computer program implementing the algebraic formula for NPV is written, except that when dollar value of future liabilities or interest rates is called for, it is replaced by random numbers drawn from appropriate probability distributions. More commonly, widely available software is used to run the Monte Carlo simulations.

The computer goes through the decision tree, drawing a sample from the relevant probability distributions at each point where an event occurs, and then applies simple logic to determine how to proceed through the tree. At each point in the tree where the computer must choose among alternatives, it will choose to minimize cost. If the decision tree has different possible events, the computer will model each event and the possible outcomes. This process is repeated until meaningful probability distributions can be established.

Many companies have applied Monte Carlo analysis to the problem of comparing the possible costs of alternative environmental remediation options. Using Monte Carlo random sampling from an option's cost-probability distribution, the probability that one option will cost more than another can be estimated and the most likely costs of each operation can be compared. Probabilities (that is, confidence levels) can be assigned to a range of possible costs, leading to more credible and defendable comparisons.

Option pricing, option assessments, and option screenings

Option pricing is a method for calculating the expected market value of an option. It models the time series interaction between investments and has been used most often in the financial markets (stock options). The value of a stock option is determined primarily by the volatility of the underlying stock. The same kind of methodology can be applied to social, environmental, and economic investment decisions. As social, environmental, and economic regulations and information change, so do options, processes, and products. The value of the strategic social, environmental, or option increases with the riskiness of the underlying cash flows.

Real options analysis provides a way to aid the framing of decisions for risk analysis. It is consistent with discounted cash flow approaches but also provides a recognition that plans often change as new information is obtained. Thus, when a static model may be inadequate, real options may be used for capital investment decisions to articulate how small early expenditures may preserve options for future investments. The calculation of the value of the investment is thus likely increased due to the value of preserving these options.

Option assessments and **option screenings** are designed to provide decision-makers with a full vision of alternative courses of action, their associated costs, and their relative attractiveness. This helps analyze the choices, options, and the value of retaining some of those options for future managerial decisions. The process consists of four steps:

1. Drawing a flow diagram
2. Identifying the major social, environmental, and economic issues
3. Defining the options
4. Selecting the most likely options for further investigation, based on cost-effectiveness, relevance for decision-makers, and sustainability impact

The EPA has proposed an option-rating weighted-sum method for screening and ranking pollution prevention options. The method involves three steps:

1. Important criteria in terms of program goals and constraints are determined and each is given a relative weight
2. Each option is rated on each criterion on a scale of 0–10

3. The rating of each option for a particular criterion is multiplied by the weight of the criterion

The option with the best overall rating is chosen and may be subject to further technical and economic analysis.[28]

Niagara Mohawk Power Company has used option screening to compare various externalities. It implemented a system to identify and measure the options related to both demand and supply sides of electric power usage. The company used option screening to determine the optimum mix of demand and supply strategies that provide electrical energy services at the lowest cost, within a set of various constraints. It used focus groups to determine the appropriate options and assign probabilities to the most likely scenarios.

Niagara Mohawk developed five separate tests of cost-effectiveness to use in the screening analysis. The results of these tests were used in the screening process to determine all relevant costs and benefits and to choose the best option. The objective was to optimize and balance economic, financial, environmental, energy and engineering, and customer service objectives to determine the best resource plan considering trade-offs relative to numerous uncertainties, constraints, and policy objectives.

Summary

The evaluation of the social, environmental, and economic impacts of an organization on society is important for management decisions. This evaluation is important to better meet the needs of the various stakeholders and usually benefits all of the stakeholders. By more broadly examining the needs of all of the stakeholders, both sustainability benefits and long-term corporate profitability are often increased. The method of evaluation of the impact of an organization's activities, products, services, and processes on society, environment, and economy is critical.

Although most managers understand the importance of measuring social, environmental, and economic impacts, it often remains difficult to implement. Through methods such as those described here, sustainability performance can be measured. Available data are gathered, assembled, and processed to provide the best available information, the likely outcomes, and the likely impacts of those outcomes. Although these methods often seem to lack precision, they can provide an estimate of how companies are performing. These methods provide guidance to managers when making difficult decisions when social, environmental, or economic interests and corporate interests are not aligned. They provide a solid academic foundation for developing measurements for sustainability performance.

The following chapter provides specific guidance on how to execute these methods and design measurements to use in improving resource allocation decisions for both operating and capital investments.

CHAPTER 7

Implementing a social, environmental, and economic impact measurement system

The identification and measurement of the costs and benefits from corporate sustainability activities is critical to the evaluation of projects within the company and the evaluation of the company and its components and members. As the previous chapter shows, there is a solid academic foundation for measuring sustainability performance. Significant improvements in the development of corporate performance measurement systems that include both financial and nonfinancial measures permit much-improved evaluation of social, environmental, and economic impacts. This is aided by vast improvements in corporate information technology capabilities that permit the collection, aggregation, and disaggregation of information for improved analysis, management, and reporting.

In the last chapter we looked at methods such as hedonic pricing, market pricing, and contingent valuation. In this chapter we translate these concepts and approaches into systems and measures that can be effectively implemented. We look at:

- The drivers of sustainability performance
- Measuring reputation
- Measuring risk
- Measuring social, environmental, and economic impacts

Many social, environmental, and economic impacts may appear to have no market consequences and no financial effect, but many of the externalities are internalized in future periods and do affect the operations and profitability of the firm in the long term. Proper evaluation of the consequences of these long-term impacts when activities are being planned and products and processes are being designed indicates a company's sensitivity to stakeholders that is essential for profitability and sustainability.

A company must develop a structure and systems that will evaluate both the impacts of sustainability initiatives on financial performance and the trade-offs that ultimately must be made when there are many competing organizational constraints and numerous barriers to implementation. The systems assist corporate executives in developing a sustainability strategy and in allocating resources to support it. The systems also assist sustainability and environmental managers as they evaluate the trade-offs and decide which sustainability projects provide the largest net benefit to both sustainability and financial performance. However, setting up the appropriate structure and systems is only one step in the pursuit of a sustainability strategy—measurement is also critical.

Only by making the business case for sustainability performance can managers truly integrate social, environmental, and economic aspects into their business strategies. The lack of a detailed business case creates additional barriers for managers trying to get support for social, environmental, and economic projects. In fact, a sustainability survey reports that the main reason respondents provide for not adopting sustainable business practices is the inability to present a clear business case.[1] So, to implement their sustainability strategy, companies face an enormous challenge: quantifying the link between corporate actions and sustainability and financial performance.

Nike recognizes how corporate financial performance is impacted by stakeholder reactions to corporate sustainability performance and incorporates it into decision-making. This impact depicts the "business case" for sustainability. Awareness of these reactions allows the company to think beyond the financial versus sustainability trade-off, and see how stakeholder reactions ultimately have a financial impact on the company, which turns many win-lose scenarios into win-win scenarios. More specifically, facilitating and recognizing the positive financial value of stakeholder impacts minimizes the magnitude of the loss in a "win/lose" scenario or, when the value of these impacts exceeds the cost of an initiative, turns it into a "win/win" scenario. For example, sustainability can create financial value for the company through enhanced revenues or lower costs. Revenues can be increased through increased sales due to improved corporate reputation, or costs can be lowered due to process improvements and a decrease in regulatory fines. Identifying the social, environmental, and economic issues that are important to key stakeholders, and improving stakeholder relationships, can foster loyalty and trust, which will impact corporate profitability. Nike Grind project is an example that was set up for doing good to the environment but that also tried to prove a business benefit (in terms of cost cutting, revenue increase, and brand awareness). The project included taking worn shoes and grinding them up for reuse in the production of other products. The initiative started as a project that is good for the community; today, it provides brand value.

Another example is P&G's distribution of Pu – R (a water purifying technology developed by P&G and the US Centers for Disease Control and Prevention) which is accomplished through a nonprofit venture to reach more people, make a bigger health impact, and help build corporate reputation.[2]

Measurement is critically important because it links performance to the principles of sustainability and facilitates continuous improvement. Managers implementing new programs can use indicators to define goals and targets to improve their sustainability performance; they can then compare these indicators to actual performance, along with various benchmarks, and measure success. Managers must constantly use feedback to challenge their assumptions about the viability of their decisions and their long-term implications for both the company and society. Appropriate measurement systems provide the proper tools for feedback and corrective actions.

Mapping the actions that drive performance

The Corporate Sustainability Model discussed in Chapter 1 (see also Fig.7.1) offers guidance to managers trying to make the business case for sustainability initiatives. The drivers of corporate sustainability performance, the actions that managers can take to affect that performance, and the consequences of those actions on corporate sustainability and financial performance rely on a thorough identification of performance metrics characterizing each component of the framework. Managers must quantify how one variable drives another until the link to profit is clear. This argues for explicitly linking corporate strategy and sustainability actions to sustainability and financial performance.[3]

The objectives, the drivers, and the metrics for sustainability success should be part of a clear articulation of the causal relationships leading from the inputs to the processes and then flowing to the desired outputs and outcomes. It is important to identify and communicate the causal links throughout the organization to guide the formulation and implementation of sustainability strategies. The causal linkage model of drivers, sometimes called a strategy map, is useful to ensure that all necessary actions are taken to achieve success, that unnecessary actions are not taken, and that all employees understand their critical roles (Fig.7.2). It provides the specific actions that lead to success in financial and sustainability performance and is supportive of and consistent with the Corporate Sustainability Model.

Causal relationships between drivers within each of the four elements (inputs, process, outputs, and outcomes) as well as between drivers in different elements are based on hypothetical assumptions of causes and effects. These hypothesized relationships may not be a perfect description of actual relationships that are underlying the sustainability strategy nor are they supposed to be constant through time. On the contrary, they need to be continually tested and revised. A clear understanding of the causal relationships underlying the primary drivers of value is one of the most important determinants of the model's effectiveness.

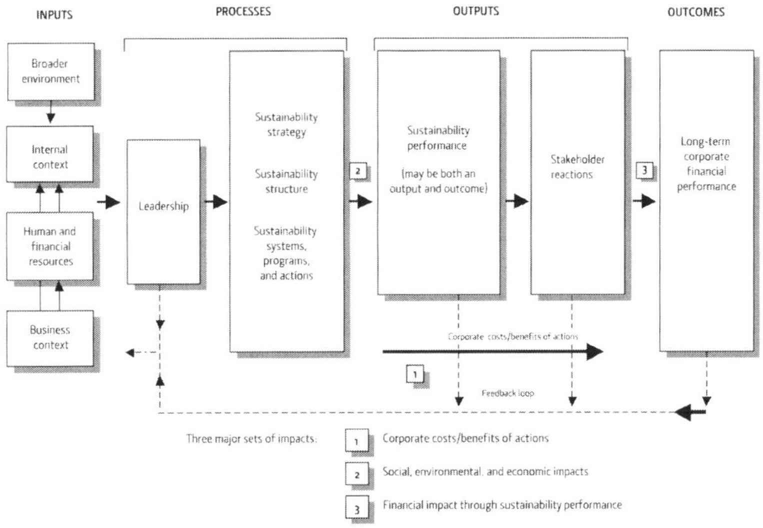

Figure 7.1 Corporate Sustainability Model

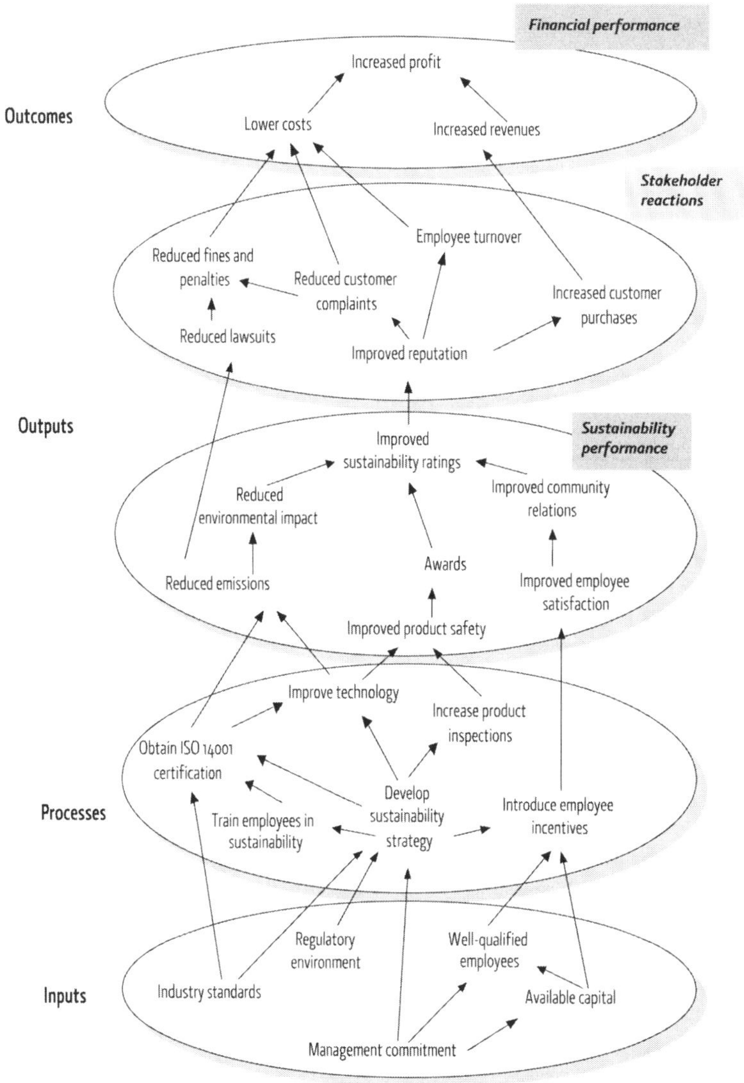

Figure 7.2 Causality of sustainability performance drivers

Figure 7.2 illustrates an example of sustainability performance drivers and the many causal relationships between them. For example, in a particular regulatory environment or industry (input), a company may choose to become ISO 14001-certified, improve technology, and increase product inspections (process). These actions should result in improved sustainability performance (output), such as improved product safety, reduced emissions, and improved employee satisfaction. Improved product safety may lead to international awards, reduced emissions may decrease the company's environmental footprint, and employee satisfaction may lead to increased voluntary activities thus improving community relations. All three outputs may improve the company's sustainability ratings. Sustainability performance drives improved stakeholder reactions (output) such as reduced lawsuits, fines and penalties, reduced employee turnover, improved reputation, and increased consumer purchases. These improvements may ultimately lead to lower costs, increased revenues, and increased profit (financial outcomes).

Companies may also consider sustainability performance as a final outcome and, in this case, would see reduced environmental impact and improved community relations as a distinctly separate successful result. Many companies have determined boundary conditions that set limits on poor sustainability behavior regardless of the financial payoff. For example, for Nissan, full compliance with environmental laws is an important boundary condition. Thus, Nissan accepts a large financial loss on sales of high-efficiency vehicles in order to meet the average gas mileage laws incorporated in the CAFE (Corporate Average Fuel Economy) standards. Some of Nissan's competitors choose to violate the law and pay a fine rather than take the financial loss. This is a trade-off that is never consciously evaluated by Nissan. The company complies without the need to address the financial/environmental calculus.

The Home Depot (THD), on the other hand, sets boundary conditions on factory conditions at its suppliers. The company never knowingly chooses to deal with a supplier that employs child labor. When THD found that its key supplier of tarps was violating these boundary standards, it cancelled further orders from this supplier and found other sources. In doing so, the company accepted large financial losses, both in costs and revenue, because alternative suppliers could not deliver the quality and volume provided by the original supplier.[4]

All four elements of the Corporate Sustainability Model connect in a chain of cause and effect. In other words, one category of measurement drives performance in the next. These drivers and subsequent measures should reinforce each other, all contributing to measuring the impact of sustainability performance on financial performance. To closely monitor these cause-and-effect relationships, metrics must be developed.

Sustainability performance metrics

Specific and appropriate measures that reflect the sustainability strategy are essential to monitor the key performance drivers (inputs and processes) and assess whether the implementation of the sustainability strategy is achieving its stated objectives (outputs) and thus contributing to the long-term success of the corporation (outcomes). Without appropriate metrics, companies often waste resources on projects or do not invest when they should because they cannot effectively evaluate the potential payoffs of sustainability initiatives.

Every component of the Corporate Sustainability Model should be associated with specific performance indicators. Impacts related to sustainability strategies can be translated into indicators of company performance in quantitative or financial terms. The inputs, processes, and outputs will be measured by evaluating various dimensions of strategies, processes, leadership, and other elements and reported quantitatively. They will be linked and converted into monetary terms as the evaluation of the impacts is summarized in the outcomes of sustainability performance and financial performance. The metrics listed in Tables 7.1–7.6 are examples of relevant measures for inputs, processes, outputs, and outcomes. It is expected that a small number from each section will be customized, adapted, and adopted as appropriate to each company.

TABLE 7.1 Corporate Sustainability Model: sample metrics for sustainability success—inputs

Inputs	Performance measures
Broader context	• Average temperature
	• Regulatory regimes

Inputs	Performance measures
	• Hazardous waste disposal regulations
	• Pollution standards
	• Existence of nondiscrimination laws
	• Local vs global standards
	• Demographic characteristics
	• Etc.
Internal context	• Number of strategic business units
	• Percentage of products, processes, and activities with life-cycle assessment
	• Results of a social audit
	• Existence of corporate code of conduct
	• Existence of a sustainability strategy, structure, processes
	• Percentage of employees with completed environmental training
	• Etc.
Business context	• Social/environmental performance of competitors
	• Number of competitors
	• Relative size of competitors
	• Breadth of competitors by geographic region, product diversity, or other
	• Number of customer channels
	• Average market size
	• Geographic diversity of production
	• Geographic diversity of sales
	• Etc.
Human and financial resources	• Funds available for employee training

Inputs	Performance measures
	• Funds committed for research and development on more effective pollution control and energy conservation efforts
	• Number of employees with environmental training
	• Cost of training per employee
	• Median or average years of schooling
	• Median or average salary
	• Etc.

TABLE 7.2 Corporate Sustainability Model: sample metrics for sustainability success—processes

Processes	Performance measures
Leadership	• Average time on top management's meetings devoted to environmental issues
	• Average years of experience of senior executives
	• Leadership turnover rate
	• Number of sustainability-related criteria for CEO evaluation
	• Number of sustainability-related criteria for board evaluation
	• Existence of a clearly articulated mission/vision
	• Number of hours of management time for volunteer work
	• Etc.
Sustainability strategy	• Planned percentage or number of suppliers certified for sustainability standards
	• Planned percentage of products undergoing life-cycle analysis
	• Planned diversity of workforce
	• Planned diversity of management
	• Planned percentage of internal promotions
	• Planned increase of facilities with screening procedures against the use of child labor (number of facilities)

Processes	Performance measures
	• Planned increase in gender diversity (percentage of workforce)
	• Planned reduction of lost workdays (number of days)
	• Planned reduction of emissions (percentage reduction)
	• Planned budget set aside for cooperatives/nonprofits
	• Number of activities intended to exceed compliance
	• Etc.
Sustainability structure	• Number of senior managers with social and environmental responsibilities
	• Number of levels of management with specific environmental responsibilities
	• Number of functions with environmental responsibilities
	• Percentage compliance with industry standards of corporate governance
	• Existence of indices of independent and active board of directors such as lead director and external director
	• Etc.
Sustainability systems, programs, and actions	• Percentage of employees with health insurance
	• Percentage of health insurance paid by organization
	• Cost of employee benefits
	• Funds donated to community
	• Funds donated to community causes chosen by employees
	• Matching funds
	• Number of hours of ethics training per employee
	• Number of hours providing technical assistance to vendors
	• Number of hours of employee time paid for volunteer work

Processes	Performance measures
	• Number of family leave days
	• Investments in cleaner technologies ($)
	• Investments in community projects ($)
	• Number of safety improvement projects
	• Safety training programs (hours)
	• Support programs for minority-owned businesses (percentage of volume of business)
	• Number of facilities with social/environmental performance evaluation systems in place
	• Number of facilities with environmental accounting systems in place
	• Number of facilities with ISO 14001 certification
	• Number of employees with financial incentives linked to environmental goals
	• Etc.

TABLE 7.3 Corporate Sustainability Model: sample metrics for sustainability success—sustainability performance

Processes	Performance measures
Sustainability performance	• Number of plant closures
	• Number of business opportunities generated locally
	• Percentage decrease in volume of hazardous waste
	• Percentage of trade with fairtrading partners
	• Percentage of product/process materials recyclable
	• Percentage of materials recycled
	• Number of products with instructions for environmentally safe use and disposal
	• Percentage decrease in volume and cost of energy use
	• Cost of fines and penalties for pollution

Processes	Performance measures
	• Percentage decrease in volume of emissions to air and water
	• Percentage decrease in vehicle fuel use
	• Percentage decrease in packaging volume
	• Percentage decrease in fresh water consumption
	• Average response time to environmental incidents
	• Number of trees planted
	• Percentage of suppliers certified/audited for sustainability compliance
	• Percentage of suppliers with sustainability lawsuits or legal actions
	• Average frequency of audits
	• Volume of landfill use
	• Workplace profile (demographics) compared to customer and community profiles
	• Percentage of workforce in volunteer programs
	• Number of volunteer hours per employee
	• Number of community causes supported through volunteer program
	• Cost of minority business purchases
	• Cost of cause-related marketing
	• Eco-efficiency of product use
	• Noise levels in community
	• Donations of products or services
	• Percentage of supplying companies owned by minority groups
	• Percentage of women in senior positions
	• Number of detected cases of bribery
	• Number of supplier violations

Processes	Performance measures
	• Number of cause-related events supported (e.g., breast cancer, Aids)
	• Cost of community support (parks, safety, recreation, etc.)
	• Number of events sponsored by organization
	• Number of accidental spills/discharges from plant
	• Remediation costs
	• Cost of community improvement
	• Number of public sponsorships
	• Number of prosecutions of criminal, antitrust, or trade violations
	• Number and type of human rights and labor violations
	• Results of ethics audit
	• Number of local jobs created
	• Number of employees with disabilities
	• Percentage of employees owning company stock
	• Salary gaps between gender/races
	• Rate of defective products
	• Average duration of product use
	• Value of products and services as measured by consumer surveys
	• Score on quality assessment
	• On-time delivery rates
	• Percentage of high-quality products on delivery
	• Product availability and back order
	• Etc.

TABLE 7.4 Corporate Sustainability Model: sample metrics for sustainability success—stakeholder reactions

Processes	Performance measures
Stakeholder reactions	• Stakeholder perceptions of corporate ethical performance
	• Percentage increase in volume of sold "green" products
	• Percentage of customer returns
	• Number of consumer complaints
	• Number of product recalls
	• Number of lawsuits and violations for inadequate disclosure related to such items as financial disclosures, product labeling, and environmental performance
	• Results of surveys of stakeholders regarding satisfaction with disclosures and meeting of their informational needs
	• Evaluation of external disclosures by external stakeholders
	• Customer satisfaction survey scores
	• Percentage of loyal customers
	• Percentage of new customers
	• Percentage of sales based on "word of mouth"
	• Employee turnover
	• Employee satisfaction scores
	• Number of applicants per job opening
	• Number of lost work days
	• Percentage of bonuses earned
	• Percentage of sick days used
	• Number of employee accidents
	• Number of days work stoppages
	• Percentage of non-paid hours overtime
	• Average work week hours
	• Average length of employment

Processes	Performance measures
	• Number of children in company-sponsored daycare
	• Percentage of employees owning company stock
	• Number of employee grievances
	• Percentage of employees using car pools
	• Number of employees participating in environmental programs
	• Employee certifications achieved
	• Percentage of internal promotions vs external hiring
	• Number of (community) awards and accolades received
	• Number of marketing or pricing practices challenged by government, by type of challenge
	• Number, type, and outcome of product liability complaints and suits, by type
	• By-product revenues
	• Improved image (survey)
	• Average time for new product development
	• Increased market share
	• Credit rating
	• Number of shareholder complaints
	• Number of social funds listing stock
	• Number of surprise inspections
	• Number of lawsuits
	• Number of community complaints
	• Number of protests
	• Number of letters to the editor
	• Percentage of favorable vs unfavorable press mentions
	• Number of plant visits
	• Number of certifications

Processes	Performance measures
	• Etc.

TABLE 7.5 Corporate Sustainability Model: sample metrics for sustainability success—outcomes

Outcomes	Performance measures
Long-term corporate financial performance	• Income and percentage of sales from "green" products
	• Income from sales of cause-related marketing affiliations
	• Income from recycled products
	• Income from recycled waste materials
	• Increased sales from improved reputation
	• Cost savings from reduction in energy costs
	• Cost savings from pollution reduction
	• Cost savings from reduction in cost of debt
	• Cost avoidance from environmental actions
	• Cost savings from employee turnover reduction
	• Workers' compensation costs
	• Cost savings from reduction in natural resource use
	• EVA (economic value added)
	• ROI (return on investment)
	• ROCE (return on capital employed)
	• Percentage of proactive vs reactive expenditures
	• Increase in relative percentage of proactive expenditures
	• Percentage of environmental costs direct-traced
	• Social/environmental costs as a percentage of sales
	• Cost of capital investments
	• Cost of operating expenditures

Outcomes	Performance measures
	• Disposal costs
	• Cost of fines and penalties
	• Percentage reduction in hiring costs
	• Cost of warranty claims
	• Etc.

TABLE 7.6 Sample metrics for measuring governance

Inputs	• Percentage of independent directors
	• Number of hours of training and education for directors
	• Existence of a code of conduct for directors
	• Number of committees
Processes	• Average overall attendance at meetings
	• Number of meetings with management other than CEO
	• Existence of an annual report on succession planning
	• Number of meetings with stakeholders
	• Number of hours spent on long-term strategic issues
	• Regular performance evaluations conducted (for CEO, board, directors)
	• Percentage of compensation linked to performance
Outputs	• Number of complaints (employees, community, customers)
	• Number of ethical/legal violations
	• Evaluation of quality of external disclosures by stakeholders (survey) or by experts
	• Percentage of projects accepted by board that met or exceeded projected ROI
	• Percentage of major projects that met operating goals
Outcomes	• Percentage change in revenue per employee
	• Percentage change in stock price

	• Average ROI
	• Percentage change in earnings (overall and per business unit)

Measuring sustainability inputs

Metrics must be developed to assess the impact that the four inputs might have on sustainability processes. Measures such as the number of employees available for sustainability programs and the dollars required to train them are examples of metrics that permit corporate and functional leadership to assess the resources available to focus on sustainability actions. Since the product characteristics have so much influence on sustainability performance, identifying and measuring those impacts is critical. Finally, companies must realize the impact that their corporate missions, strategies, structures, and systems might have on sustainability strategies and performance. Table 7.1 provides examples of metrics that measure the inputs.

Measuring sustainability processes

Companies need to develop performance indicators to monitor and assess the value of sustainability actions. Each element of sustainability actions must be translated into a metric that will eventually be linked to sustainability performance (Table 7.2). First, the sustainability strategy must be translated into measurable goals such as planned percentage of products undergoing life-cycle analysis. Metrics to describe programs must also be developed and often include the level of expenditures in sustainability programs and technology. Measures of leading indicators of performance around the structure of the sustainability function, management systems, programs, and actions including performance evaluations and rewards should also be monitored. Metrics such as the number of certified suppliers or the percentage of facilities certified to the ISO 14001 standard are examples of metrics that permit managers to assess the impacts of these initiatives on a specific aspect of sustainability performance. Other useful sustainability process measures include: number of hours of employee time paid for volunteer work, number of family leave days, dollar investments in cleaner technologies, etc.

Measuring sustainability performance

Every sustainability initiative undertaken should be associated with a specific sustainability performance indicator. Sustainability performance can be an intermediate output, an outcome, or both. Companies can develop strategies with the ultimate goal of improving society, the environment, or the economy with no link to improving profitability. Alternatively, companies might need to prove the business case for sustainability strategies by linking an improvement in social, environmental, and economic impacts to an improvement in corporate profitability. In this case, sustainability performance is an output, rather than the final outcome.

As managers implement new programs or invest in new technologies to improve their sustainability performance, they must clearly define goals and targets and compare these to actual performance. Actual changes in the production of waste, the recycling of the waste, and the changes in both environmental and financial impacts of these changes must be monitored. Both leading and lagging indicators should be included. Lagging indicators, like most financial measures, record the effect or results of prior actions. In contrast, nonfinancial indicators are also included because they are presumed to be predictors of future performance. Investments made in recycling equipment are an example of a leading indicator of hazardous waste. The rate of work-related injury (sustainability performance element) is a lagging measure of health and safety program efficiency (sustainability action element), and also a leading indicator of employee satisfaction (stakeholder reaction element).[5] In addition, evaluations of the number of human rights and labor violations, number of employee grievances, number of product recalls, number of social funds listing company stock, number of fines, and number of awards received are all potential measures of sustainability performance. Table 7.3 provides metrics that can be used to measure sustainability performance.

Measuring stakeholders' reactions

Stakeholders' reactions are an important component of the Corporate Sustainability Model as they may significantly affect short-term revenues and costs and long-term corporate performance on many levels. Employees choose whether to work for the company, customers choose whether to buy products, investors choose whether to buy shares, and government officials choose whether to increase or decrease regulation and enforcement. Research has shown that stakeholders do react to a company's reputation for corporate sustainability.[6] A study of the failed WTO (World Trade Organization) talks in Seattle found that investors drove the market capitalization of companies without reputations for social responsibility down by an average of US$378 million but did not penalize firms with a reputation for social responsibility.[7]

Companies are now gaining lasting advantage through stakeholder relationships uniquely structured to provide strategic advantage. Customers provide this advantage through loyalty and a long-term stream of purchases. Employees do the same when they commit to great service, innovation, and reliability. Shareholders provide a lasting advantage when they provide long-term, patient capital. Because gaining advantage through stakeholders has been recognized as a driver of strategic success, companies must identify the key stakeholder groups that are the primary drivers of their strategy including shareholders, customers, suppliers, employees, and communities. They should develop metrics for each of these stakeholder groups to gauge reactions to the company's sustainability performance (Table 7.4).

Measuring corporate financial performance

Costs and benefits associated with sustainability strategy must be measured and incorporated into management decisions (Table 7.5). As stated earlier, corporate financial performance can be an outcome of sustainability performance. Benefits often come from positive and improved relations with regulators and other stakeholders. For example, regulators may ease the permitting process for companies who have consistently demonstrated a strong sustainability performance record, thus reducing the time and investment required to bring new products and services to market. Better access to capital is another benefit as the financial community pays greater attention to environmental, economic, and social performance and gives preference to companies with favorable records.

Sustainability actions can also lead to cost reductions, perhaps from material substitution or less packaging, lower energy consumption during the production process, reduced material storage and handling costs, or reduced waste disposal. As well as generating cost reductions through improved efficiency, they may also create a positive reaction from customers who benefit from these savings or product improvements or who value the social, environmental, and economic contribution. They may contribute positively to a company's reputation for excellent sustainability performance and to shareholder value. They may also send a positive message to financial analysts and investors about the company's manufacturing performance. In these ways, these actions can simultaneously impact both sustainability performance and financial performance.

Companies should also include other impacts such as projected costs for compliance with legislation that is on the horizon but not yet enacted. While these costs may not affect current financial performance, companies must make current decisions with these future costs in mind. For example, some local and national governments set minimum requirements for labor practices or require the take-back of some products. These regulations attach market implications to the social, environmental, and economic impacts and make the cost of these impacts clearer, essentially shifting the boundary between external and internal costs—that is, internalizing an externality.

Though this book has primarily focused on the social, environmental, and economic aspects of sustainability, all of the principles of sustainability discussed in Chapter 1 (Table 1.1) can be measured using the same model and techniques. For example, in the case of governance, there are inputs, processes, outputs, and outcomes for board activities. Corporations make important choices in board composition (inputs) that have a significant impact on board performance. Board structure and systems (processes) also significantly affect its decisions and performance. The composition of the board affects how it prepares for, deliberates on, and makes important decisions and affects its success at fulfilling its roles and responsibilities and improving its performance (outputs), and ultimately improving corporate performance (outcomes). Continuous feedback provides a basis for improvement for the directors, the board, and the corporation. There are also metrics associated with the inputs, processes, outputs, and outcomes that can be considered in board evaluation (Table 7.6).[8] The balanced scorecard format discussed in Chapter 5 can also be used.[9]

Engage with your stakeholders

In recent years, many companies have dramatically increased the quality and quantity of interaction they have with stakeholder groups on a regular basis. Investment in stakeholder engagement pays off as a preventive strategy and can mitigate sustainability shocks when they occur. This is particularly important because of the rise of social media which speeds up the entire news cycle on a global scale. Proactive interaction may be with employees, customers, community activists, environmental groups, human rights groups, or product safety associations. Some companies have established community panels to learn about public concerns but, too often, companies have been taken by surprise by these sustainability issues, and organizational crises and related costs occur.

Sberbank: A crowdsourcing platform for engaging with stakeholders

In 2012, Sberbank, the largest bank in Russia and Eastern Europe, launched a crowdsourcing platform http://sberbank21.ru, which was used to engage with stakeholders on various projects including public discussion of *Corporate Social Responsibility Report 2011.* A total of 5,368 users registered for the CSR project; 2,447 took part in the discussion, 1,792 proposals were submitted, and 35,244 comments were made. The discussion involved clients, employees, representatives of public organizations, educational institutions, etc. Bank managers and heads of divisions can now use the project results in their work. The outcome of the crowdsourcing project is included in *Corporate Social Responsibility Report 2012* and will be taken into account in developing the bank's CSR strategy.

One of the key issues identified by project participants was raising financial literacy. In 2012, the bank implemented a wide range of initiatives aimed at raising public financial literacy:

- Awareness campaigns in mass media: The bank used federal periodicals, radio shows, popular Internet resources, and social networks bringing information to the widest possible audience. A total of 25 awareness campaigns were held that covered nearly 78 million people

- Special long-term web-based projects designed to raise financial literacy, such as Sberbank Our Home—the bank's YouTube channel. Their total coverage exceeded 8 million users

- Game and training applications in social media: The bank launched 18 game and training applications which attracted over 20 million users

- Financial literacy classes for school and university students: Classes were held in a game-like live format in 16 of the largest Russian cities, covering 20,000 senior school pupils and 1–2-year university students

- Children's training literature: The bank has prepared and issued a unique training book for children above 6 years of age *When I Grow Up I'll Become a Sberbanker.*

- Raising financial literacy among pensioners[10]

A company's engagement with its stakeholders may initially be based on a more careful understanding of the social, environmental, and economic impacts of corporate activities, products, services and process (such as described in the gas-drilling case study in section entitled "Cost of control"). This would be part of a more comprehensive discussion with stakeholder groups. But how much engagement is required?

The amount of engagement will be determined in part by the company impacts and the products, geography, industry, and customer characteristics. It will also be affected by the existing trust or distrust of the product, company, or industry.[11] When stakeholders have significant distrust of a product, company, or industry, it will be particularly challenging to persuade them that the company is effectively managing its sustainability impacts. When companies develop a reputation for corporate social, environmental, and economic responsibility, it helps to protect them when crises occur. There are many company examples where a lack of community trust caused increased costs, along with empirical data to support the value of building community trust and reputation.[12]

The level of stakeholder trust can range from negative to positive. Stakeholders may believe the company:

- Is trying or willing to do harm to the stakeholders
- Wishes to do no harm but is unwilling to expend any resources to protect the stakeholders
- Wishes to do good for society but has a lack of capacity or systems in sustainability, is uncommitted to sustainability, or has a lack of competence to execute

- Is committed to sustainability and willing to expend resources to establish organizational systems to effectively manage sustainability

The actions that will be necessary for effective stakeholder engagement and response to stakeholder needs will depend, in part, on the level of trust or distrust that already exists. Excellence in sustainability performance is important in any case. But this level of trust will impact on stakeholder perceptions of risk, corporate performance, and future company actions. The well-known examples of Nike (child labor) and Shell (Brent Spar) show how sustainability and financial performance are closely linked. These incidents were caused, in part, by ineffective stakeholder engagement, perceived community risks, and ineffective crisis management. The detailed approaches described in this book can mitigate many of these corporate risks while improving both financial and sustainability performance.

Indeed, consumers concerned with working conditions, environmental issues, and outsourcing are increasingly demanding accountability in retail clothing. Customers, even those who were focused on discount prices, are not only willing to pay more, but are actually paying more, for clothes that carry signs about fair labor practices. As a consequence, some retailers are disclosing information about exactly how, and where, their products were made.[13] There are several channels for engaging with stakeholders, ranging from focus groups and opinion polls to formal progress meetings with government and NGOs. The choice of which to use usually depends on the relationship with the stakeholder.

Focus groups uncover issues, uncertainties, vulnerabilities, and concerns and are recommended when assessing customers, employees, and societal stakeholders. In a focus group interview session, eight to ten participants are asked to actively discuss among themselves the issue at hand. The format is as follows:

- Participants are asked for feedback on ideas, insights, issues, and experiences
- A discussion leader moderates and structures the debate
- Sessions are repeated a number of times with different participants

The benefits of focus groups are:

- A wide range of stakeholders are involved
- Points of view are shared and hidden issues can be uncovered
- "Local knowledge" becomes available to identify threats and opportunities

The limitations of focus groups are:

- Dominant group members can bias the discussion in favor of their interests or points of view

- Discussion can lead to prejudiced reactions, rather than brainstorming[14]

Opinion polls are useful when measuring the effectiveness of campaigns or actions but responses are usually influenced by present circumstances, rather than looking at long time horizons. Polling firms, a public relations team, and information from other companies can identify which stakeholders may view certain issues negatively and generate reputation risk:

- **Third-party polling.** Polling by opinion research firms can provide a sense of how important these issues are to the company's stakeholders, and how they might react. A company's public relations team can calculate the impact of reactions like opting for another brand, boycotting, or a negative media campaign, based on numbers provided by polls and using methods similar to measuring brand value
- **Surveys.** Surveys can give some insight into the intensity of stakeholder reaction to certain issues. These "intensity of feeling" polls can be translated into monetary terms by asking questions about the impact of certain issues on buying products. The resulting numbers can then be calculated as lost sales
- **Other companies.** Researching the impacts on other companies that have experienced a similar issue can give insight into the monetary impact they experienced

Panels and surveys are increasingly used to measure and monitor stakeholder reactions and provide valuable feedback. These surveys assess opinion on the company's performance in such areas as air pollution, water pollution, labor practices, and community involvement. Survey results may be communicated to stakeholders through various means including social and environmental, or sustainability annual reports. The surveys are important because they assist the company in identifying and measuring the impacts, and in improving internal management decisions. (The use of surveys will be discussed later in this chapter.)

Dow Chemical, like many other companies, has established community advisory panels in most of the communities in which it has facilities. The company began its CAP (Community Advisory Panel) program with the goal of building trust, cooperation, and mutual respect between Dow and the community. For more than a decade, the CAP has served as a voice of the community, representing a cross-section of the community in terms of cultural diversity, age, education, and employment. The CAP has suggested a variety of efforts such as emergency response education for residents, community projects, and local hiring.

BHP Billiton, the world's largest mining company, identifies its key stakeholders and considers their expectations and concerns for all activities throughout the lifecycle of the operations, in accordance with the Community Group Level Document. Operations specifically consider any minority groups (such as Indigenous groups) and any social and cultural factors that may be critical to stakeholder engagement. BHP Billiton engages with the Indigenous groups through community consultation and engagement groups, participation in BHP Billiton activities, community perception surveys, newsletters and target communications, and support through local foundations.[15]

Nonprofits offer significant opportunities to companies because of the impact they have on communities. They have trust, knowledge of social issues, credible channels of communication, opportunities for employee engagement and volunteerism, differentiation of their brands, and proven record of positively contributing to community.[16] Finding a right strategic partnership with an NGO or nonprofit will support the sustainability strategy and increase credibility in communication with stakeholders.

Another approach to engaging stakeholders is the use of a **stakeholder network.** A stakeholder network is a group of organizations and individuals who voluntarily come together to address an issue. In this approach, the company is not at the center of the stakeholder relationships; instead, each stakeholder is equally involved and responsible. GlaxoSmithKline, a global pharmaceutical company, is involved in a stakeholder network to improve hospice care in Canada.[17] GlaxoSmithKline and leaders of the hospice community realized that none of the problems in hospice care could be addressed by any one organization. GlaxoSmithKline held a forum, which included caregivers, physicians, nurses, the clergy, media, activists, and other associations, to share information and develop strategies to address hospice care. The result of the forum was a strategy, implemented by the 650 members of the network that focused on encouraging public dialog about hospice care, educating and supporting caregivers and healthcare providers, and changing public policy.

Other nontraditional channels include internet and hotlines to allow constituents a convenient and open forum to communicate globally. BP began OpenTalk, a system that allows employees, contractors, and others with whom BP comes into contact, to anonymously raise concerns on unethical business conducts such as discrimination, bribery, or environmental accidents. There is a confidential 24-hour telephone line where concerns can be raised through fax, email, or letter. In 2012, 1,295 cases were raised through OpenTalk, with the most common issues relating to the people section of the BP's Code of Conduct.[18]

Measuring reputation

Reputation risk is considered a cost resulting from, and therefore a secondary effect of, social, environmental, economic, and political risk. A company's reputation depends partly on its reputation among its stakeholders on specific issues.[19] Stakeholders' opinions are based on their perceptions and expectations of what companies are doing. In some cases the perception will be an accurate reflection of reality and in others the perception may not reflect reality; but, whether reputation is based on real negative actions or perceived negative actions, the effect on company costs can be significant.

One way to place a value on reputation is through use of a **reputation quotient.** The reputation quotient evaluates stakeholder perceptions across 20 attributes that are grouped into six dimensions of reputation:

1. Emotional appeal
2. Products and services
3. Vision and leadership
4. Workplace environment
5. Social and environmental responsibility
6. Financial performance[20]

Use of the quotient as a measure of reputation can help inform the effect of sustainability issues because it looks beyond financial success. It assesses not only a corporation's overall reputation but also the factors that led to that reputation. Therefore, using the quotient as a measure of reputation can help inform corporate executives about the effect of sustainability issues on stakeholders.

Alternatively, corporate reputation can be determined by assessing the company's **reputational capital.** A company's reputational capital is the excess market value of its shares—the amount by which the company's market value exceeds the liquidation value of its assets. The use of reputational capital as a way to measure reputation has several advantages:

- It is simple to derive
- It enables comparisons of companies across industries and over time
- It recognizes the reputations of companies involved in both the manufacturing and the service sectors
- It takes into account the value of a company's brands and its intangible assets
- It enables comparisons of companies with more than one product line or business[21]

Reputation is often viewed as the personality or image of the company. A **corporate personality scale,** based on seven areas, is another approach to measure the internal and external perspective of reputation. The seven areas or pillars are: agreeableness, enterprise, competence, ruthlessness, chic, machismo, and informality. These are measured through surveys of customers and employees with a score for each pillar rated on a scale from 1 to 5. Companies can use the personality scale to estimate their brand image and the loyalty of customers and employees.

Reputational audits have also been used to help identify and manage reputation and the risks that a damaged reputation poses. A reputational audit may begin with a review of the company's current identity, image, and reputation, followed by an analysis of the trends, plans, and competitive positioning of the company. A careful identification and measurement of the likely reputational impacts of company activities, products, and processes is then completed. Lastly, companies must take appropriate actions to manage the transition.

Reputation costs can also be measured through lost sales minus the cost of producing those goods, or the **lost net profit. Share price** and **market share decline** are two other potential issues to consider. Perrier was once the leading sparkling water brand in the US, holding 80% of the US imported bottled-water market and close to 6% of the total bottled-water market. In 1990, benzene was found in the bottled water sold in South Carolina and the company recalled 70 million bottles in the US and Canada while claiming that it was an isolated incident. When similar contamination was discovered by Danish and Dutch officials, the company did a worldwide recall and claimed that benzene naturally occurred in the CO_2 that made its water "sparkling" and was usually filtered out. It lost substantial market share. Six years later, Perrier's sales was still at only one-half of its 1989 peak, and the company had to spend large amounts of money on increased advertising, free samples, and other marketing and promotional expenditures in an attempt to recover its market share.[22] We discuss additional ways to measure reputation later in this chapter.

Measuring the value of brand name

Although the benefits of having a solid reputation are often intangible assets, there is undoubtedly a financial benefit from following a sustainable strategy that can build stakeholders' confidence in the company. A good proxy for valuing reputation capital is to calculate the value of a brand name. Interbrand, a global brand consulting company, provides a methodology that goes beyond opinion polls or the budgets for ad spending.

The process begins with quantifying what the brand's overall sales are and, with the participation of financial analysts, projects net earnings for the brand. Next it deducts a charge for the cost of having tangible assets and the residual income is the value added by intangibles such as patents, customer lists, and the brand name. It then isolates the contribution of the brand name from the other intangibles, with the use of market research and interviews with industry executives. Finally, Interbrand analyzes the resilience of the brand name by looking at seven factors, including the brand's market leadership, its stability, and its global reach, and determines a discount rate that reflects the risks of the future brand earnings.

With this methodology, Interbrand publishes a ranking of the world's 100 most valuable brands. Heading the list is Coca-Cola, with a brand name of estimated worth over US$77 billion in 2012. Other companies in the top ten include Apple, IBM, Google, Microsoft, General Electric, McDonald's, Intel, Samsung, and Toyota.[23]

By estimating the impact of a damaged reputation, companies can design the right reputation risk management system to identify early warnings and anticipate stakeholders' reactions. By incorporating this process into management systems and decision-making, companies can improve corporate responsibility, develop trust with stakeholders, and, ultimately, enhance their corporate reputation.

Measuring risk

Conducting a risk analysis is one method to help organizations measure inputs and develop processes to mitigate any negative affect that taking a risk might have on the company.[24] Additionally, often it is a company's ability to identify and manage risks that others cannot that leads to innovation, opportunity, and market success. After identifying all of the possible social, environmental, economic, and political issues that could affect the organization and compiling them in a comprehensive risk profile, as discussed in Chapters 4 and 6, a company must develop metrics for each issue to assess its potential impact. This will enable managers to integrate the project-or company-related social, environmental, economic (sustainability), and political risks into ROI calculations.

Seven stages for measuring sustainability, and political risks

After identifying the various potential risks, measuring them is a seven-stage process:

1. Calculate the benefit associated with each issue that may generate risk
2. Calculate the potential costs associated with each risk, including reputation costs
3. Estimate the probability that each risk will materialize
4. Multiply the potential cost of each risk by its expected probability of materializing to calculate the expected value of each risk
5. Estimate when, over time, the risk may emerge. Calculate the NPV (net present value) of the risk
6. Aggregate the NPVs of all risks. Insert as a line item in ROI calculations
7. Calculate the expected value of the ROI

Step 1: Calculate issue benefit

Measuring the cost of social, environmental, economic, and political risks involves monetizing the savings and costs associated with each issue that could generate risk. For example, corporations commonly consider operating in a region where child labor is employed. In considering this option, the savings from either employing children or using contractors that may employ children should be calculated by measuring the difference in the wage rates between paying an adult and a child. The savings of using child labor would represent the issue benefit, which is generally assigned a positive value.

Step 2: Calculate risk costs including reputation

Although some industries such as clothing and shoe manufacturing have been seriously damaged by the use of child labor, and have therefore attempted to stop the practice, others, such as the chocolate industry, for a long time did not consider this a risk. Children working as cocoa bean pickers were employed in the supply chain. Chocolate and candy manufacturers largely ignored the issue, until newspapers began publishing stories of kidnappings and forced child labor on cocoa plantations in West Africa. If a company carefully considers this outcome, it should calculate each potential cost associated with employing (or contracting with others that employ) this labor force, and the public discovering it. These costs could include:

- Lost sales and other reputation impacts (measuring reputation is addressed in a later section)
- Managing a consumer boycott by hiring a public relations firm, creating a new advertising campaign, hiring a stakeholder relations manager, communicating internally with employees, and senior management's time devoted to dealing with the issues
- Diminished brand value
- Negative impact on recruiting potential hires
- Damage to company culture and morale

Each of these costs is assigned a value to calculate the risk costs of employing child labor.

The biggest cost of social, environmental, economic, and political risk is usually a reputation cost, typically as a result of lost sales due to consumer boycotts or protests. In 2000, the Rainforest Action Network began a campaign against Citigroup for financing projects that destroy rainforests. The campaign included protests at bank branches and television commercials. As a result, Citigroup lost about 20,000 customers.[25] The major consumer protests that have plagued many companies like Nike and Shell illustrate the significance of the events and the related reputational and financial costs.

Where possible, the impacts of share price and market share decline, as discussed earlier, should be included in calculations as potential long-term losses. The costs of managing stakeholders in the medium to long term, either through additional personnel or other strategies, should also be included. Marsh & McLennan, a professional services and insurance firm, experienced a 40% drop in its stock price when accusations of bid-rigging activity made the news in November 2004. In addition to a downgrading of its debt by credit-rating agencies because of its deteriorating reputation, Marsh & McLennan cut 5% of its workforce on predictions of a 94% decline in its third-quarter profits.[26]

Step 3: Estimate probability

After the potential costs of each risk to the company have been calculated, the potential likelihood, as a percentage, that each risk will occur and cause damage to the company, is approximated. This number is the estimated probability. (Later the impact on the company in expected value will be calculated.)

However, a footnote can be included in the ROI analysis that indicates that these numbers are midpoints or point estimates within a range, and the range can also be included. An estimated probability should be assigned to each identified risk. For example, the estimated probability of the emergence of social, environmental, economic, and political risks for a fictitious coffee processing plant in Colombia might include:

- Workers being kidnapped by the local militia: 60%

- Being "taxed" by local militias or cartels: 25%
- Being found guilty and paying fines under the US Foreign Corrupt Practices Act or other home-country laws that regulate bribery and payoffs: 6%
- Supplier's coffee plantation destroyed by crop dusters as part of local government-led cocaine eradication scheme creating a supply gap: 27%

Step 4: Calculate expected value of each risk
After approximating the estimated probability, the expected value for each risk is calculated, by multiplying the estimated cost of the risk by the percentage estimated probability of its occurrence. For example, if the costs of a reaction to use of child labor are estimated to be US$100,000, and the likelihood that this risk would materialize is estimated 10%, then:

Child Labor Risk Expected
Value=($100,000)þ(10%)=$10,000

Step 5: Calculate NPV of each risk
After Steps 1–4 have been completed, the NPV of each issue is calculated. Note that each issue has risks that emerge at different times. NPV is calculated on the outcome of:

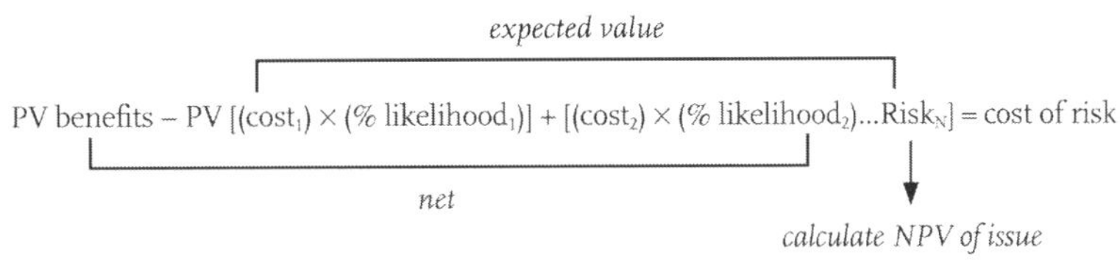

NPV calculations for social, environmental, economic, and political risk are completed in the same way as traditional NPV calculations. Discounting back, using a set discount rate, is also done in the traditional manner. These calculations are carried out for each identified social, environmental, economic, and political risk.

Step 6: Aggregate NPVs of social, environmental, economic, and political risks

Once all NPVs for social, environmental, economic, and political risks have been calculated, the NPVs should be added together. The aggregate sustainability risk NPV and the aggregate political risk NPV should then be inserted as line items in the normal ROI calculation. Schedules should be provided that show the calculations of benefit, expected value, likelihood, and cost of sustainability and political risk, as illustrated in Figure 7.3. It is critical that senior management see both the process and the output of doing these calculations.

CALCULATE THE MONETARY BENEFITS OF THE PROJECT

Outputs	Benefits	NPV
New product	Added revenue stream	$..................
	Labor cost savings	$..................
	New customer base	$..................
	Total benefits	$..................

↓

CALCULATE THE TOTAL COSTS OF THE PROJECT

Costs		NPV
Shipping	Transport rates, import duty, transporting goods to and from factory	$..................
Raw materials		$..................
Total sustainability risk costs	See Schedule A	$..................
Total political risk costs	See Schedule B	$..................
	Total costs	$..................

↓

CALCULATE THE PROJECT ROI

$$\text{ROI} = \frac{\text{Total benefits} - \text{Total costs}}{\text{Capital costs (investments)}} \times 100$$

Figure 7.3 Integrating sustainability and political risk costs in ROI calculations

Schedule A Costs of social risks

Risk	Benefit	Cost types	Costs	Likelihood	Expected value
Civil unrest surrounding site	$	• Costs of engaging employers skilled in negotiating with protesters • Cost of engaging extra security personnel *Reputation-related:* • Cost of hiring community relations manager • Cost of managing activist NGO relations	$ ...... $ ...... $ ...... $	.... %	$ ....
Prostitution near site	$	• Costs of implementing health education for workers to teach about sexually transmitted diseases (to avoid costs related to HIV infection)	$	.... %	$ ....
Child labor	$	*Reputation-related:* • Costs of reputation damage • Cost of managing boycotts when information reaches activist consumers • Cost of NGO relations manager	$ ...... $ ...... $	.... %	$ ....
Infringement of indigenous lands	$	• Costs of litigation in international courts • Cost of remunerating population • Cost of work stoppages due to local strike, reputation damage, community protests, work stoppages *Reputation-related:* • Cost of hiring community relations manager • Cost of managing activist NGO relations	$ ...... $ ...... $ ...... $ ...... $	.... %	$ ....
				Reputation costs, including lost sales and profits	$ ....
				NPV	$ ....

Schedule B Costs of political risks

Risk	Benefit	Cost types	Costs	Likelihood	Expected value
Changes in legislation that change the rules of the game	$	• Lost revenues • Increased taxes and tariffs	$ .... $	.... %	$ ....
Forced contract negotiation with host government	$	• Lost profits • Lost investment	$ .... $	.... %	$ ....
Armed insurrection	$	• Cost of hiring private security • Cost of training local police/military to prevent human rights abuses (if required to use these forces by contract)	$ .... $	.... %	$ ....
Associated reputation risk	$	• Costs of incentive packages to attract workers to location • Cost of protests, etc. due to potential linkages with human rights abuses	$ .... $	.... %	$ ....
Endemic corruption	$	• Costs of payoffs and bribes • Costs of potential lawsuits for that activity	$ .... $	.... %	$ ....
Targeted criminal activity	$	• Costs of protecting personnel, including extra security, reinforcing security at private homes, providing security training to employees and families • Costs of attracting workers, including increased pay, time off, and hardship bonuses • Costs of increased security to protect facility • Costs of potential work stoppages	$ .... $ .... $ .... $	.... %	$ ....
Terrorism	$	• Costs of reinforcing infrastructure • Costs of hiring additional security personnel • Costs of rebuilding	$ .... $ .... $	.... %	$ ....
				Reputation costs, including lost sales and profits	$ ....
				NPV	$ ....

Schedules A and B in Figure 7.3 list examples of potential social (sustainability) and political risks. Schedule A lists risks that could emerge for a company—for instance, in the extractive industry—that operates in an unstable region. Although some issues that emerge, such as civil unrest near the site, would probably not present any benefits, others, such as establishing operations on Indigenous lands, could produce short-term savings because of low land prices. However, costs associated with these sustainability risks that might be incurred include:

- Remuneration for indigenous land
- Hiring someone to negotiate with protesters or assigning some of current employees' time to those negotiations;
- The cost of extra security to protect the site
- Hiring a community relations manager
- Executive time spend strategizing on managing NGO relations
- Work stoppages due to community protests
- Reputation damage
- The potential for litigation fees and fines if the issue goes to court

Most underlying causes of political risks (unlike some social, environmental, and economic risks) do not present any savings to a company. Although entering a country with political instability can bring both benefits and costs, antibusiness legislative changes, policy changes or contract renegotiation that would be considered risks offer a company little or no benefit. Favorable policy or legislation changes, however, would not be considered political risk as defined here. Schedule B lists various costs the company would incur if the risks mentioned were to materialize. For instance, if there were an armed insurrection targeting the company site, costs could include:

- Hiring private security to protect executives and their homes
- Training personnel in self-defense (defensive driving, home invasion protection, etc.)
- Extra training of local police who protect the company site on the level of force company standards allow (where they go beyond local laws), to protect the company from litigation for human rights abuses

If the company overseas faced endemic corruption, costs associated with this risk could include:

- Dollars (or equivalent) paid directly in bribes, or other methods of payment to facilitate transactions
- Legal fees and fines if found guilty of bribery practices in a lawsuit filed under the Foreign Corrupt Practices Act or similar legislation
- Reputation damage sustained by the company for being associated with a corrupt regime

Reputation costs have been included as a separate line item in each schedule because they represent a large component of social, environmental, economic, and political risk. In addition to how reputation costs were treated in the previous reputation risk section, they can be listed as lost sales and profits.

Step 7: Calculate expected value of ROI
Once Schedules A and B have been calculated, their results can be integrated into traditional ROI calculations, as illustrated in Figure 7.3. This process can also be used to calculate opportunities for organizational innovation.

Integrating social, environmental, economic, and political risks into ROI calculations enables managers to better understand: (1) the full range of risks their operations face; and (2) their costs. Although the output of the analysis is useful, the analysis process itself also provides the opportunity to strategize for risk management—either to develop ways to avoid the risk, to create risk mitigation plans, and to capitalize on opportunities.

In an increasingly globalized world, a company needs to integrate sustainability and political risks to manage its risks effectively, to improve resource allocation, and identify opportunities. This demands the quantification of social, environmental, economic, and political risks. To account for these risks, they must be identified, measured, monetized, and included in ROI calculations. And then they can be more effectively managed to improve both sustainability and financial performance.

Measuring social, environmental, and economic impacts

Once metrics have been specified, a methodology for measuring social, environmental, and economic impacts based on the concepts discussed in the previous chapter is critical.[27] The first step in measuring impacts is to identify the impact to be valued and the population, or affected group, whose values will be measured. Next, the choice of method (revealed preference or stated preference) needs to be determined. A single method or multiple methods can be used to measure impact. For instance, travel cost and hedonic pricing can both be used to estimate the benefits of cleaning up a polluted river.

If using a revealed preference method, secondary data sources should be identified and evaluated. Multiple data sources may be necessary to gather all needed information. Some methods, such as travel cost and contingent valuation, require the collection of primary data through surveys. Finally, an estimate of WTP or WTA is derived from the data collected from the primary and secondary sources.[28]

Measuring impact at The Co-operative Bank

In 2001, UK-based The Co-operative Bank developed a methodology to measure how its sustainability practices affect revenue and growth. The methodology uses several calculations and survey questions. From the survey, the bank determined that 53% of personal current account customers state that sustainability is one of a number of important factors in why they opened or maintained an account at the bank, while 31% cite sustainability as the most important factor. The bank has estimated the profitability of each product, including all direct costs and indirect costs attributable to the product. The profitability is then multiplied by the sustainability factor, and then aggregated to produce a sustainability profitability contribution range (which ranges from those customers for whom sustainability is the most important determining factor to those customers for whom sustainability is one of a number of factors).[29] The bank concluded that 26% of profits could be assigned to customers who cite sustainability as an important factor, and 14% to customers who cite sustainability as the most important factor.

Surveying nonmarket externalities

Several methods for valuing externalities were introduced in the previous chapter. Contingent valuation, along with other methods, uses surveys of relevant populations to elicit values placed on goods and services. Surveys can aid companies in fully measuring outputs and outcomes. There are currently at least six main methods to gather information through surveys, but all fall into a general approach consisting of the following steps:

1. A sample of the population is questioned about its value for a specified good
2. The responses are documented and form the basis for estimating WTP or another relevant method
3. Results are extrapolated to the entire population

Here is a description of each of the six methods related to WTP and the general appropriateness of their application.

1. **Open-ended WTP.** In this method respondents are asked to state their maximum WTP for the good or resource under evaluation. Questions could be framed in the form of "What is the most that you would be willing to pay to guarantee that the wilderness area will remain closed to development?" The method is not favored much because of the potential for unrealistic responses
2. **Close-ended iterative bidding.** Respondents are asked first if they would be willing to pay a specified amount for the good or service. If the response is affirmative, then the number is raised in increments until the respondent answers negatively. If the initial response is negative, the amount is lowered until a value is agreed. This method is commonly used, but there is evidence that the responses are biased depending on the

initial price set for the good (this is known as starting-point bias)

3. **Contingent ranking.** Respondents rank specific combinations of quantities of the good and willingness to pay for each segment. For example, combinations could range from low water quality at low price to high quality at higher prices. The combinations are ranked from most preferred to least preferred. This method is popular because it is easier for respondents to answer and for surveyors to analyze

4. **Dichotomous choice.** In this method, respondents are given randomly assigned prices and asked whether they would be willing to pay that price. As a result, analysts construct a distribution of the responses and calculate the probability of respondents answering positively to a set amount. This method requires a large sample size to be able to determine with high-percentage probability if the population is willing to pay for a specified value

5. **Payment card with comparative tax prices.** Individuals are asked to value a good after seeing a card showing the tax-prices for a range of other publicly provided goods. The card shows, for example, the dollar amount that an individual with a particular annual income pays for national parks. The idea is to provide a reference point so that individuals make an informed decision, thus

reducing the occurrence of outliers from open-ended surveys

6. **Payment card with a range of prices.** Respondents are asked the maximum price they would be willing to pay for a good from a range of dollar values. This survey could be conducted anonymously and by mail, reducing bias resulting from personal interviews

Given the high cost of surveys, it is possible to use information from existing or related databases, provided that differences in the population sample are controlled statistically. Also, conductors of surveys are encouraged to document the characteristics of the samples so that they can be useful in future analyses. Where extensive surveys are not practical, the thought process and discussion of the relevant issues can at least provide recognition of alternative views that may be relevant for the decisions. Furthermore, a well-executed stakeholder engagement process, though not a substitute for extensive surveys and analysis, can be a significant aid to understanding alternative views and the intensity of stakeholder attitudes.

Survey bias

There are at least three main sources of bias that arise due to poorly designed surveys.

1. **Sample bias.** Most surveys rely on a small number of people to represent the target population. In almost all cases, samples are selected by a probability method, which produces

simple random samples that give each individual in the target population an equal probability of being selected. A broader issue with samples is how to select the target population. For CV purposes, the relevant target population is all constituents affected by the project or the investment being analyzed. A stakeholder analysis will help identify who is affected and will help in the definition of the target population. If the sample is appropriately selected, then sample bias can be avoided

2. **Nonresponse bias.** Even if sample bias is eliminated through a proper survey design, there will still be individuals who do not respond. Nonresponse can be either a voluntary refusal or simply that the individual was unavailable. Refusals can be solved by expressing the legitimacy of the survey or by offering incentives. But for unavailable responses the extrapolation of results may misrepresent the target population. If nonresponse is purely random, it can be offset by increasing the size of the sample population and testing for nonresponse bias

3. **Interviewer bias.** This form of bias occurs when respondents perceive that the interviewer prefers a particular answer. To avoid this source of bias, surveys can be conducted by mail. However, doing this could decrease the response. Alternatively, careful training of the interviewers and design of the instruments can reduce this bias

Besides the systematic biases described above, the contingent valuation methodology has some particular sources of bias related to quantifying willingness to pay. They can be categorized into four types of behaviors:

1. **Strategic behavior** is when respondents deliberately bias their response to serve their personal interest

2. **Compliance bias** occurs when respondents try to please either the interviewer or the organization sponsoring the survey

3. **Free-riding behavior** is when respondents undervalue a good or service because they are confident that somebody else would pay for it and they can still benefit from using it. Free-riding is generally related to public goods and services where the use by one individual does not exclude others from also benefiting

4. **Embedding bias** results when responses are influenced by the amount of information provided. Given the hypothetical nature of surveys, contingent valuation issues often involve complex scenarios that require clarity and objectivity. Delivering a clear message often depends on the logistics, as discussed below

Survey methods

Another consideration in researching for contingent valuation is the type of instrument to be used in surveying. The first, personal surveys, requires the presence of an interviewer who can explain the issues at hand and motivate the respondent to cooperate in a more objective manner. This personal touch is lost when performing the second type of research, namely telephone surveys, which could lead to the respondent's misunderstanding and lack of interest.

The third instrument, mail, or email, can improve the explanation of issues by using visual aids, and also eliminate interviewer's bias, but it is handicapped by leaving the respondent "alone" to understand the issues being analyzed. If the survey is intended to be sequential, another limitation of written research is the inability to pace the survey. Respondents could also be tempted to answer or browse through sections in a disorderly manner, affecting the final results. Mail surveys are also a source of nonresponse bias.

When designing a research survey, these logistic considerations should be balanced to determine if the potential cost savings in using telephone or mail/email surveys outweigh the limitations.

The information collected from questions such as those asked in the survey in Figure 7.4 can aid the energy companies in their sustainability decisions. The energy companies wanted to know how stakeholders make trade-offs on energy and development. The nonprofit advocacy organizations and governmental organizations surveyed represent important corporate stakeholders and the measurements of sustainability performance, and the subsequent stakeholder reactions provide important inputs to the corporate decision-making process. Corporate decision-makers should integrate these stakeholder reactions and the estimated impacts on corporate profitability into their sustainability decisions. The information can also be effectively used by other stakeholders to better understand the perception of these impacts by various stakeholder groups.[30]

In the previous chapter, the Wyoming gas drilling case study (page 150) introduced an overview of the issues, setting, and context. This research study used CV to measure a cross-section of stakeholders' willingness to pay to offset the effects of development on the environment. To elicit data on stakeholder reactions, a survey was administered to both the local population and to a national audience. By conducting both a local and a national survey, a very broad set of stakeholders was represented. This was supplemented by archival data from many government, business, and community sources, and extensive interviews.

Willingness to pay

Q-9 Conservationist groups may pool money in order to purchase or lease land to help provide contiguous habitats and ensure free passage of animals through migration routes. This may help mitigate impacts from *residential expansion*. Please CIRCLE the value of the HIGHEST amount you would agree to donate each year for the next ten years above what you currently donate to conservationist groups to counteract the effects of *residential expansion* on wildlife.

$0	$1	$5	$10	$20	$30
$50	$75	$100	$150	$200	OVER $200

Q-12 It is important to know how much protecting wildlife is worth to you. Please think about:

- Your current annual income
- Your current annual expenses
- Other possible uses for your income

Keeping these factors in mind, CIRCLE the value of the HIGHEST amount you would agree to pay each year for the next ten years in higher energy prices for programs and technologies to mitigate potential impacts on wildlife?

$0	$1	$5	$10	$20	$30
$50	$75	$100	$150	$200	OVER $200

Trade-offs

Q-4 Reducing residential development in Wyoming may help mitigate potential impacts on wildlife preservation. Please indicate your preference for the trade-off between residential development and wildlife preservation in Wyoming: (Circle one) (1 = residential development/no wildlife preservation, 7 = no residential development/wildlife preservation)

1 2 3 4 5 6 7

Q-5 Energy companies can invest in programs and technologies to mitigate potential impacts on wildlife. The cost of energy is partially affected by the preservation efforts from energy companies. Please indicate your preference for the trade-off between the cost of energy and wildlife preservation in Wyoming: (Circle one) (1 = lower cost energy/no wildlife preservation, 7 = higher cost energy/wildlife preservation)

1 2 3 4 5 6 7

Follow-up

Q-13 If you answered anything but $0 in Question 12, please read through the entire list below and then put a 1 by the statement that best matches your most important reason. (If you have more than one reason, put a 2 by your second most important reason, and so on).

- ☐ I want to preserve wildlife for future generations.
- ☐ I want to preserve wildlife for the enjoyment of all citizens.
- ☐ I want to support the protection of undisturbed wildlife.
- ☐ I want to preserve wildlife for my personal enjoyment, such as hunting, fishing, and viewing.

Importance

Q-1 When thinking about Wyoming, how important do you believe are the following: (Circle one per part) (1 = not important, 7 = very important)

a. Maintaining stable wildlife populations

1 2 3 4 5 6 7

b. Adequate supply of natural gas

1 2 3 4 5 6 7

c. Clean air

1 2 3 4 5 6 7

d. Affordable residential housing

1 2 3 4 5 6 7

e. Clean lakes and streams

1 2 3 4 5 6 7

FIGURE 7.4 **Measuring the impact of natural gas drilling: abbreviated survey questions**

Source: Epstein and Widener (2007) *Measuring Multiple Stakeholder Costs and Benefits*

To design a survey, an organization should begin by performing interviews with important stakeholders to gain knowledge of the decision being studied. The survey should then be pre-tested to gauge the clarity and understandability of the survey questions. At this point, a survey method such as mail or person-to-person interviews, as discussed earlier, should be determined and surveys conducted.

Summary

After examining the academic literature and the conceptual foundations for measuring sustainability impacts in the previous chapter, in this chapter we demonstrated how to measure and execute a measurement system for management decision-making and managerial actions. As companies assess the choice of appropriate measures to evaluate sustainability investments, numerous potential issues arise. Since the choices are different for each company, substantial customization is necessary. Here are six initial questions for senior managers that can lead to the development of appropriate measures for improved evaluation of the social, environmental, and economic impacts of current operations, a new initiative to improve corporate sustainability, or a new corporate initiative or investment:

- What measurement systems are currently in place and being utilized within the organization?
- What are the important criteria to the company and its constituencies and stakeholders?
- What does the company wish to accomplish with this sustainability initiative or corporate investment?
- What is the anticipated time-frame associated with this initiative or investment?
- Who are the parties involved in implementing this initiative or investment, and who will be affected by the results?
- What critical processes are associated with the successful execution of the project?

To answer these questions, companies must not only customize their sustainability measurement approach but also use multiple measures to fully analyze their situations. Different measurement criteria are important for companies that have different strategies or may be in a different stage of their life-cycle or the development and implementation of their sustainability strategy. The multiple measures will typically include both financial and nonfinancial measures that are leading and lagging indicators of performance. The measures should be linked to strategy, and include a combination of input, process, output, and outcome measures. They may be used in a balanced scorecard or other approach and can be developed specifically for sustainability or as a part of an overall corporate performance system.

Though challenging, measurement of sustainability impacts can be done and is needed in corporate decision-making. Currently, most companies do not include extensive measures of social, environmental, and economic impacts in their decision-making processes and ignore what are potentially significant effects. They acknowledge the importance but decline to include them in ROI calculations claiming that the measurement is too difficult.

Although measurement may be imprecise, it is still relevant. Social, environmental, and economic impacts must be included in ROI calculations and managerial decision-making at all levels. Proper measurement systems evaluate the impacts of sustainability initiatives on financial performance and the trade-offs that ultimately must be made when there are many competing organizational constraints and numerous barriers to implementation.

In the following chapter we look at how managers can use the information gathered from measurement systems to improve their organizations' products and processes for improved sustainability performance.

CHAPTER 8

Improving corporate processes, products, and projects for corporate sustainability

Analysis of sustainability performance, as discussed throughout this book, is important for improved performance. The organization's measurement system will provide important information to aid in management decision-making, but improvements will occur only if managers and organizations learn and redesign processes, products, services, projects, and other activities to achieve improved sustainability impacts and performance.

The feedback process is an important aspect of sustainability performance and will probably challenge and change strategies and assumptions. Various mechanisms at different levels in the organization can provide feedback to top management to promote knowledge sharing and to enhance capabilities for improved sustainability performance. The performance evaluation systems and performance indicators discussed earlier are critical in providing relevant information to managers as they improve processes, products, and projects. And stakeholder engagement is effective only if the organization uses information gathered to affect changes and improve decision-making.[1]

The potential for learning associated with appropriate information is significant. Companies implementing sustainability actions should develop mechanisms to access and share good practices and initiatives across the organization. Feedback mechanisms and continuous learning are important parts of any learning organization and in the implementation of systems to improve corporate sustainability. Constantly using feedback to challenge assumptions about the viability of various decisions and their longterm implications for both the company and society will help improve organizations and their sustainability performance.

In this chapter we will discuss how the following actions can aid in improving performance:

- Organizational learning
- Life-cycle analysis
- Redesigning products and processes
- Rethinking markets, including bottom of the pyramid
- Integrating sustainability in the supply chain
- Internal reporting

Looking again at the Corporate Sustainability Model (Fig.1.4), the dotted part of the arrows shows that the feedback process does not rely exclusively on data relating to financial performance. Sustainability performance and stakeholder reactions are important elements in the feedback loop affecting the decisions of managers. When managers see the impact of their activities, products, and services on sustainability, stakeholder reactions, and financial performance, they can make changes in corporate and business unit strategy, structure, and systems. Then, increased attention to sustainability may change the strategy, structure, and systems of the sustainability function to drive improvements in sustainability performance.

Organizational learning: the new battleground?

The concept of "learning" organizations has grown in popularity and interest in recent years.[2] It seems to be the new battleground as the ability of an organization to "learn" faster than its competitors holds the promise of sustainable competitive advantage. To evaluate how sustainability implementation affects the organization knowledge base, both the process (learning mechanisms) and the outcome (capabilities) of the process must be examined. The development of capabilities is indeed connected to the learning process. A company's knowledge assets (core capabilities) are embodied in four dimensions:

1. **Skills and knowledge** relate to the organization's employees and their expertise and qualifications. This includes both company-specific and general knowledge and skills
2. **Physical technical systems** reflect the skills and knowledge that are embedded in hard data and codified procedures over time. Such systems include databases, software, and machinery
3. Integrating sustainability matters into its decision-making process should also impact on a company's **managerial systems.** These systems guide the organization's accumulation of knowledge. Organizations may create knowledge through training, and encourage and control knowledge through performance evaluation systems and reporting structures
4. The fourth dimension, **values and norms,** determines and controls the type of knowledge that is sought and nurtured in the three previous dimensions. Values serve as screening and control mechanisms. Effectively communicating knowledge, values, and norms inspires employees to contribute and support the organization's overall strategy and are critical to its implementation[3]

To promote knowledge transfer and transparency, Johnson & Johnson, the multinational manufacturer of healthcare products, established a Healthy Future 2015 goal in 2010 to have all 23 major Johnson & Johnson brand websites share product sustainability information.

Ten of these major brands and/or companies, or just over 43%, are now sharing sustainability information on their websites, including Neutrogena and Aveeno.[4]

MillerCoors uses benchmarking for organizational learning

In 2011, a team from MillerCoors, one of the world's leading brewers, visited three of its own breweries in South America that have introduced innovative solutions to reduce water and energy use in their operations, improve process efficiency, and decrease beer loss. The team observed first-hand the South American breweries' sustainable environmental practices. For example, to improve energy efficiency, the breweries recover and reuse heat energy and recover and reuse steam energy from brew-house operations. The breweries are filled with windows that take advantage of natural light, thereby reducing demand for electric lighting. To reduce water usage, the breweries optimize beer pipe and beer tank cleaning systems and, where possible, use their last rinse cycle to start the next cleaning cycle. Brewery grounds are landscaped with drought-tolerant plants to reduce demand for irrigation. To reduce waste, the breweries carefully maintain their equipment for maximum operational efficiency with a focus on avoiding spilling or wasting product. The MillerCoors team was particularly impressed by the South American breweries' use of short-interval controls, a process by which employees monitor brewery performance in real time at several intervals within a work shift. Short-interval controls give brewery managers a chance to quickly correct problems, or take advantage of opportunities as they arise, and directly engage employees in day-to-day brewery operations. By using benchmarking, MillerCoors developed a plan to incorporate the South American breweries' short-interval control and other best practices into its US operations.[5]

Organizations committed to social, environmental, and economic objectives should ensure that an appropriate managerial infrastructure, aligned with set social, environmental, and economic objectives, exists to support and promote desirable behavior.

Organizations must develop learning mechanisms to develop and maintain capabilities that will meet present and future challenges of sustainability management. A company's ability to learn (its absorptive capacity) affects its actual learning process, and absorptive capacity is an important determinant of a company's ability to exploit new or outside knowledge. For example, does a company have the skills and expertise to exploit a new and cleaner technology? Does it have a clear understanding of all of the options, technical and managerial, available to it? Companies should recognize the importance of building their absorptive capacity through technical training and extensive monitoring of technical literature in the field. These capabilities will determine the company's ability to compete on the traditional competitive priorities: cost, quality, dependability, flexibility, and innovation.

Learning can be divided into single-loop and double-loop. **Single-loop learning** occurs when members of an organization make corrections to errors to maintain the features, strategy, or culture already in place. In **double-loop learning,** on the other hand, assumptions and strategies may be challenged and changed; feedback is used to question the basic assumptions about the strategy and whether it remains viable.[6]

Figure 8.1 shows the main elements of an effective learning process:

- A shared vision that facilitates and promotes systematic team problem-solving
- A feedback process that efficiently transfers knowledge and information about the organization's own experiences, experimentation, and others' experiences and best practices
- A review process that evaluates, challenges, and adapts prevailing practices and strategies in light of new information

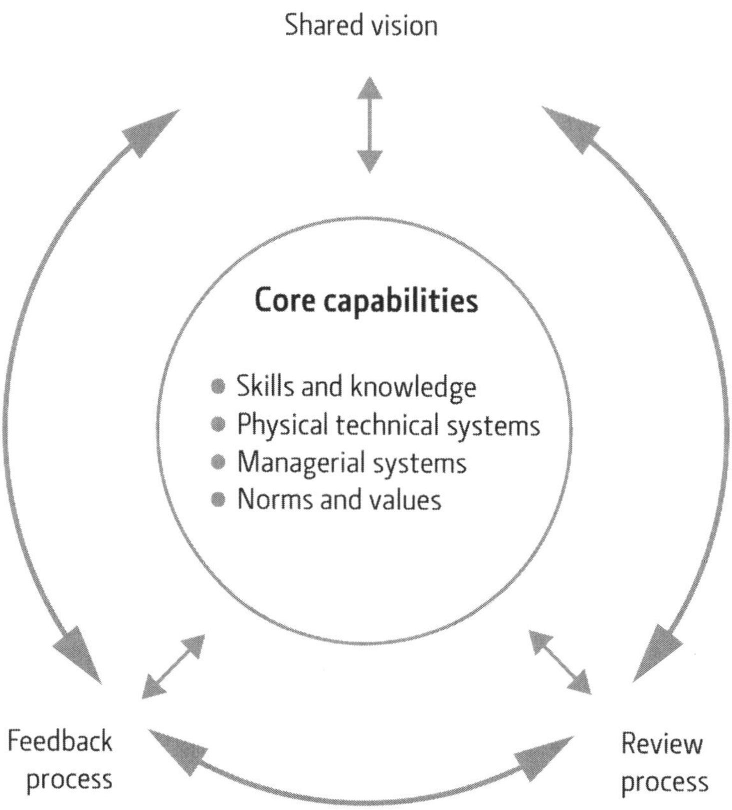

FIGURE 8.1 **Capabilities enhancement and learning activities**

Source: Epstein and Roy (1997) "Using ISO 14000 for Improved Organizational Learning and Environmental Management"

Several industry and voluntary standards cover capability-building activities. Through many of their requirements, standards such as ISO 14001 contribute to learning activities and the following features of most voluntary standards provide valuable learning mechanisms.

- **Documentation requirements/records.** For learning to have an impact, knowledge must be spread quickly and efficiently through the organization. Most standards require that the sustainability process be effectively documented through written procedures and information tracking. Documentation-control requirements also promote easy access and availability of these procedures
- **Identification of environmental aspects/legal and other requirements.** Procedures to access and track information on legal and other requirements (e.g., industry associations) and environmental aspects should be established to keep the organization aware of new developments. This will also enable the company to learn from the experiences and best practices of others
- **Communication requirements.** Opening up boundaries and stimulating the exchange of ideas are also important learning mechanisms. Appropriate internal and external channels of communication will contribute to a fresh flow of ideas and learning from shared experiences

Plan, Do, Check, Act

The **PDCA cycle** is a valuable tool for learning and promoting change in organizations and provides a valuable framework for continuous improvement. The

PDCA cycle consists of four phases, each involving key activities:

1. **Plan**

    –Conduct initial sustainability reviews

    –Define sustainability strategy

    –Design sustainability programs

    –Set objectives and targets

2. **Do**

    –Develop structure

    –Provide training

    –Introduce programs

3. **Check**

    –Conduct internal audit

    –Monitor and measure performance

4. **Act**

    –Management review

1. **Plan.** This phase includes all the activities that will guide the organization to a better understanding of

the issues at stake before it commits itself to its sustainability strategy. Managers should focus on the company's current impacts and profile, its likely future sustainability impacts and profile, and the likely impact of regulations on activities. Then they need to focus on future sustainability performance, set goals, and establish an implementation plan.

2. **Do.** During this phase, the actual sustainability programs are introduced. This phase may vary considerably according to the type of programs being introduced. There are many ways an organization may choose to achieve their objectives. Among these are product, process, and managerial system modifications. They can be minor changes or radical new ways of doing business.

3. **Check.** The purpose of this phase is to help the organization assess its situation against the initial plan. Tools and procedures are required to ensure proper feedback and corrective capabilities for the sustainability systems. This phase promotes activities that are essential to the organizational learning process, such as knowledge transfer throughout the organization and systematic problem-solving.

4. **Act.** The final phase of the PDCA cycle is the management review. It addresses the effectiveness of the sustainability systems and possible need for changes to policies, objectives, and targets, and other elements. This review is made in light of information

such as audit results and any organizational or external changes.

This process can help create business opportunities and create value for both organizations and the environment. There are numerous examples of companies that have discovered win–win opportunities through careful investigation of their operations and transferring of technologies and techniques throughout their organizations. They have been forced to question some of the basic assumptions of their business decisions and found substantial profits through redesigning products and processes. Effective environmental management systems designed within this framework can thus be used to drive concerns for issues of production yield, waste reduction, marketing opportunities, and many others through the organization. Both the implementation of organizational strategies and the strategies themselves must be constantly re-examined. Improvements to the organization can then be driven through all of the systems in the organization so that organizational learning occurs.

Improving sustainability performance

Implementing systems to improve sustainability performance is quite different than implementing systems aimed at maximizing financial performance. To achieve goals such as revenue increases most companies start with mission statements, and then define their strategy. To implement their strategy, companies define performance measures and tie incentives and rewards to these performance measures. As employees strive to achieve the performance measures, the company succeeds in implementing its strategy.

Improving sustainability performance begins by communicating to all employees the importance of social, environmental, and economic performance to the corporation, to their individual welfare, and to their jobs. Correspondingly, the message should include the identification of corporate stakeholders (employees, suppliers, customers, community, etc.) and the importance of treating all stakeholders well. This message can be communicated in internal communications to employees and through training programs that sensitize employees to the social, environmental, and economic impacts of various activities, processes, and products.

Through a partnership between the World Business Council for Sustainable Development (WBCSD) and Cambridge Programme for Sustainability Leadership (CPSL), an e-learning tool for business education in sustainability literacy was developed, called Chronos. Accompanied by simulations, which may be used in addition to the tutorial as guides and materials, Chronos can be customised to reflect the needs of individual companies. Chronos is already being used by almost 200 organizations across the world, many of them Fortune 500 companies. Over 15 companies have customized Chronos to their own needs; for example, Philips and Heineken have introduced Chronos to their induction programs to ensure all new company entrants understand basic sustainability principles. And at Rio Tinto, employees are asked to evaluate company sustainable development policies using a customised version of Chronos.[7]

Using life-cycle assessment to improve performance.

To manage the learning process more effectively, organizations must create systems and processes that support these learning activities and integrate them into daily operations. One approach to helping organizations better understand their long-term social, environmental, and economic impact is through LCA. LCA provides a valuable framework for identifying the total impact of a corporation's activities, processes, and products. By examining the impact of products, processes, services, and other activities over the complete life-cycle, managers can redesign these activities to improve sustainability and financial performance.

Producer responsibility

The concept of product take-back continues to gain popularity and is causing more companies to think in life-cycle terms. Governments are experimenting with new forms of regulation focused on inducing "producer responsibility" or "extended producer responsibility" through both mandatory and voluntary product take-back schemes. A growing number of large food and beverage companies in the US are assuming the cost of recycling their packaging after consumers are finished with it, a responsibility long imposed on packaged goods companies in Europe and, more recently, in parts of Asia, Latin America, and Canada. Several factors are converging to make what is known as "extended producer responsibility" more attractive. On the one hand, more environmentally conscious consumers are demanding that companies share their values. On the other hand, it has become cheaper to recycle an aluminum can into a new can than it is to make one from virgin material. But also, many local governments are struggling with losses and are looking for ways to shift the costs of recycling onto someone else. Companies that make the packaging are logical candidates. Many states in the US have laws requiring companies to take responsibility for spent products such as batteries and mercury switches. Maine, however, has a law that might shift the cost of discarded packaging to business.[8]

Such plans have their roots in 1991 German regulations requiring "take-back" of packaging waste. Since then, other countries have considered a broad range of take-back legislation, from regulation of waste generation during manufacturing to mandating the actual physical recovery of products at the end of their lifecycle. The EU's End-of-Life Vehicles Directive, which became effective January 1, 2007, requires auto manufacturers to take back any of their vehicles, at no cost to the owner. The vehicle is then disassembled, the pieces are disposed of in compliance with other environmental regulations, and what's left is shredded. At least 85% of the vehicle's weight must be recycled.[9]

The assignment of responsibility for the end of the product life-cycle to the producer forces a more complete transition to life-cycle thinking. It goes beyond simply improving accounting for internal costs and demands attention to what were previously considered external costs: the costs of final disposition. This reframing of the producer's domain makes life-cycle costing a tool of even greater importance, as identification of costs throughout the life-cycle becomes an essential ingredient in decisions about corporate strategy.

A growing number of companies are using life-cycle assessment to better understand the costs and benefits of various actions and to improve management decisions. Sony Ericsson first completed a full life-cycle analysis (LCA) in 2008 on the W890 phone. That work resulted in an LCA model which Sony Ericsson still uses internally to measure and keep track of the carbon footprint of its products. The LCA that Sony Ericsson conducts on its products is based on a three-year life expectancy. For W890, the biggest impact area is the electronic component manufacturing because it is very energy intensive. The second largest impact is the user phase which includes the energy that the end user consumes to charge the phone, while the transportation of the components and the phones has the third biggest impact. High-end phones generally have a higher carbon footprint than low-end phones, such as the Sony Ericsson W890, because smartphones have greater functionality and this increases their energy consumption.[10]

Involving managers in a discussion of the environmental impacts of product development, manufacturing, delivery, use, and disposal adds a new dimension to decision-making at each link in the value chain. One of the first companies to use product life-cycle review as an integral part of improving social, environmental, and economic performance was Bristol-Myers Squibb, a global pharmaceutical company. At Bristol-Myers Squibb, multifunctional product life-cycle review teams brainstorm ideas and identify particularly salient opportunities to reduce environmental, health, and safety (EHS) impacts from design and development through manufacturing, distribution, sales, use, and ultimate fate. In addition to EHS benefits, Bristol-Myers Squibb product life-cycle (PLC) reviews of products produce economic benefits at the company's operations around the world.[11]

Baxter International, a global healthcare company, addresses environmental and social issues across the entire product life-cycle. These range from sustainable design and bioethics during R&D, to energy and materials efficiency during manufacturing and transport, to responsible advertising and promotion and, finally, to product repair, refurbishment, and recycling for electronic products, as appropriate, at end-of-life. Lifecycle assessments (LCA) are used to evaluate the environmental performance of products and determine ways to reduce their environmental footprint. This may include decreasing the presence of chemicals of concern and reducing life-cycle water or energy consumption, GHG (greenhouse gas) emissions and waste generation. For example, during 2011, Baxter used LCA to inform the development of its next-generation home hemodialysis system. In 2010, Baxter undertook a streamlined LCA that compared two generations of dialyzer products to evaluate how material changes affect environmental performance. The company's family of XENIUM+ dialyzers is 13–22% lighter than earlier versions, which offers the potential for reduced fuel consumption in shipping and biohazard waste removal. XENIUM+ dialyzers also use approximately 25% less cardboard in their packaging. To further these efforts, Baxter launched a new Product Sustainability Program in 2012, building on extensive partnership with the R&D, marketing, and supply chain groups. The program team initiated new projects to define life-cycle environmental impacts for both individual products and entire therapies.[12]

LCA helps managers to understand how the inputs and outflows of the life-cycle inventory translate into social, environmental, and economic impacts and to develop a more complete risk profile for products and processes. By carefully completing the process of identifying and measuring impacts and better understanding the social, environmental, and economic consequences of their products and processes, managers can more effectively manage their sustainability performance. Business units that have traditionally viewed social, environmental, and economic activities as separate functions from operating activities can develop a new model of integrated decision-making that accounts for the ways social, environmental, and economic concerns impact all of a company's operations.

Reducing social, environmental, and economic impacts

Often, the focus of various feedback mechanisms that are such a critical part of the Corporate Sustainability Model and in managing sustainability is the development of methods to reduce the negative social, environmental, and economic impacts produced by processes and products. Feedback provides information to aid in decision-making and managerial actions. At least four methods have been identified that provide opportunities for companies to become more socially, environmentally, and economically efficient.[13]

Redesign the product or service

First, companies can redesign their products or services. For years, Nike shoes contained a GHG called sulfur hexafluoride or SF6. In 2006, Nike developed a technology that uses nitrogen instead of SF6 to create the air pocket in its Nike Air sneakers. The nitrogen breaks up more easily and is not harmful to the environment. Although it took Nike nearly 14 years to develop this technology, it sees the effort as an important element in its sustainability performance.[14]

There are important benefits to society, the environment, the economy, and the corporate bottom line from product and service improvements motivated by a thorough review of the social, environmental, and economic impacts of products and processes. Product and service redesign often leads simultaneously to sustainability and financial benefits. United Parcel Service of America (UPS) has focused for decades on using the most fuel-efficient transport mode or combination of modes to meet service requirements, and on being able to fluidly shift modes in real time to reduce energy intensity whenever possible. The various transport modes used in the transportation sector have different energy intensities (energy required per unit of volume transported), ranging from aircraft at the high-end to ships at the low-end. UPS expertise in this area enabled the company to avoid approximately 2.4 million metric tons of GHG emissions in 2012 by shifting delivery volume from air to ground, and nearly 0.9 million metric tons of emissions by shifting volume from ground to rail while keeping service commitments to customers.[15]

Examples like this are becoming more common, yet in most companies similar opportunities for both cost savings and reduction in social, environmental, and economic impacts continue to exist. Long time horizons and challenges of measurement and incentive systems often remain as barriers to improvements in both sustainability and financial performance.

Many companies are also innovating their packaging to cut down on their environmental impact. For example, Dell was the first major company to pilot the use of mushroom-based packaging materials. This packaging is grown rather than manufactured, by injecting mushroom spores into agricultural waste to create an organic form that is as strong and safe as foam, but can be easily composted. Dell used it to cushion Dell PowerEdge R710 servers shipped in Multipack.[16]

Re-engineer the process

Companies can re-engineer their processes to reduce consumption, reduce pollution, and avoid risks. China National Petroleum Corporation, the world's fourth largest oil company, developed a boiler technology in which wastewater from thermal recovery of heavy oil can be recycled without the need to remove SiO_2 (silicon dioxide), thereby greatly reducing chemical costs and sludge generation.[17] Many companies have achieved substantial benefits by analyzing processes and products to determine ways in which waste and toxicity can be reduced. Pfizer, the world's largest research-based pharmaceutical corporation, has a manufacturing plant in Kalamazoo, Michigan, which is implementing a process that uses waste to clean equipment. The Kalamazoo API (Active Pharmaceutical Ingredients) Cleaning Solvent Project eliminates the need for expensive high-purity solvents, such as methanol or acetone, to clean equipment. Instead, it uses a high-quality recycled solvent mixture created from distilled solvents, which can be used to dissolve leftover product residue from the equipment as part of cleaning. This process is reducing the amount of chemical byproducts that are typically discarded by more than 25%, and is on target to save over US$1 million in 2013.[18]

Apple's data center in North Carolina employs an innovative cooling system that reuses water 35 times, resulting in a 20% reduction in overall water consumption. The data center also uses a rainwater-supplied system for onsite landscape irrigation, further reducing overall water consumption. By installing sophisticated irrigation systems that monitor local weather conditions and soil moisture, Apple is able to adjust landscape irrigation schedules and avoid unnecessary watering, resulting in a 40% reduction in landscape watering.[19] The Gazprom Group, the Russian energy giant, works on intensifying reusing waste in order to reduce the waste accumulation rate on production sites. By the end of 2012 the amount of accumulated waste was 47% less than in 2011. Gazprom Group increased the amount of waste transferred to other organizations for further use, neutralization, landfilling, and disposal by 48.5%. The amount of waste reused at its own facilities grew by 18.8%, which made 8% of the total waste.[20]

Create more but use less

Third, broader sustainability initiatives have also been introduced that are primarily focused on creating more goods and services while using fewer resources and producing less waste and pollution. Some have proposed that companies should base their products and processes on a cradle-to-cradle model. This model argues that products should be designed for eventual reuse in another product or be used by nature. Many companies are buying-in to the cradle-to-cradle or zero-waste concept.[21]

Coca-Cola has a whole subsidiary, Coca-Cola Recycling L.L.C, devoted to its stated goal of ensuring 100% recycling of its cans and bottles in North America by 2015 and 50% in the rest of the world.

Starbucks has bins in which customers can deposit their cups at 18% of its stores in the US and Canada. The company has a goal of 100% by 2015. In Chicago, Starbucks has developed a closed-loop system in which cups from its stores are trucked to a recycling facility where they are mixed with other recycled material and turned into paper napkins to be used in its stores.

Stonyfield Farm, the Vermont yogurt maker, put collection bins in Whole Foods stores. Customers can take any polypropylene plastic cup or container—margarine tubs, other brands' yogurt containers—which most municipalities do not recycle, to Whole Foods stores. They are collected, taken to a plant for processing, and then turned into toothbrushes and razors by Preserve, a company that creates products out of recycled materials.[22]

H&M (Hennes & Mauritz) was the first fashion company in the world that offered its customers around the globe the opportunity to hand-in clothes from any brand that they no longer want. In return, they receive a small reward in thanks. H&M then helps to reuse and recycle them.[23]

Vodafone, a British multinational telecommunications company with over 85,000 employees, for years encouraged customers to return their unwanted handsets and accessories to Vodafone for reuse (where possible) and recycling. In 2012, Vodafone changed its approach to proactively ask customers what they are doing with their old handsets when they upgrade. The new Vodafone BuyBack scheme offers both consumers and business customers attractive incentives to return used phones and tablets, including discounts on new handsets, charity donations, or store credit.

Returned handsets are refurbished and resold. In cases where equipment cannot be refurbished or resold, Vodafone works with specialist partners to separate and recycle the components. BuyBack is not only helping Vodafone to increase the number of phones collected for recycling (from 1.37 million in 2011/12 to 1.62 million in 2012/13), but also has a significant commercial benefit for Vodafone.[24] These innovations are all examples of companies striving to fully recycle their products or achieve zero waste rather than a reduction in current waste. In many cases, these also produce clear financial benefits.

For example, Baxter has been steadily increasing its recycling rate since 2007. Of the 58,000 metric tons of non-hazardous waste generated in 2012, Baxter recycled approximately 40,000 metric tons, or 69%. Baxter also recycled 1,600 metric tons of regulated waste in 2012, for an overall recycling rate of 66%. Recycling activities at Baxter generated nearly US$6.6 million in revenue in 2012. Although some recycled waste streams do not generate revenue, even in those cases recycling typically costs less than disposal.[25]

Companies can also certify their products with a third party, multi-attribute eco-label Cradle to Cradle Certified [CM] administered by the Cradle to Cradle Products Innovation Institute. Products or materials from any industry or country are eligible to apply for certification. Since the program began in 2005, more than 150 companies from over 15 countries have participated in the Cradle to Cradle Certified [CM] program. Currently there are over 425 certified products, which include building materials, interior design products, textiles and fabrics, paper and packaging, and personal and homecare products. Participating companies include UPS, Shaw Industries, Herman Miller and Steelcase.[26]

Rethink the market

Companies can find new ways to meet customer needs by rethinking their markets. In order to gain advantages over competitors, companies can differentiate their products and processes to reduce social, environmental, and economic costs or to improve the benefits provided to its stakeholders. The success of a differentiation strategy depends on the existence of three conditions:

1. The product or service reaches customers who are willing to pay a premium for something that is either environmentally friendly or socially and economically responsible
2. The benefits provided can be effectively communicated to its customers and stakeholders
3. The company is able to protect its product design or processes from competitors[27]

Using the methods for measuring social, environmental, and economic impacts and surveys of consumer preferences discussed in Chapters 6 and 7, companies are able to assess and quantify the first condition—people's WTP or WTA compensation for an environmentally friendly product. The second condition requires transparency achieved through disclosures of sustainability performance, backed by proper auditing of social and environmental, or sustainability reports. The last condition requires the efforts of internalizing the social, environmental, and economic impacts into the corporate structure, systems, and culture. By doing so, companies' know-how can be safeguarded and internal processes become more difficult to imitate. New product developments should also be protected by patents where appropriate.

Star-Kist, a tuna manufacturer and Heinz subsidiary, faced a product development challenge when the public became aware of tuna-fishing practices. Following the release of video footage showing dolphins dying in the course of tuna-fishing operations, the company announced that it would sell only dolphin-friendly tuna, fished in the western Pacific, where the fish do not swim with dolphins. This became a very popular program and market share rose.[28] However, it faced challenges because the three conditions were not met. The strategy backfired because consumers did not value the effort required to provide a dolphin-safe tuna. Contrary to what the company had surveyed, people wanted a cheap source of protein. Also, despite publicly announcing the efforts to protect dolphins, the fishing techniques sacrificed other animal species existing in the western Pacific. And, to worsen matters, since no proprietorship exists over fishing methods and international waters, competitors could follow Star-Kist's steps almost immediately.[29]

Many companies have been successful in making a shift from selling products to selling services. For instance, Interface, the world's largest carpet manufacturer, traditionally sold its carpets to clients who would have to purchase new carpet when it needed to be replaced. Now, under its Evergreen Service Contract, Interface leases carpets, rather than selling them. Monthly inspections detect worn carpet tiles and the tiles are replaced as needed. This method is better for the environment, saves Interface money, and saves its customers money because it is cheaper to produce and replace small tiles than entire carpeting.[30]

Bottom of the pyramid

One differentiation strategy used by companies is marketing products toward the "bottom of the pyramid" (BOP). The people at the bottom of the economic pyramid, the world's poorest, represent two-thirds of the world's population.[31] Almost half the world—over three billion people—live on less than US$2.50 a day.[32] Most have little formal education, live in rural villages, and do not have the usual distribution or communication systems found in more developed countries. This market is still being underserved by large firms and multinational corporations. Not only can effective innovation strategies aid in alleviating global poverty, they also give corporations markets for future growth since the BOP is an untapped market.

To seize opportunities at the BOP, companies usually need to refocus their business models to include the appropriate structures, systems, and strategies suited for the existing conditions at the BOP. For instance, most poor people are paid in cash on a daily basis and therefore tend to purchase only what they need at that particular time. To meet this need, in India Procter & Gamble has begun to offer its Pantene shampoo, a high-end product, in single-serve packets priced at two cents each. Unilever offers Close-Up toothpaste for less than ten cents a package in sub-Saharan Africa. These products are of the same quality as those offered in wealthy markets but are affordable and packaged to meet the needs of those at the BOP. Also, companies such as Dow Chemical and Cargill are currently experimenting with totally biodegradable plastic packaging so that this trend will not result in problems of waste.

GE is commercializing small-scale, distributed technologies such as distributed solar, point-of-use water treatment, and portable, small-scale health devices. The vast, underserved market at the base of the pyramid is an ideal place for the incubation of new, sustainable technologies through a bottom-up form of innovation.[33]

Amanco Guatemala, a subsidiary company of the Brazilian GrupoNueva, developed a business model addressed specifically at low-income markets. The project involved a drip irrigation system adapted to the requirements of small farmers. The majority of small farmers in Guatemala make exclusive use of the natural rainfall cycle, which limits agricultural production to the rainy season and therefore to two annual harvests. The Amanco system was called the 4þ4 Irrigation Module because it made it possible to produce during all four seasons of the year and obtain four annual harvests on any type of land. In the first year, small producers were able to harvest twice as much, with savings of 33% in terms of labor costs, yield increases of over 22%, and significant water efficiencies. With the same amount of water they could irrigate an extra 50% of land.[34]

To understand the non-users in the BOP, ethnographic process involving direct observation is often most appropriate (spending days with villagers, learning about their food habits, and observing the patterns).[35] As mentioned earlier, CEMEX designed Patrimonio Hoy (Property Now), one of the most successful social programs in Latin America, providing 453,538 low-income families with access to building materials such as cement, concrete blocks, and steel. Patrimonio Hoy provides these products at average market prices as well as microfinancing, technical advice, and logistical support to assist participants in building their own homes. To explore this segment of customers and build a new business model, a team of CEMEX employees lived in the shantytowns for six months.[36]

Working with local partners to develop and distribute new technologies and products is another key to being successful in BOP economies. For example, Roche, a Swiss global healthcare company, is the main external sponsor of Transnet-Phelophepa Healthcare Train. Described by locals as a "miracle train," the Transnet-Phelophepa Healthcare Train is a clinic on rails that brings primary healthcare to remote regions of South Africa where there is only one doctor for every 5,000 residents. Since its launch in 1994, Roche has been funding its primary healthcare, community outreach, and other ancillary services. From modest beginnings as a three-car train, the service was expanded in 2012 to two 18-coach trains. With both trains running 36 weeks a year, Phelophepa can deliver medical services and healthcare education to reach over 550,000 people annually, including more than 90,000 patients in need of care. Train staff also visit schools, providing vital health checks, medicines, and education to children.[37]

Innovation in product distribution is also critical. Grameen Telecom, a spin-off of Grameen Bank, developed an innovative approach to provide a wireless phone service to rural communities in Bangladesh. The company lent up to US$175 to women entrepreneurs to cover the cost of a mobile phone, a solar recharger, and basic training. The entrepreneurs then provided phone usage to inhabitants in rural villages for a reasonable service fee. Grameen was able to offer a product that otherwise would not have reached poor communities.

A decade ago, ICICI Bank, India's largest private sector bank, launched several microfinance programs, including running a network of village internet kiosks, partnering with MFIs (microfinance institutions) which acted as loan service agents, and collaborating with social entrepreneurs. By using nontraditional methods, ICICI has grown to be the second largest bank in India.[38]

The move to less developed economies is supported by the same potential benefits as from observing the guidelines for social, ethical, and environmental leadership: it is also a matter of sustainability. To pursue these opportunities, corporations should follow the same analysis, processes, and actions discussed throughout this book. Companies need to begin to integrate the BOP mentality into their strategic definition and implementation, and promote internal interest toward underdeveloped economies.

Both the board and the CEO will need to be involved in developing a strategy for opening the company to developing nations. These four broad strategies can help companies find success in BOP markets:

- Focus on the BOP with unique products, services, or technologies that are appropriate for BOP needs

- Localize value creation through franchising and using members of the community as vendors or suppliers
- Enable access to goods and services through innovative distribution and packaging strategies
- Partner with governments, nonprofits, or other organizations when necessary[39]

As is the case with the implementation of sustainability, the strategy needs to be institutionalized, measured, and evaluated before it leads to long-term increases in sustainability and financial performance. If developed and managed properly, working with the poor can improve both sustainability and financial performance. It can provide increased access to what is currently a very large, and rapidly growing, market. It can also be a setting to develop innovations at low cost that can often then be transferred to the developed economies. And it can be a major contribution to alleviation of poverty.

Involve the supply chain

Another way to minimize social, environmental, and economic impacts is to put pressure on suppliers to reduce the negative impacts of the components of the products or services they provide. By pushing these concerns throughout the supply chain, companies can reduce their sustainability impacts and their costs. It is also an opportunity to stimulate markets for socially, environmentally, and economically sensitive products and materials.[40]

Foxconn Technology Group, Taiwan, the world's largest electronics contract manufacturer, established a supplier GHG management portal where more than 300 suppliers' carbon data are on file. Foxconn can readily manage the GHG inventory by accessing the data through e-management. By the end of 2011, 230 suppliers had finished their GHG inventory based on the ISO 14064 international standard. Among these suppliers, 141 have received verification statements from third parties, with the suppliers having combined carbon emissions of approximately 50 million tons of CO_2e in the base year.[41]

Reputation risk can become a factor when deciding on suppliers and contract facilities. Many companies have endured protests from social activists and environmental groups because of their supply practices. From 1997 to 1999, environmentalists protested against THD (The Home Depot) stores around the country stating that the company purchased its wood from endangered forests. In August 1999, THD stated that it would not purchase from endangered forests, and by 2003 the company was working with environmental groups to protect forests in Chile.

Responding to these claims is one step; however, being proactive about supplier systems can help companies face these difficult challenges. The many benefits of socially, environmentally, and economically sensitive purchasing systems include:

- Cost avoidance: lower waste management fees and hazardous material management fees
- Savings from conserving energy, water, fuel, and other resources
- Easier compliance with regulations
- Reduced risk of accidents, reduced liability, and lower health and safety costs
- Improved image

These efforts should include ensuring that labor sourcing is done in a socially responsible way and cover not only the purchase of parts and materials to manufacture products but also the purchase of office supplies. After achieving 100% green purchasing in the areas of office supplies and computers in 2002, Toyota now focuses on creating relationships in which suppliers in various countries and regions and Toyota do business on an equal footing based on mutual respect, thus building firm bonds of trust and promoting mutual growth and development. Toyota's global purchasing activities revolve around three policies: fair competition based on an open-door policy; mutual benefit based on mutual trust; and contribution to local economic vitality through localization.[42]

Companies have developed several methods of instituting sustainable purchasing initiatives into their systems, including:

- **Written policies and communication.** Companies are establishing sustainability policies that their suppliers must follow. For example, Dell's Restricted Materials Program requires suppliers to restrict or eliminate certain materials in components supplied to Dell. Adherence to the program is a requirement of any contractor doing business with Dell[43]
- **Questionnaires and audits.** Questionnaires are used by companies to screen new suppliers or evaluate existing suppliers. Some companies conduct sustainability audits of suppliers to ensure that minimum standards are met. British American Tobacco created a Business Enabler Survey Tool (BEST) to evaluate suppliers.[44] BEST focuses on the sustainability of supply and sets the standards for suppliers. It currently uses 102 criteria to measure performance, including:

  –Employee working relationships that promote a stable and productive workforce

  –Effective controls on environmental impacts

  –Process control procedures to ensure quality production

  –Effective cost control programs

- **Supplier meetings.** These meetings are held to communicate expectations to suppliers and share information
- **Training and technical assistance.** Some companies provide training and assistance to help suppliers develop sustainability management systems
- **Collaborative research and development.** Suppliers are involved in the design process to help develop more innovative, socially, environmentally, and economically friendly products[45]

Nike has developed a single evaluation system across all types of products and all Nike brands that redefines what "good" looks like for factories that supply to Nike. A new manufacturing index (MI) has been implemented in 2012 with four equally weighted areas: quality (product issues, defective rates, manufacturing processes); delivery (delivery precision, planning precision); cost (cost competitiveness, cost accuracy, cost timeliness); and sustainability performance management (lean implementation, environment/energy, labor, health and safety). It integrates scores from these key performance areas into a single dashboard rating. Factories are evaluated as Gold, Silver, Bronze, Yellow or Red. Contract factories that are able to consistently exceed Nike requirements in the equally weighted areas of Quality, Cost, Delivery, and Sustainability performance management, and show consistent performance leadership in the industry, will achieve a Silver rating in the MI. Contract factories that go beyond the industry and are demonstrating innovation and benchmark performance within the broader manufacturing landscape will achieve Gold. At a minimum, factories in Nike's supply chain will be expected to achieve and sustain a Bronze rating, indicating that they meet Nike baseline standards and can self-govern through integrated systems and a lean approach.[46]

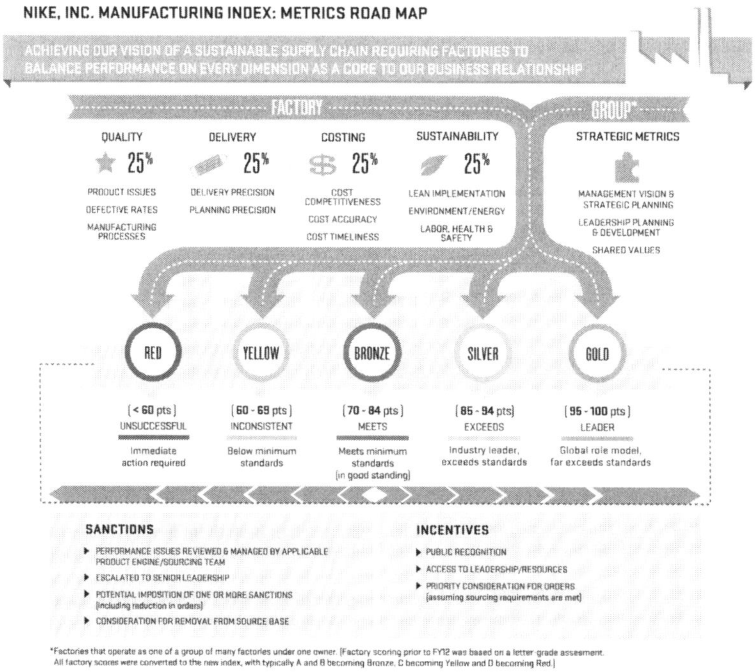

FIGURE 8.2 **Nike's MI for suppliers**

Source: Nike (2012) *Nike Inc., FY10/11 – Sustainable Business Report Summary*

To do business with L'Oréal, suppliers' operations must meet the same standards as L'Oréal's own sites. As a signatory of the United Nations Global Compact, L'Oréal expects its suppliers to adhere to the basic conventions of the International Labor Organization and to local legislation, especially in matters concerning minimum wage, working time, and occupational health and safety. Social audits are carried out by independent external companies on the suppliers' premises, thus making close monitoring of their adherence possible. Audits are also systematically carried out on subcontractors, wherever in the world they may be based, as well as on suppliers of raw materials, packaging, and point-of-sale advertising in sensitive countries. Social audits are paid for by L'Oréal and carried out by independent assessors on L'Oréal's behalf. In 2010, L'Oréal assessed working conditions and labor standards at more than 560 supplier sites through its social audit program. The high volume of purchases made by L'Oréal gives the group considerable economic and social leverage. This allowed the group to launch the Solidarity Sourcing Program in 2010, which is neither a sponsorship program nor a charity. Rather, it enables and promotes inclusion of suppliers with better access to major customers and those who employ those generally excluded from employment (e.g., the disabled or socially disadvantaged, people in rural communities). This approach is also followed by existing L'Oréal suppliers when it comes to their own purchasing policies.[47]

It is important that companies use their supplier audits not only to find deficiencies but also to improve performance.

adidas Group has invested more resources than ever before in 2012 on training and building capacity with its suppliers. Overall, in 2012, adidas Group conducted 172 training sessions on different topics for more than 3,000 people. One aspect of this focus was the group's effort to drive a more common approach and methodology in developing and executing the training program for suppliers. This included defining the target suppliers and the training subjects more accurately, strengthening the sharing of training materials between regions, using external experts and consultants for certain subjects, such as HR management systems or safety officer training, and best practice sharing. Factories that already operate at the self-governance level are invited to share their experiences, knowledge, and success stories. To review progress, suppliers receive assignments after the training to ensure they put into practice what they have learnt.[48]

In 2007, Walmart introduced a packaging scorecard to evaluate suppliers on the sustainability of their product packaging. The scorecard is based on GHG emissions related to production (15%), material value (15%), product-to-packaging ratio (15%), cube utilization (15%), recycled content usage (10%), innovation (5%), the amount of renewable energy used to manufacture the packaging (5%), the recovery value of the raw materials (10%), and emissions related to transportation of the packaging materials (10%). By 2013, the initiative reduced the overall GHG impact of packaging by an average of 9.8% in Walmart US stores, 9.1% in Walmart Sam's Clubs in the US, and 16% in Walmart Canada stores.[49]

Unilever has committed to purchasing all of its tea from sustainable sources. Being the world's largest tea company, this strategy has the potential to have substantial impact on the industry. To aid in this strategy, Unilever is collaborating with the Rainforest Alliance to audit the estates where Unilever purchases its tea. Unilever CEO, Patrick Cescau, views this as a win-win situation for many of the company's stakeholders: "Consumers will have the reassurance that the tea they enjoy is both sustainably grown and traded fairly. Subsistence farmers will get a better price. Tea pluckers will be better off. The environment will be protected. And we expect to sell more tea."[50]

The full integration of suppliers into a social and environmental management system requires recognition of the interrelationships through the supply chain. Particularly for those in the toy, footwear, and apparel industries, the social issues, including labor, are significant since most of their manufacturing is in developing countries. Companies affect their suppliers and, through their policies, can improve social, environmental, and economic sensitivity. Likewise, a supplier's use of recycled and recyclable materials and sensitivity to social, environmental, and economic impacts can significantly reduce a company's impacts and financial costs. Process and product redesigns are often encouraged through careful monitoring of supplier relationships.

Internal reporting

In order to make decisions to improve processes and products, managers and employees need information about sustainability performance.[51] Though companies have significantly increased their external disclosures of sustainability (discussed in Chapter 9), the development of systems focused on internal managerial decisions is even more critical for improving sustainability performance. Internal reporting provides important feedback for effective decision-making and strategic planning, and also helps employees to see how their individual contributions add to the successful performance of the company. By properly disclosing social, environmental, and economic performance metrics for internal users, leading companies are empowering their employees to provide both a horizontal and a vertical analysis of their functions. The former will set benchmarking opportunities across companies in the same industry, while the latter will provide an opportunity for continuous improvement and long-term sustainability.

The interests of the various internal constituents vary both in scope and the detail of required information. Organizations must obviously disclose information to internal audiences as required by regulation or face detrimental costs of noncompliance. Companies must identify and profile internal stakeholders who have an interest in their sustainability performance. The content and placement, distribution and communication of internal reports are also important factors in obtaining information for effective decision-making.

Profiling the sustainability report audience

Reporting sustainability outputs and outcomes should operate on multiple levels to address the needs of diverse audiences, each with its own specific needs, requirements, expectations, agendas, and levels of expertise. Figure 8.3 presents the most important internal and external audiences for corporate disclosures.

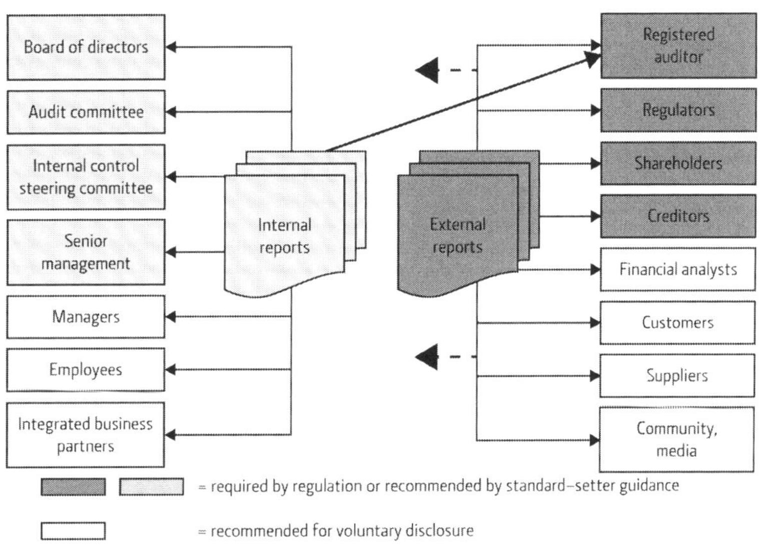

FIGURE 8.3 **Internal and external audiences interested in corporate decisions**

Source: Epstein and Rejc Buhovac (2006) *The Reporting of Organizational Risks for Internal and External Decision Making*

Although internal reports are aimed exclusively at internal audiences, there may also be interested internal audiences for some areas of external reporting, including corporate annual reports and sustainability reports (see the two broken arrows in Figure 8.3).

As Figure 8.3 shows, internal and external audiences can be further divided into two subgroups. Some audiences (audit committees, internal control steering committees, boards of directors, and senior management among internal audiences, and registered auditors, regulators, shareholders, and creditors among external audiences) must be informed about sustainability outputs and outcomes because of regulation or recommendations in standard-setter guidance. Voluntary disclosure to other internal audiences (managers, employees, and integrated business partners), and external stakeholders (financial analysts, customers, suppliers, community, and media), is recommended because of anticipated benefits for improved decision-making.

Responsibilities of internal audience members

The board of directors has the primary oversight responsibility for developing and implementing the organization's mission, values, and strategy, and must carefully review corporate processes of sustainability-issue identification, monitoring, and management. Specific reviews of financial objectives, plans, major capital expenditures, and other significant material transactions also typically fall within a board's responsibility.

Senior management has specific needs for information on sustainability. They need relevant, accurate, and reliable reports on a real-time and periodic basis for effective decision-making and control. Only by generating a wealth of sustainability-related information can organizations inform senior management with facts, not intuition, so that they can then appropriately integrate that information into management decisions and make more effective decisions to implement company strategy and goals.

Similarly, managers need relevant and accurate real-time and periodic reports. Without effective internal reporting related to sustainability, business unit and functional managers cannot (a) make optimal strategic and tactical decisions, (b) evaluate the payoffs of specific management initiatives, or (c) make new capital project decisions explicitly acknowledging the potential risks and their costs on organizational profitability.

Employees, for example, prefer to work for companies with safe and healthy working conditions. Companies also need to manage the social, environmental, and economic impacts at all levels of the organization. Machine operators and others throughout the organization need to know the sustainability impacts of their activities so adjustments can be made to improve performance. In addition, as members of the community, they are often interested in the company's impact beyond the workplace.

Organizations are increasingly beginning to consider integrated supply chain partners as internal rather than external participants. Interdependence of partners in an extended supply chain requires cooperation and collaboration in sustainability management. Integrated supply chain partners need real-time information on various sustainability outputs, particularly those related to integrated processes and technologies, so that they can contribute to improved customer satisfaction and achieve performance excellence for the supply chain as a whole.

The content of internal sustainability reports

What information needs to be provided for improved management decisions? More specifically, how detailed should the reports be? When thinking about what to include in an internal sustainability report, managers should, at a minimum, cover targets, accountability, and recommendations and decide what type of data to provide, what metrics to include, and how to explain the context of the information reported.

Set and report on targets

Previous plans and goals should be disclosed with the sustainability outputs, to permit comparisons between actual achievements and planned results. Internal audiences must be given enough clear information to allow them to understand the potential or existing operational and financial impact of sustainability on the organization. In addition, an explanation of the impact of combined outputs on the organization as a whole may be provided. Managers need to explain the link between sustainability events and response activities, and their financial consequences as an understanding of these links and financial impact is critical for improved decision-making. The internal report's ability to look across the organization will allow internal users to identify sustainability outputs in the aggregate, and determine gaps in the sustainability management strategy.

Demonstrate accountabilities

Showing where accountability lies may be specifically important for boards of directors, audit committees, and steering committees, all of whom have responsibility for oversight, and for senior management and other managers who are responsible for decision-making. Stakeholders will be interested in who is responsible for the sustainability strategy and performance of the organization. Accountability can be shown through a discussion of corporate governance and structure.

Make recommendations
Sustainability reports must also include recommendations for the intended internal audiences. Sustainability reports cannot determine how the CEO, CFO, and other senior managers should respond to individual findings. However, the recommendations should be precise, business-focused, and pragmatic, so that the recipients of reports feel sufficiently informed to act. For example, an organization may face a human-resource-related risk within a process that is found to be dependent upon the skills of one individual. The sustainability report recommendations might suggest an additional hire, cross-training, or improving documentation so that a non-specialist could operate the process.

Include different types of data
Different aspects of sustainability performance and decisions call for different types of data—qualitative or quantitative, different metrics, and other tools (such as graphs, figures, or scenarios). Graphs and other figures are specifically useful. However, the report must include sufficient relevant technical detail needed by those responsible for taking action.

Include metrics
Sustainability reports should explain presented metrics in sufficient detail. In periodic reports, metrics must be disclosed consistently from period to period, to the extent they are still relevant. However, there is no need to continue reporting on a specific output with a specific metric in one period if it is no longer

relevant, or if a more relevant metric becomes available.

Executives drive financial accountability by choosing or devising the financial measures that provide appropriate decision-making information internally, at every step in the management planning and control process. As with financials, there must also be operational, social, environmental, and economic measures that will help improve decision-making. These measures add tremendous insight to people's thinking at each step of the accountability cycle, from strategy and planning to reporting and pay.

Explain context

The context of reported impacts must be appropriately explained. Managers seeing only facts without context may react inappropriately. In addition, the reporting of specific outputs must include sufficient evidence to influence proper decisions. For example, some managers may require overwhelming evidence before they accept a problem's existence; others may simply need sufficient evidence to understand the nature of the problem. Managers may therefore decide to include information on strategy, actions, and performance in addition to information specifically focused on sustainability outputs. This broader description should be narrative, and accompany a quantitative presentation of the outputs. Alternatively, the report should clearly describe the status of the organization's processes and activities related to sustainability management initiatives.

Ben & Jerry's environmental coordinators compile monthly reports covering progress toward annual goals, status of key indicators, and any other issues regarding environmental performance. However, internal audiences (and many external audiences) will be interested not only in disclosure of specific outputs but also in the sustainability management process. A well-established and properly managed process will assure internal audiences about the reliability of sustainability reports. Organizations must therefore include information on the quality of their management process.

Distributing internal reports

For the board and committees, sustainability reporting should happen at least quarterly. As the rate of change in business activities accelerates, and information technology reduces the cost of collecting and providing updated information, reporting frequency is likely to speed up. The following means can be used for general communication of sustainability-based information across business units, processes, or functions:

- Broadcast emails, email discussion groups, or conference calls
- Corporate newsletters, letters from the CEO, or newsletters from the senior sustainability manager
- Databases supporting specific sustainability issues

- Intranet sites capturing information on sustainability management for easy access by personnel
- Messages integrated into ongoing corporate communications
- Posters or signs reinforcing key aspects of sustainability strategy

ABN AMRO, an international banking group, issues a biweekly e-newsletter on sustainability to every business unit globally. The newsletter communicates strategy, discusses performance, and covers global trends and developments. By publishing this newsletter regularly, the company communicates the importance of sustainability and ensures employees are aware of corporate performance.

Summary

Managers should not underestimate the importance of the underlying learning process associated with measuring social, environmental, and economic impacts. Through use of the various feedback systems discussed here, organizations can develop new capabilities that will enable them to achieve competitive advantage from improved sustainability performance. Sustainability programs that are designed only from a compliance perspective and that are reactive rather than proactive will not provide adequate productive learning and capability-building possibilities.

Managers should consider this an important dimension as they define their sustainability policy and objectives.

The feedback and internal reporting process should provide managers with information to help reduce social, environmental, and economic impacts substantially through:

- Process and product redesigns
- Zero-waste strategies
- Product differentiation
- Supply chain relationships

Product quality, production yields, and profitability can be increased, and waste can be reduced or eliminated. Striving for continuous sustainability improvement usually causes social, environmental, and economic impacts *as well as* corporate costs to decrease.

In the next chapter we look at external reporting.

CHAPTER 9

External sustainability reporting and verification

It is critical to collect and analyze information on sustainability for improved resource allocation decisions. This information should then be included in internal sustainability reports to improve managerial decision-making regarding processes and products. How companies perform on sustainability is also an important factor for external stakeholders since they are affected by corporate strategies and actions. Sustainability disclosure is valuable because it helps a company demonstrate that it is managing its risks and has a track record of paying attention to its sustainability performance. Surveys reveal that a positive sustainability reputation adds an extra layer of protection, leads to higher total returns, and such companies are more likely to enjoy a lower cost of capital. The 2011 global survey by the CDP (Carbon Disclosure Project), for example, reported that companies in its CDLI (Carbon Disclosure Leadership Index) and CPLI (Carbon Performance Leadership Index) provided double the average total return of the Global 500 between January 2005 and May 2011.[1]

The empirical evidence to date most strongly suggests that:

- The average investor is paying attention to sustainability when things go wrong and the company is in the limelight and usually under duress
- It is likely that the investor reaction to negative sustainability events will continue to increase as more investors pay attention and increasingly understand what these events can mean for a company
- Disclosure of sustainability performance can partially protect against drops in shareholder value when things do go wrong[2]

For this reason, all elements of the Corporate Sustainability Model (inputs, processes, outputs, and outcomes) should be measured and reported for improved management decisions and actions and for improved accountability to stakeholders.

Massey Energy: stock price reaction to poor safety

On April 5, 2010, there was an explosion at Massey Energy's Upper Big Branch mine in Montcoal, West Virginia, killing 29 miners. At that time, Massey Energy was the fourth largest producer of coal in the US and the largest coal producer in Central Appalachia. The stock fell 11% on April 6, and by the end of April was down by 33%. Facing estimated costs of US$80–150 million to compensate the families of the fallen miners and pay insurance deductibles and possible legal fees, as well as about US$62 million worth of equipment and mineral rights impacted by the disaster, Massey's financial performance deteriorated rapidly. As evidence of a poor safety record was unearthed, pressure on senior management increased, leading the CEO Don Blankenship to retire in December 2010. On January 28, 2011, shareholders of Alpha Natural Resources—known for a focus on safety—agreed to buy Massey Energy for US$7.1 billion, and the stock price jumped 10%. The Massey accident brought the consequences of shortchanging safety for the sake of profit to the fore in the coal mining industry, which has since been in decline due to cheap natural gas and increased regulation.[3]

Various pressures have caused companies to increase their social, environmental, and economic disclosures in corporate annual reports, and the quantity and quality of disclosure in separate environmental, social responsibility, or sustainability reports. Corporate responses to increased stakeholder demands for information on corporate sustainability performance vary widely. Some companies have issued social and environmental reports for each operating division or geographic area, some for the entire corporation only, and some have included this discussion in corporate annual reports. In the *Fortune* Global 250 study, 20% of the companies included a sustainability section in their annual reports, while 54% published a separate sustainability report.[4] Another study by SIRAN (Social Investment Research Analysts Network) found that:

- 79% of the S&P 100 companies have sections on their websites for sharing sustainability policy and performance information
- In 2005, a dozen new companies issued sustainability reports for the first time
- Forty-three of the S&P 100 companies issue annual sustainability reports[5]

Many reports began as only environmental reports; however, more companies have broadened their reports to include social and economic issues as well. Also, more companies are including governance and legal aspects in their reports. One of the first sustainability reports was published in 1998 by Royal Dutch Shell—*Profits and Principles: Does There Have to Be a Choice?* —with an unprecedented level of information on environmental, social, and governance issues. Along with the sustainability reporting and the GRI (Global Reporting Initiative), the mid-1990s were also characterized with another movement—integrated reporting. The integrated reporting movement emphasizes integrating sustainability information in annual reports showing their impact on each other. More than 80 of the world's largest global companies are piloting integrated reporting, including Coca-Cola,

Microsoft, Volvo, Philips, and Unilever. Novo Nordisk, the Danish pharmaceutical company, has been issuing integrated reports since 2005, and the US-based United Technologies Corporation, a US$60 billion revenue diversified company that provides high-technology products and services to the building and aerospace industries, has done so since 2009.[6] Unlike GRI, which developed into global sustainability reporting standards adopted by the majority of S&P 500 companies, actual standards for integrated reporting are under development (for more on the standards for sustainability reporting, see the subsequent section).

The rise in reporting of sustainability performance goes hand-in-hand with stakeholders' demands for reliable and credible information from management. Managers and external stakeholders must have the information they need to make better decisions, and it is important that the information is of high quality, reliable, relevant, and intelligible to likely readers. To provide confidence among stakeholders, companies should demonstrate that the sustainability performance metrics disclosed are integral and representative of actual efforts and achievements. General Mills, the American food giant, explained in its 2010 CSR report why the company did not meet its kilowatt hours of energy per metric ton of production. Its energy consumption has incrementally decreased over the past few years, but General Mills experienced a surge in demand for breakfast cereals and granola bars. The products

General Mills makes today are cooked or toasted, which requires more energy than when the company first started measuring its overall energy usage. And they are also less dense, which skews energy use per metric ton of production upward.[7] A 2005 survey of stakeholders indicated that formal external verification was the most important factor contributing to credibility.[8] It can improve the reliability of the information and the accountability to stakeholders.

In this chapter we will look at:

- The standards for sustainability reporting
- The content, format, and distribution of reports
- External disclosure of sustainability measures
- Verification
- Internal and external sustainability auditing

Standards for sustainability reporting

GRI

The most prominent approach to standardized environmental reporting began with Ceres (originally Coalition for Environmentally Responsible Economies). Ceres is a nonprofit organization composed primarily of public-interest groups, social investment professionals, and environmental groups promoting responsible activity. The Ceres Principles were an attempt to standardize information and emphasized the importance of both internal and external evaluations of sustainability performance.

Spearheaded by Ceres in partnership with UNEP (United Nations Environment Program), the GRI was established in 1997 with the mission of developing globally applicable guidelines for reporting on the economic, environmental, and social performance of corporations, governments, and NGOs. GRI incorporates the active participation of corporations, NGOs, accountancy organizations, business associations, and other stakeholders from around the world.

First released in 2000, the GRI's Sustainability Reporting Framework provides guidance for disclosure about sustainability performance, and gives stakeholders a framework to understand disclosed information. The GRI's Sustainability Reporting Guidelines represent the first global framework for comprehensive sustainability reporting. Launched in 2011, G3.1 completed the content of the G3 guidelines released in 2006. G3.1 featured expanded guidance on local community impacts, human rights, and gender. In May 2013, GRI launched its latest version of reporting standards, the G4 guidelines. The main features of the G4 guidelines include an emphasis on what is material, which encourages organizations to provide only information that is critical to their business and stakeholders. This means organizations and report users can concentrate on the sustainability impacts that matter, resulting in reports that are more strategic, more focused, more credible, and easier for stakeholders to navigate. Other important features include up-to-date disclosures on governance, ethics and integrity, supply chain, anticorruption and GHG emissions, and a generic format for disclosures on management approach. GRI will continue to recognize reports based on the G3 and G3.1 guidelines for up to two full reporting cycles. However, reports published after December 31, 2015 should be prepared in accordance with the G4 guidelines.

The GRI's framework consists of the Sustainability Reporting Guidelines and Sector Guidance. The guidelines assist in the preparation of sustainability reports by organizations, regardless of their size, sector, or location. The guidelines offer an international reference for all those interested in the disclosure of governance approach and of the environmental, social, and economic performance and impacts of organizations. The guidelines are useful in the preparation of any type of document which requires such disclosure.[9]

Although it is critical to report performance to stakeholders, companies must first develop a strategy to implement sustainability. External reporting can provide important feedback but should be seen as part of credible accountability rather than merely a public relations exercise. The primary focus should be on improving sustainability and financial performance and then reporting on progress to various internal and external stakeholders.

IRIS, GIIRS, and B Lab

A growing community of impact investors, who deliberately invest for social and environmental impact, cannot fully evaluate impact investment with GRI. The NCIF (National Community Investment Fund) and the IRIS (Impact Reporting and Investment Standards) initiative managed by the GIIN (Global Impact Investing Network), a US-based nonprofit organization, have harmonized their respective metrics to increase impact investors' use of standardized social metrics. The resulting IRIS provide a common reporting language for impact-related terms and metrics. By defining terms, impact reporting is consistent (for example, it stipulates how to calculate a metric ton of carbon), and it also serves as a repository of aggregated IRIS-compliant data to enable benchmarking across companies. Funds and direct investors can use these standards to credibly track and report social and environmental performance, and companies raising capital can attract investors by measuring and reporting both financial and nonfinancial performance by IRIS performance measures.[10]

Impact investors increasingly require an independent third-party impact rating to make investment decisions. IRIS does not provide an overall impact rating for companies or funds, or comparability among alternative investment opportunities. GIIRS (Global Impact Investing Rating System), on the other hand, is a comprehensive and transparent system for assessing the social and environmental impact of market companies and funds with a ratings and analytics approach. GIIRS uses IRIS definitions whenever a metric in the IRIS taxonomy corresponds to the metric being assessed in a GIIRS rating. GIIRS offers a company seeking investment capital a rating of its social and environmental impact, including: (1) an overall rating; (2) ratings in approximately 15 subcategories; (3) key performance indicators (KPIs) relevant to the company's industry, geography, size, and social mission; and (4) benchmark data highlighting a company's performance as compared to its peers.[11]

GIIRS is powered by B Lab, a US-based nonprofit that leads the initiative of building a community of Certified B (Benefit) Corporations. Certified B Corporation is a certification conferred by B Lab and declares that a company has met a high standard of overall social and environmental performance.[12] Through the leadership of the community of certified B Corporations, laws have been passed in 19 states (as of July 2013) creating a new type of corporation—the Benefit Corporation—that best meets the needs of entrepreneurs and investors seeking to use business to solve social and environmental problems. Benefit Corporations operate in the same way as traditional corporations but with higher standards of corporate purpose, accountability, and transparency. They give business leaders legal protection to pursue a higher purpose than profit, and they offer investors and the public greater transparency to protect against pretenders. Benefit Corporation is thus a legal status administered by the state. Benefit Corporation legislation requires officers and directors to consider all stakeholders in major business decisions, and it provides increased accountability. Each state's legislation differs somewhat from the model Benefit Corporation draft, but all of them meet the bar of providing for an overarching general public benefit obligation, accountability to all stakeholders, and impact transparency. While certified B Corporations have access to a portfolio of services and support from B Lab, Benefit Corporations do not. Benefit Corporations, however, do not need to be certified.

California's B Corp legislation took effect alongside a new law creating the FlexC (Flexible Purpose Company), which allows a firm to adopt a specific social or environmental goal, rather than the broader obligations of a B Corp. Another option in North America is the low-profit limited-liability (L_3C) company, which can raise money for socially beneficial purposes while making little or no profit.

In South America, Sistema B has been established in 2012 in partnership with B Lab to certify South American social enterprises called Empresa B.[13]

B Lab also has partners setting up their own B Lab offices in Canada (MaRS Discovery District), Australia (Small Giants and Net Balance), and in Europe.

SASB and IIRC

While GRI is global and provides hundreds of generally applicable indicators, for selection by the reporter in preparing a sustainability report, sustainability accounting standards are US-focused and industry specific, designed for use in integrated disclosure in the Form 10-K and 20-F. These standards are being developed by the SASB (Sustainability Accounting Standards Board), a US-based nonprofit, and will enable comparison of peer performance and benchmarking within an industry. SASB is currently developing standards for 88 industries in ten sectors.

The key objective of integrated reporting is to demonstrate the linkages between an organization's strategy, governance, and financial performance and the social, environmental, and economic context within which it operates. The IIRC (International Integrated Reporting Council), a global coalition of regulators, investors, companies, standard setters, the accounting profession, and NGOs, is leading the creation of the globally accepted International (Integrated Reporting) Framework that elicits from organizations material information about their strategy, governance, performance, and prospects in a clear, concise, and comparable format.[14]

Industry guidance on sustainability reporting

GRI's sector supplements are versions of the Sustainability Reporting Guidelines tailored for specific industry sectors which make reporting more relevant and user-friendly for organizations in diverse sectors. Sector supplements have been developed for ten different sectors: airport operators; construction and real estate; electric utilities; event organizers; financial services; food processing; media; mining and metals; NGOs; and oil and gas. GRI's recommendation is to use sector guidance when preparing a sustainability report, if available. The contents of the ten GRI sector supplements available by May 2013 have been reorganized to fit the G4 guidelines' content, structure, and requirements. It is presented in the May 2013 GRI sector disclosures documents, in a new format, to facilitate its use in combination with the G4 guidelines.[15]

In addition to GRI's sector supplements, there is also industry-specific guidance on sustainability reporting, such as *Oil and Gas Industry Guidance on Voluntary Sustainability Reporting,* which does not establish an industry standard but rather serves as a resource for interested companies. It was established by IPIECA (the former International Petroleum Industry Environmental Conservation Association, now the global oil and gas industry association for environmental and social issues), API (American Petroleum Institute), and OGP (International Association of Oil & Gas Producers) in 2005 and revised in 2010. The guidance is designed as a "stand-alone" reference tool and differs from GRI G3 guidelines in four ways. First, it is designed to provide flexibility to meet the reporting needs of a variety of organizations in the petroleum industry, such as multinational majors, national oil companies to smaller international and domestic companies. Second, the guidance captures industry consensus on the material sustainability issues faced by oil and gas companies, accompanied by appropriate indicators and reporting elements for these significant issues. Third, the guidance shares oil and gas industry-specific good practice, including greater technical depth on quantitative performance tracking, particularly on environmental, health and safety issues, with practical options for qualitative reporting, especially on economic and social issues. Fourth, it is aligned with recommendations from other good practice and guidance documents published by IPIECA, API, and OGP for their members.[16]

Let everyone know how you're doing

For years, reporting was often based on mistrust, as senior management questioned the willingness of outsiders to handle corporate information responsibly.[17] Today, the premise is not just that senior management should base their reporting communication policy on trust in order to be more accountable; organizations can also expect tangible benefits from fair and broad disclosure of sustainability outputs and outcomes. Different groups have interests in disclosure:

- Owners primarily rely on financial reporting to assess the current financial condition of the organization, its financial performance over time, and its prospects. However, current and prospective owners have interests beyond the relative transparency of an entity's material costs and liabilities, and expect information on all organizational issues, including sustainability
- Creditors have a particular vested interest in complete and timely disclosure of organizational risks, to assess credit risks and potential joint liability for loans secured by, for example, contaminated properties
- The list of external audiences for sustainability reporting also includes customers, suppliers, and communities (interest groups, media, the scientific community, and the general public)

Communication with stakeholders or investor relations is one of the most important corporate governance aspects investors monitor before making an investment.[18] Public-interest groups and customers have also gained senior managers' attention. Organizations see increasing pressure for greater transparency, mandatory or voluntary, and a better alignment of externally reported information with the information that is reported internally to senior management for decision-making. Stakeholders expect and demand increased sustainability disclosure to improve both monitoring and decisions. This requires effective external reporting of the social, environmental, and economic issues the organization is facing, and of the management team's plans to capitalize on emerging opportunities or to minimize the risk of failures.

There are corporate accounting methods that can be used to hide social, environmental, and economic liabilities in reports. They include hiding big issues in the footnotes, delaying the quantification of liabilities, avoiding meaningful qualitative disclosure, disaggregating social, environmental, and economic liabilities, and employing artificial time horizons.[19] Each of these methods is legal but can be used to keep important and material information from stakeholders. However, leadership companies will want to disclose information on social, environmental, and economic liabilities in a way that is accessible and comprehensible to stakeholders.

To increase transparency and stakeholder trust, organizations may want to disclose broader organizational outputs and outcomes to external audiences. This approach may be especially important, because external constituents expect disclosure of how the organization is prepared for and manages sustainability. With appropriate sustainability structures and processes, organizations can enhance corporate image and win the trust and loyalty of those outside the organization: customers, shareholders, suppliers, and others they depend on to conduct business. The content, format, placement, distribution, and communication of external reports are important considerations in an organization's reporting framework.

The content of external reports

When deciding what to report externally, managers should choose from the data that it has already collected for its internal reports. A study found that stakeholders, particularly social investors, prefer indicators that are simple, easily collected, and readily available.[20] The UN Global Compact recommends that companies use measurement and reporting systems already in place and report only some of that information to stakeholders. The ISO 14031 standard makes no recommendations about which metrics a company should use or report. However, it lists almost 200 topics from which companies can select metrics that comprehensively describe their sustainability impacts. But companies must take care to be selective and balance a desire for more complete information with a need to keep it understandable and useful. In many cases, the presented data is so extensive that it is difficult to get a clear understanding of sustainability performance.

Generally, senior management must assure stakeholders that sustainability processes and impacts are well managed. External information users recognize that leading nonfinancial performance measures should link strongly to the organization's future performance. Key input, process, and output measures are leading indicators, and can be used to forecast future results. For example, fines and penalties may be a leading indicator of corporate reputation, the amount of a company's toxic emissions suggests future environmental costs, and employee turnover is a leading measure of future recruitment and training costs.

Some companies are reluctant to report internal performance indicators, especially if the news is not entirely favorable. However, just as the disclosure of information in corporate reports can signal good performance, it can also be used to soften the impact of poor performance. Companies reporting a deficiency can use the opportunity to discuss steps they have undertaken to improve performance. And disclosures should reflect the results of past sustainability performance as well as the strategies and systems in place to improve future performance. Once they begin to increase their voluntary disclosure, companies are acknowledging their acceptance of greater responsibility and accountability on an ongoing basis, engendering trust and building credibility with stakeholders—whether the news is good or bad. That credibility is important to all stakeholders, including investors who value improved information for decision-making. For example, Bayer, a global healthcare enterprise headquartered in Germany with over 100,000 employees, disclosed in its sustainability report not only environmental incidents and transport accidents that are classed as major ones, but also incidents that came to the attention of stakeholders (Table 9.1).[21]

TABLE 9.1 Bayer discloses environmental incidents and transport accidents observed by stakeholders

	Location of the incident	Description	Explanation
1	Bayer MaterialScience, Tarragona, Spain	January 11, 2012 Leak in gas pipeline	During roadworks at the production site, a digger damaged a gas pipeline and caused a leak. This was plugged by an emergency team. No one was hurt and there was no environmental damage.
2	Bayer CropScience, Haelen, Netherlands	March 31, 2012 Fire in operations room of Research & Development (R&D) Department	In an operations room in the R&D building, a faulty freezer cabinet caused a fire. The fire protection measures were initiated right away. The fire department extinguished the fire immediately. No one was hurt and there was no environmental damage.
3	Bayer MaterialScience, Krefeld-Uerdingen, Germany	April 3, 2012 Burst waste gas line	For reasons still unknown, a waste gas line in the nitro-benzene wastewater treatment plant burst under pressure. No one was hurt and there was no environmental damage.
4	Bayer MaterialScience, Dormagen, Germany	May 14, 2012 Deflagration	A deflagration occurred in a hydrogen drying tower. The building sustained material damage, but no one was hurt. No substances were released into the environment.

Source: Excerpt from Bayer (2012) Sustainable Development Report

There is growing consensus that external sustainability reports should contain more comprehensive information than just that required by regulatory agencies. Stakeholders say they want to see human rights, energy-and eco-efficiency, and health and safety in reports (see Fig.9.1).

A five-part test devised by Zadek and Merme can help to decide what information a company should disclose.[22] The five areas cover the following areas for disclosure:

1. The report covers the traditional direct short-term financial impacts of sustainability performance, such as carbon emissions
2. The company discloses performance associated with declared policies, regardless of short-term financial consequences
3. The company discloses similar information of its market peers
4. Stakeholder concerns are addressed. Are companies disclosing information that is likely to impact stakeholder behavior?
5. Aspects of performance that might not be currently regulated but could be regulated in the future are discussed

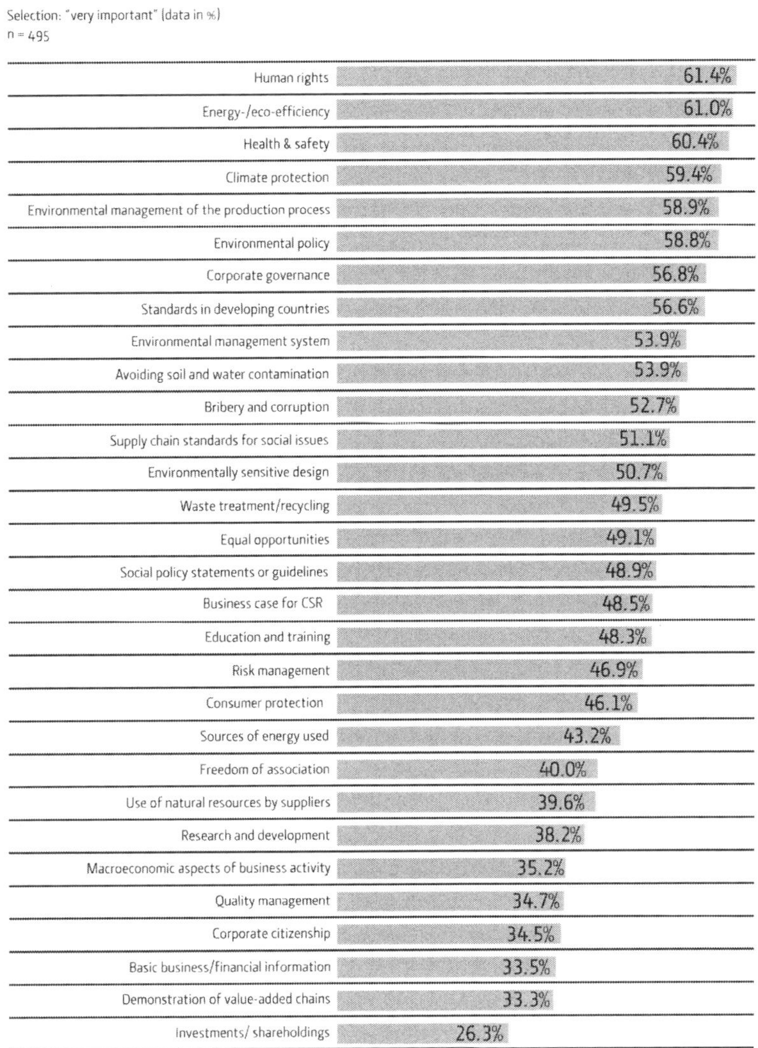

FIGURE 9.1 **CSR issues in reports: what stakeholders want to see**

Source: Pleon (2005) *Accounting for Good*

This five-stage test provides a basis for determining what information is material. Stakeholders and corporate managers can apply it to see how the company's reporting is evolving and where it needs to go in the future.

The content of external sustainability disclosures should be customized to the company context and issues. For example, the supply chain and employer relations are intently watched in the apparel and sporting goods industry. Nike was the first company in its industry, in 2005, to provide a complete list of the factories that it contracts with to make Nike brand products, detailing the number of employees, workforce information, as well as addresses and contact information.[23] Similarly, adidas Group reports results from its factory audits, including number of audits divided by region and type (Table 9.2), number of warning letters issued (Table 9.3), and number of business relationship terminations (Table 9.4). The company also reports the number of training sessions given to suppliers to help them comply with adidas health, safety, and labor standards, and key labor noncompliance findings.[24]

Companies often provide information that enables analysis against targets or other benchmarks such as industry standards. They can also provide data that compares performance over time. In both cases, and especially in the latter, it is particularly important to provide some guidance to the reader to aid in evaluating the impact and relevance of the disclosure (for example, the effect of emissions on the community).

Other information reported by companies includes:

- Inputs, such as material, energy, and other natural resource use
- Processes, such as management systems and policies, including goals, targets, and accountability systems; risk management methods, accident and safety data, and stewardship practices; product data such as life-cycle analyses, product-packaging changes, and remanufactured products
- Outputs, such as waste and emissions, stakeholder identification, social, environmental, and economic impacts and concerns, and stakeholder reactions
- Outcomes, such as financial data on reactive versus proactive spending, capital and operational expenditures, charitable contributions, and costs avoided

Increasingly, companies are indicating specifically when they are reporting GRI indicators. With the GRI Content Index included at the end of the report, a company lists every G4 guidelines' disclosure addressed in its report. It communicates which GRI disclosures have been reported, and the reason why certain disclosures have not been reported. The Index is a gateway for finding all reported sustainability information. It should have clear and direct referencing and, if used online, can be an interactive navigation tool. By using the Index, report users should easily find the specific GRI data of interest. In its 2012

Sustainability Report, Volkswagen Group takes full account of the reporting guidelines of the GRI. Selected indicators and the degree to which they are reported are disclosed as shown in Table 9.5.

Number of audits divided by region and type

Region	Initial Assessments[1]			Performance Audits[2]			Environmental Audits			Total		
	2010	2011	2012	2010	2011	2012	(2010)[4]	2011	2012	2010	2011	2012
Asia	355	379	**270**	788	790	**712**	(118)	114	**190**	1,142	1,283	**1,172**
Americas	59	50	**38**	78	75	**73**	0	0	**0**	137	125	**111**
EMEA	33	47	**30**	37	46	**47**	0	0	**4**	71	93	**81**
Total[3]	**447**	**476**	**338**	**903**	**911**	**832**	**(118)**	**114**	**194**	**1,350**	**1,501**	**1,364**

[1] Every new supplier factory has to pass an Initial Assessment to prove compliance with the Workplace Standards prior to order placement.
[2] Audits conducted in approved supplier factories.
[3] Includes audits done in licensee factories. In addition, there was a considerable number of full environmental assessments conducted for selected suppliers in Asia.
[4] The number of Environmental Audits was recorded and disclosed separately in 2010; therefore it is not included in the total number of audits for 2010.

TABLE 9.2 **Number of supplier audits in adidas Group in 2012**

Source: adidas Group (2012) *Sustainability Progress Report: Performance Counts*

Number of warning letters issued to adidas Group suppliers by region*

Region	1st Warning			2nd Warning			3rd and Final Warning -> recommended termination			Total Warning Letters		
	2010	2011	2012	2010	2011	2012	2010	2011	2012	2010[1]	2011[1]	2012[1]
ASIA	43	39	**34**	2	7	**9**	1	0	**8**	46	46	**51**
AMERICAS	3	2	**4**	0	0	**1**	0	0	**0**	3	2	**5**
EMEA	1	0	**1**	0	0	**0**	0	0	**0**	1	0	**1**
Total	**47**	**41**	**39**	**2**	**7**	**10**	**1**	**0**	**8**	**50**	**48**	**57**

* Including warning letters issued by licensees and agents. But excluding warnings to main suppliers for the non-disclosure of subcontractors, which is issued either directly through business entities, or by the adidas Group legal department where there is a breach of contract obligations under a manufacturing agreement.

TABLE 9.3 **Number of warning letters issued to adidas Group suppliers in 2012**

Source: adidas Group (2012) *Sustainability Progress Report: Performance Counts*

Number of business relationship terminations due to compliance problems

Region	2010	2011	2012
Asia	0	9	**10**
Americas	0	1	**0**
EMEA	0	3	**0**
Global	**0**	**13**	**10**

Worldwide rejections after Initial Assessment due to compliance problems

	2010	2011	2012
Total number of first time rejections[1]	149	171	113
First time rejection rate	44%	48%	44%
Total number of final rejections[2]	27	34	21
Final rejection rate	6%	7%	6%

[1] Factories that were directly rejected after first visit, i.e. with no chance being visited a second time, and factories that were rejected after Initial Assessments but which were given a chance for a second visit.
[2] Factories that were directly rejected after first visit, i.e. with no chance being visited a second time, and factories that were rejected after being visited a second time.

TABLE 9.4 **Number of business relationship terminations in adidas Group in 2012**

Source: adidas Group (2012) *Sustainability Progress Report: Performance Counts*

GRI Standard Disclosure		Reference	Status	UNGC	GSC
EN27	Packaging materials	118, 131, OSR 131	◍	8, 9	
EN28	Sanctions for non-compliance with environmental regulations	D	○	8	
Social Performance Indicators: Labor Practices and Decent Work					
	Disclosure on management approach	8, 9, 14-21, 58-66, 69-72, 74-80, OSR 66	○	1, 3, 6	
LA1	Workforce by employment type and region	60, 138-139	○		
LA2	Employee turnover	141	◍	6	
LA4	Employees with collective bargaining agreements	18-20, 42, 66, 69	○	1,3	
LA5	Minimum notice period(s) regarding operational changes	69, OSR 17	○	3	
LA7	Occupational diseases, lost days, and number of fatalities	77, 141, OSR 66	◍	1	15
LA8	Training on serious diseases	74-76, OSR 66	○	1	16
LA10	Training per employee	65	◍		16
LA13	Composition of governance bodies	73, 77-78, 139-140, AR 214-215, 133, 147-150	○	1, 6	16
LA14	Gender pay disparity	66, OSR 17, 18	○	1, 6	
Social Performance Indicators: Human Rights					
	Disclosure on management approach	18-20, 34-46, AR 201, OSR 15, 17, 18, 66	○	1-6	
HR1	Investment agreements	18, 34, 36, 39, 151, OSR 15, 17, 18	○	1-6	17
HR2	Supplier screening on human rights	34-39, 151	○	1-6	17
HR4	Incidents of discrimination	18-20, 45-46, OSR 15	○	1, 2, 6	16, 17
HR5	Freedom of association and collective bargaining	18-20, 36, 42-46, 69-70, OSR 17, 18	○	1-3	
HR6	Child labor	18-20, 36, 42-46, AR 193, OSR 17, 18	○	1, 2, 5	17
HR7	Forced labor	18-20, 36, 42-46, AR 193, OSR 17, 18	○	1, 2, 4	17
Social Performance Indicators: Society					
	Disclosure on management approach	16-27, 42-46, 81-89, OSR 15, 18	○	10	
SO1	Impacts on communities	18, 42-47, 86-89, 98, 126, 151	○		18
SO2	Corruption risks	42-46, AR 134-136	○	10	
SO3	Anti-corruption training	44, AR 134-136	○	10	
SO4	Actions taken in response to incidents of corruption	45, 46, AR 134-136	○	10	20
SO5	Lobbying	17, 19-20, 42, 45, 100-101, AR 131	○	1-10	
SO8	Sanctions for non-compliance with laws and regulations	AR 233-234	○		20
Social Performance indicators: Product Responsibility					
	Disclosure on management approach	8, 16, 18, 32, 33, 42-46, 103-110 AR 208-210, 222, 223	○	1, 8	
PR1	Health and safety impacts along product life cycle	32-33, 102-105, OSR 90	○	1	
PR3	Product information	103-111, 158, 159, OSR 2	○	8	
PR6	Marketing communication standards	16-21, 32-33, 107, OSR 15, 114, 137, BP	◍		
PR9	Sanctions for non-compliance with product and service related regulations	AR 233-234	◍		

Status: ○ fully reported
◍ partly reported
● not reported

AR = Annual Report
GP = Group Portal www.volkswagenag.com
BP = Brand Portal www.volkswagen.com
OSR = Online Sustainability Report
UNGC = United Nations Global Compact
GSC = German Sustainability Code
D = Direct Answer (online-index)

TABLE 9.5 **Volkswagen Group's Sustainability Report 2012 includes GRI Content Index**

Source: Excerpt from Volkswagen Group (2012) *Sustainability Report*

To confirm the amount of GRI standard disclosures a company has addressed in its sustainability report, companies can apply for GRI Application Level Check. GRI thus confirms that a sustainability report has the required set and number of disclosures to meet the organization's self-declared Application Level. Table 9.6 shows a Statement of the GRI Application Level Check for Volkswagen Group's *Sustainability Report* 2012. It states that Volkswagen Group's *Sustainability Report* 2012 fulfils the requirement of Application Level A+. Application Level A, intended for advanced reporting organizations, applies to sustainability reports of companies that have executed a thorough materiality process in consultation with their stakeholders. As a result, these reporters are able to "report or explain" to the fullest extent (all profile disclosures, all disclosures on management approach, and all core performance indicators). "+" reveals that Volkswagen Group has had its reporting externally assured (Fig 9.2).

Global Reporting Initiative™

Statement
GRI Application Level Check

GRI hereby states that **Volkswagen Group** has presented its report "Sustainability Report 2012" to GRI's Report Services which have concluded that the report fulfills the requirement of Application Level A+.

GRI Application Levels communicate the extent to which the content of the G3 Guidelines has been used in the submitted sustainability reporting. The Check confirms that the required set and number of disclosures for that Application Level have been addressed in the reporting and that the GRI Content Index demonstrates a valid representation of the required disclosures, as described in the GRI G3 Guidelines. For methodology, see www.globalreporting.org/SiteCollectionDocuments/ALC-Methodology.pdf

Application Levels do not provide an opinion on the sustainability performance of the reporter nor the quality of the information in the report.

Amsterdam, 26 March 2013

A+
GRI REPORT
GRI CHECKED

Nelmara Arbex
Deputy Chief Executive
Global Reporting Initiative

The "+" has been added to this Application Level because Volkswagen Group has submitted (part of) this report for external assurance. GRI accepts the reporter's own criteria for choosing the relevant assurance

The Global Reporting Initiative (GRI) is a network-based organization that has pioneered the development of the world's most widely used sustainability reporting framework and is committed to its continuous improvement and application worldwide. The GRI Guidelines set out the principles and indicators that organizations can use to measure and report their economic, environmental, and social performance. www.globalreporting.org

Disclaimer: Where the relevant sustainability reporting includes external links, including to audio visual material, this statement only concerns material submitted to GRI at the time of the Check on 19 March 2013. GRI explicitly excludes the statement being applied to any later changes to such material.

FIGURE 9.2 **GRI Statement of the GRI Application Level Check for Volkswagen Group's Sustainability Report 2012**

Source: Volkswagen Group (2012) *Sustainability Report*

The format of external reports

How long should a sustainability report be? The most important thing is that companies should provide stakeholders with all pertinent information. A survey of stakeholders (Fig.9.3) found that, if a sustainability report has the appropriate content, then length doesn't matter. However, more than 60% of the respondents disliked reports of more than 50 pages. The survey also found that two-thirds of respondents spend no more than 30 minutes and most read only selected parts, so companies should highlight the parts that they really want stakeholders to pay attention to. This would probably be a clear and concise discussion of sustainability performance and important processes and outputs in language written for a general audience. Additional detail can be included where appropriate.

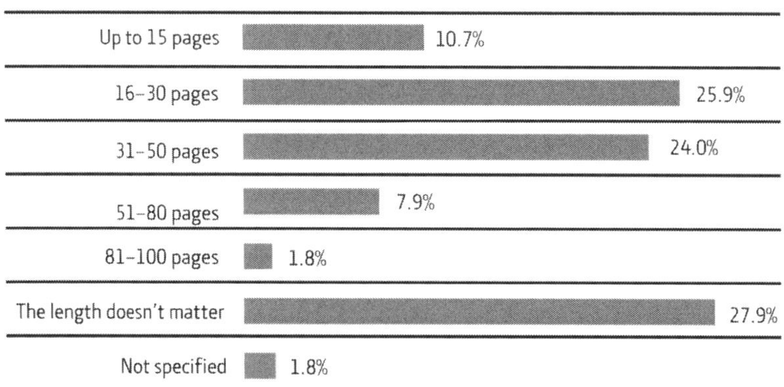

FIGURE 9.3 **Length of sustainability report**

Source: Pleon (2005) *Accounting for Good*

Because of the rise of the internet and the trend toward electronic dissemination of financial and other information on websites, concerns about the organization of information may change. Users of corporate websites have greater control over which portions of the report to review and which to disregard. As these technologies develop, the sequence of information in a traditional paper annual report and the length of the report might become increasingly less important, but providing the information in an easily accessible format and language remains critical.

Distributing external reports

Large companies now increasingly produce separate sustainability reports in addition to their corporate annual reports. However, the president's letter in annual reports, along with other voluntary disclosures, should offer a brief overview of the organization's performance on key sustainability issues.

Generally, the communication strategy may include analyst meetings, press conferences, formal documents, and other channels of communication, such as the internet or websites. Some may access the information in electronic form, but others will continue to want information on paper.

Dell maintains a robust, multifaceted approach to reporting on its corporate responsibility priorities, goals and impact. In addition to its annual *Corporate Responsibility Report* that complements the company's

Annual Report, it provides a link within the corporate responsibility section on www.dell.com to Dell's comprehensive GRI A-level Report produced using the GRI Sustainability Reporting Guidelines. It also maintains extensive content within the corporate responsibility section of its website to provide stakeholders with current, detailed information. This includes case studies, policies, interactive tools, opportunities for feedback, and learning resources—on a wide range of Dell's corporate responsibility activities, all organized by corporate responsibility action areas. This ongoing reporting on Dell's website complements the more focused "snapshot in time"—such as content presented in Dell's annual corporate responsibility reports. Moreover, since 2003, Dell has provided a detailed report each year on its carbon emissions to the CDP—the largest database of primary corporate climate change information in the world. Dell requires its Tier 1 suppliers to report their emissions to the CDP. Dell also reports and participates in the Investor CDP, CDP Supply Chain and CDP Water.[25]

For its 2012 *Sustainability Progress Report,* DuPont shortened the printed (pdf) report but expanded the content on the website. The printed report makes references to where additional information can be found on the website. ScottishPower has chosen to use the internet as its sole method of communicating in its 2005/06 *Corporate Responsibility Report.* The new format is intended to be more user-friendly and to enable the company to make more frequent updates. Avon enables its internet users to build their own Avon Corporate Responsibility Online Report by checking the boxes for the pages the user would like to include in his or her very own pdf report. Whichever method is practiced, the reporting objective should be to provide a sound basis for external audiences to assess sustainability performance and actions.

External disclosure of sustainability performance measures

Companies that adopt a broader set of performance measures for internal decisions—measures that flow from strategy and point to profitability—and that integrate internal and external reporting, can be more accountable to their constituents.[26] Beyond the benefits of greater accountability, such as enhancing the corporate image and engendering trust among stakeholders, disclosing sustainability performance measures to external stakeholders has been shown to boost company valuation as it reduces investor uncertainty.

Demand for more disclosure is growing, from Wall Street to Main Street. Financial analysts indicate they want more nonfinancial data from corporate external reports, including the annual report.[27] More complete information could help them diagnose company problems and make more accurate valuations. Just as managers need broader sets of indicators to better understand their company's performance and to improve their decision-making, analysts and other "outsiders" need similar information to evaluate prospects for future earnings and share value.

Many companies now report broader financial performance metrics than the traditional metrics of the income statement, balance sheet, and cash flow statement. Among them are measures of economic profit or shareholder value, and, increasingly, leading measures of financial performance.

Since 1994, Baxter Healthcare has been issuing its annual *Environmental Financial Statement* that details environmental revenues, control costs, and the financial impact on the company of its environmental actions from preceding years (Table 9.6). This is one of the more creative and detailed disclosures of environmental impacts and an example of how companies can improve accountability through voluntary external disclosures. In calculating savings and cost avoidance for resource reduction activities, Baxter assumes that production and distribution grow at the same rate as the company's cost of goods sold, and that resource use and waste generation increase at that same rate in the absence of reduction initiatives. Baxter determines this rate by calculating the average annual increase in the company's published cost of goods sold over the past six years. It then adjusts this number for new acquisitions and changes in inventory, and subtracts inflation, which is calculated as an average of three major, relevant US producer price indexes. The company then rounds the resulting growth rate down to the nearest whole number to conservatively report performance.[28]

Novo Nordisk, the Danish insulin producer, used the IIRF (International Integrated Report Framework) proposed by the IIRC to report on its financial, social, and environmental performance for 2011. By following this format, Novo Nordisk is trying to bring together information in a way that reflects the commercial, social, and environmental context of a business. It believes that this approach allows the company to better understand, manage, and report on multiple dimensions of value, and helps managers make better decisions and manage in a way that creates shared value. The company reports additional information online.[29] Key figures for 2012 are provided in Table 9.7.

Understanding and creating customer value is also a priority for top management and a key performance driver at many companies. A number of companies use and report a customer satisfaction metric, but others go further, releasing leading measures of customer satisfaction and performance. Ford Motor Company focuses on a long-term view of customer satisfaction, and has measured and reported customer satisfaction after three months, and then after three years, of product use. Allstate Insurance has reported marketing and advertising expense data, internet and phone accessibility for customers, and sales to specific demographic groups, such as working-and retirement-age customers.

The sustainability processes companies must excel at to deliver value to customers are wide-ranging, encompassing purchasing, manufacturing, distribution, and social and environmental management processes and actions. Equally wide-ranging are the process-related disclosures companies make, including information on supplier relationships, material usage and disposal, operational performance, productivity, workplace safety, waste generation and disposal, and social investments.

Companies also report metrics about activities that typically stress reskilling, systems development, change procedures, and the development of personal and organizational capabilities. Table 9.8 lists several innovative measures disclosed in external reports. Some of these measures relate directly to sustainability, while others are nonfinancial measures that would be of use to some stakeholders. Companies can look at those examples and develop a list of additional measures they could report that would be of value to their stakeholders.

Baxter 2012 Environmental Financial Statement

Estimated Environmental Costs, Income, Savings and Cost Avoidance Worldwide[1]

ENVIRONMENTAL COSTS (dollars in millions)	2012	2011	2010	2009	2008
Basic Program					
Corporate Environmental – General and Shared Business Unit Costs[2]	$2.2	$2.0	$1.9	$2.0	$1.9
Auditor and Attorney Fees	0.4	0.3	0.4	0.4	0.3
Energy Professionals and Energy Reduction Programs	1.2	1.2	1.2	1.2	1.2
Corporate Environmental – Information Technology	0.3	0.3	0.4	0.4	0.4
Business Unit/Regional/Facility Environmental Professionals and Programs	12.1	11.8	11.9	11.5	11.2
Pollution Controls – Operation and Maintenance	4.0	3.9	3.7	3.6	3.4
Pollution Controls – Depreciation	2.3	2.4	2.5	3.1	2.4
Basic Program Total	**$22.5**	**$21.9**	**$22.0**	**$22.2**	**$20.8**
Remediation, Waste and Other Response (proactive environmental action will minimize these costs)					
Attorney Fees for Cleanup Claims and Notices of Violation	$0.3	$0.2	$0.1	$0.1	$0.1
Settlement of Government Claims	0.0	0.0	0.0	0.0	0.0
Waste Disposal	8.6	8.3	7.3	7.5	8.4
Carbon Taxes, Credits and Offsets[3]	1.2	1.1	1.2	0.2	0.2
Environmental Fees for Packaging[4]	0.9	0.9	0.9	1.0	0.9
Environmental Fees for Electronic Goods and Batteries	0.1	0.1	0.1	0.1	0.1
Remediation/Cleanup – On-site	0.2	0.2	0.4	0.1	0.2
Remediation/Cleanup – Off-site	1.5	0.4	0.8	0.4	0.1
Remediation, Waste and Other Response Total	**$12.8**	**$11.2**	**$10.8**	**$9.4**	**$10.0**
Total Environmental Costs	**$35.3**	**$33.1**	**$32.8**	**$31.6**	**$30.8**

ENVIRONMENTAL INCOME, SAVINGS AND COST AVOIDANCE (dollars in millions; see Detail on Income, Savings and Cost Avoidance from 2012 Activities below)	2012	2011	2010	2009	2008
From Initiatives in Stated Year					
Regulated Waste Disposal	$0.5	$0.4	$1.1	($0.4)	($0.2)
Regulated Materials[5]	(1.8)	(0.1)	2.8	(0.6)	(2.0)
Non-hazardous Waste Disposal	1.1	(0.9)	0.7	0.2	0.5
Non-hazardous Materials[5]	9.2	(11.0)	3.6	2.5	1.3
Recycling (net income)	6.6	5.0	5.9	3.5	5.5
Energy Conservation	3.9	2.4	0.3	4.5	5.2
Water Conservation	0.4	(0.2)	0.4	0.6	0.8
From Initiatives in Stated Year Total[6]	**$19.9**	**($4.4)**	**$14.8**	**$10.3**	**$11.1**
As a Percentage of Basic Program Costs	88%	-20%	67%	46%	53%
Cost Avoidance from Initiatives Started in the Six Years Prior to and Realized in Stated Year[6,7]	**$36.4**	**$40.2**	**$81.2**	**$92.6**	**$98.1**
Total Environmental Income, Savings and Cost Avoidance in Stated Year	**$56.3**	**$35.8**	**$96.0**	**$102.9**	**$109.2**

DETAIL ON INCOME, SAVINGS AND COST AVOIDANCE FROM 2012 ACTIVITIES (dollars in millions)	Income and Savings	Cost Avoidance	Total Financial Benefit
Regulated Waste Disposal Cost Reduction	$0.1	$0.4	$0.5
Regulated Waste Materials Cost Reduction	(2.9)	1.1	(1.8)
Non-hazardous Waste Disposal Cost Reduction	(0.4)	1.5	1.1
Non-hazardous Waste Materials Cost Reduction	6.1	3.1	9.2
Recycling Income	6.6	0.0	6.6
Energy Consumption Cost Reduction	(8.1)	12.0	3.9
Water Consumption Cost Reduction	(0.8)	1.2	0.4
Total	**$0.6**	**$19.3**	**$19.9**

COST AVOIDANCE DETAIL FROM EFFORTS INITIATED IN THE SIX YEARS PRIOR TO REPORT YEAR (dollars in millions)	2012	2011	2010	2009	2008
Regulated Waste Disposal	$0.8	$0.4	$0.7	$0.1	$0.2
Regulated Waste Materials	(5.3)	(2.4)	0.3	0.1	(1.6)
Non-hazardous Waste Disposal	2.3	1.5	3.4	2.5	3.9
Non-hazardous Waste Materials	5.6	3.6	19.4	23.8	23.6
Energy Consumption	27.8	32.7	50.1	59.2	65.5
Water Consumption	5.2	4.4	7.3	6.9	6.5
Total	**$36.4**	**$40.2**	**$81.2**	**$92.6**	**$98.1**

[1]Financial numbers rounded to nearest US$100,000 to reflect appropriate degree of data accuracy. [2]Corporate environmental costs comprise total environmental costs related to operating corporate environmental programs that report into Baxter manufacturing and legal groups. While corporate Environment, Health and Safety (EHS) and certain business unit EHS groups were integrated in 2003, total business unit program costs remain in the Business Unit/Regional/Facility Environmental Professionals and Programs line, as those environmental costs more directly support facility programs. [3]Cost of carbon taxes, credits and offsets includes expenses associated with purchasing renewable energy certificates and Chicago Climate Exchange (CCX) carbon credits purchased through the IntercontinentalExchange. [4]Following completion of the 1996-2005 packaging-reduction goal, Baxter discontinued tracking program costs and financial savings associated with packaging-reduction initiatives at the corporate level. Baxter may reinstitute this line item in future financial statements. [5]Reflects change (positive for decrease and negative for increase) for purchases of raw materials due to changes in material use efficiency and associated generation of waste. [6]In calculating savings and cost avoidance for waste-, energy- and water-reduction activities, it is assumed that production and distribution activities grew proportionately with Baxter's publicly stated cost of goods sold, adjusted for changes in inventory and inflation. Baxter uses a three-year rolling average of the annual percentage change in growth in the cost of goods sold to determine the financial values for each stated year. For 2012, the three-year rolling average was 3%; for 2011, 3%; for 2010, 2%; for 2009, 3%; and for 2008, 5%. This rolling average helps avoid distortions due to certain acquisitions/divestitures and the delayed environmental effects from changes in production. [7]To be conservative, the accumulation of reported cost avoidance from conservation activities in prior years is terminated after seven years, the approximate duration of many facility conservation and process-improvement projects, after which additional process improvements and changes are possible.

TABLE 9.6 **Baxter Environmental Financial Statement**

Source: Excerpt from Baxter (2012) *Environmental Financial Statement*

DKK million	2008	2009	2010	2011	**2012**	2011–2012
Financial performance						**Change**
Sales						
Modern insulins (insulin analogues)	17,317	21,471	26,601	28,765	**34,821**	21%
Human insulins	11,804	11,315	11,827	10,785	**11,302**	5%
Victoza®	–	87	2,317	5,991	**9,495**	58%
Protein-related products	1,844	1,977	2,214	2,309	**2,511**	9%
Oral antidiabetic products (OAD)	2,391	2,652	2,751	2,575	**2,758**	7%
Diabetes care total	33,356	37,502	45,710	50,425	**60,887**	21%
NovoSeven®	6,396	7,072	8,030	8,347	**8,933**	7%
Norditropin®	3,865	4,401	4,803	5,047	**5,698**	13%
Hormone replacement therapy	1,612	1,744	1,892	2,054	**2,163**	5%
Other products	324	359	341	473	**345**	(27%)
Biopharmaceuticals total	12,197	13,576	15,066	15,921	**17,139**	8%
Total sales by business segment	45,553	51,078	60,776	66,346	**78,026**	18%
North America	15,154	18,279	23,609	26,586	**34,220**	29%
Europe	17,219	17,540	18,664	19,168	**19,707**	3%
International Operations	6,353	6,835	8,335	9,367	**11,080**	18%
Japan & Korea	4,196	4,888	5,660	6,223	**6,617**	6%
Region China	2,631	3,536	4,508	5,002	**6,402**	28%
Total sales by geographical segment	45,553	51,078	60,776	66,346	**78,026**	18%
Underlying sales growth in local currencies	12%	11%	13%	11%	**12%**	
Currency effect (local currency impact)	(3%)	1%	6%	(2%)	**6%**	
Total sales growth as reported	9%	12%	19%	9%	**18%**	
Depreciation, amortisation and impairment losses	2,442	2,551	2,467	2,737	**2,693**	(2%)
Operating profit	12,373	14,933	18,891	22,374	**29,474**	32%
Net financials	322	(945)	(605)	(449)	**(1,663)**	270%
Profit before income taxes	12,695	13,988	18,286	21,925	**27,811**	27%
Net profit for the year	9,645	10,768	14,403	17,097	**21,432**	25%
Total assets	50,603	54,742	61,402	64,698	**65,669**	2%
Equity	32,979	35,734	36,965	37,448	**40,632**	9%
Capital expenditure, net	1,754	2,631	3,308	3,003	**3,319**	11%
Free cash flow[1]	11,015	12,332	17,013	18,112	**18,645**	3%
Financial ratios						
Percentage of sales						
Sales outside Denmark	99.2%	99.2%	99.4%	99.3%	**99.4%**	
Sales and distribution costs	28.2%	30.2%	29.9%	28.6%	**27.6%**	
Research and development costs	17.2%	15.4%	15.8%	14.5%	**14.0%**	
Administrative costs	5.8%	5.4%	5.0%	4.9%	**4.2%**	
Gross margin[1]	77.8%	79.6%	80.8%	81.0%	**82.7%**	
Net profit margin[1]	21.2%	21.1%	23.7%	25.8%	**27.5%**	
Effective tax rate[1]	24.0%	23.0%	21.2%	22.0%	**22.9%**	
Equity ratio[1]	65.2%	65.3%	60.2%	57.9%	**61.9%**	
Return on equity (ROE)[1]	29.6%	31.3%	39.6%	46.0%	**54.9%**	
Cash to earnings[1]	114.2%	114.5%	118.1%	105.9%	**87.0%**	
Payout ratio[1]	37.8%	40.9%	39.6%	45.3%	**45.3%**	
Payout ratio excl non-recurring events[2]	36.6%	40.9%	42.8%	45.3%	**45.3%**	
Long-term financial targets						**Targets[3]**
Operating profit margin[1]	27.2%	29.2%	31.1%	33.7%	**37.8%**	40%
Operating profit growth	38.4%	20.7%	26.5%	18.4%	**31.7%**	15%
Operating profit after tax to net operating assets[1]	37.4%	47.3%	63.6%	77.9%	**99.0%**	125%
Cash to earnings, (three-year average)	97.6%	111.5%	115.6%	112.8%	**103.7%**	90%

	2008	2009	2010	2011	**2012**	2011–2012
Social performance						**Change**
Patients:						
Least developed countries where Novo Nordisk sells insulin according to the differential pricing policy	32	36	33	36	**35**	(3%)
Healthcare professionals trained or educated in diabetes (1,000)	N/A	425	373	835	**1,274**	53%
People with diabetes trained (1,000)	N/A	416	494	626	**836**	34%
Donations (DKK million)	78	83	84	81	**84**	4%
New patent families (first filings)	71	55	62	80	**65**	(19%)
Employees:						
Employees (total)	27,068	29,329	30,483	32,632	**34,731**	6%
Employees (average FTEs)	26,069	27,985	29,423	31,499	**33,061**	5%
Employee turnover	12.1%	8.3%	9.1%	9.8%	**9.1%**	
Assurance:						
Relevant employees trained in business ethics	N/A	N/A	98%	99%	**99%**	
Business ethics assurance activities	25	30	35	46	**48**	4%
Fulfilment of action points from facilitations of the Novo Nordisk Way	92%	93%	93%	93%	**94%**	
Product recalls	2	2	5	5	**6**	20%
Warning Letters and re-inspections	0	0	0	0	**1**	–
Company reputation with external key stakeholders (scale 1–7)	N/A	N/A	N/A	5.6	**5.7**	2%
Long-term social targets						**Targets**
Patients reached with diabetes care products (million) (estimate)[4]	N/A	N/A	N/A	21	**23**	40 million by 2020
Working the Novo Nordisk Way (employee assessment) (scale 1–5)	N/A	N/A	N/A	4.3	**4.3**	4.0
Diverse senior management teams[5]	43%	50%	54%	62%	**66%**	100% by 2014
Environmental performance						**Change**
Resources:						
Energy consumption (1,000 GJ)	2,533	2,246	2,234	2,187	**2,433**	11%
Water consumption (1,000 m^3)	2,684	2,149	2,047	2,136	**2,475**	16%
Emissions and waste:						
CO_2 emissions from energy consumption (1,000 tons)	217	166	95	94	**122**	30%
Wastewater (1,000 m^3)	2,542	2,062	1,935	2,036	**2,272**	12%
Waste (tons)	24,314	26,362	25,627	41,376	**82,802**	100%
Long-term environmental targets						**Targets**
Energy consumption (change compared with prior year)	(9%)	(11%)	(1%)	(2%)	**11%**	3% annual growth[6]
Water consumption (change compared with prior year)	(17%)	(20%)	(5%)	4%	**16%**	5% annual growth[6]
CO_2 emissions from energy consumption (change compared with 2004)[7]	(1%)	(24%)	(56%)	(57%)	**(44%)**	10% reduction by 2014
Share performance						**Change**
Basic earnings per share/ADR in DKK[1]	15.66	17.97	24.81	30.24	**39.09**	29%
Diluted earnings per share/ADR in DKK[1]	15.54	17.82	24.60	29.99	**38.85**	30%
Dividend per share in DKK	6.00	7.50	10.00	14.00	**18.00**	29%
Total dividend (DKK million)	3,650	4,400	5,700	7,742	**9,715**[8]	25%

1. For definitions, please refer to p 93.
2. Impact of Zymogenetics, Inc. share divestment, discontinuation of all pulmonary diabetes projects and impact of DAKO A/S share divestment.
3. The long-term financial targets were updated in February 2013. Please refer to p 10.
4. The accounting policy has been updated in line with WHO definition, and historical data have been restated accordingly. Please refer to p 97.
5. By the end of 2014 all senior management teams must comply with the target to be diverse in terms of both gender and nationality or explain why this is not achievable.
6. For target definition, please refer to p 14.
7. The accounting policy has been updated and historical data have been restated accordingly. The target remains unchanged. Please refer to p 102.
8. Proposed dividend for the year (not yet declared).

TABLE 9.7 **Novo Nordisk 2012 integrated report: performance highlights**

Source: Excerpt from Novo Nordisk (2012) *Annual Report*

TABLE 9.8 Disclosed sustainability measures of nonfinancial measures in external reports

Measures	Company
Economic profit	Coca-Cola
Market value of real-estate assets	Rouse Company
Recycling income	Baxter International

Measures	Company
Purchases from minority businesses	Procter & Gamble
Number of customer complaints	The Co-operative Bank
Global image survey results	BP Amoco
Consumption per capita	Coca-Cola
On-time delivery performance	Analog Devices
Packaging reduction	Baxter International
Number of sites with environmental certification	Unilever
Sources of energy	BC Hydro
IT expense as a percentage of administrative expense	American Skandia
Employee turnover–voluntary and involuntary	Dow Chemical
Number of jobs posted and filled internally	Dow Chemical
Training expense per employee	Milliken
Technology coverage	Allstate Insurance
Environmental, health, & safety capital expenditures	Alcoa
Transportation incidents	Dow Chemical
Recycled materials	ABB
Marketing and advertising dollars	Allstate Insurance

Source: Adapted from Epstein and Wisner (2001) "Increasing Corporate Accountability"

Verifying sustainability performance and reporting

Independent verification is an important component of external reporting. In 2005, 30% of sustainability reports of the Global 250 included assurance statements. Major accounting firms issued 60% of the statements, with various other consulting and specialized verification firms issuing the balance.[30] One of the major challenges in auditing social, environmental, and economic performance is that there is little standardization of sustainability management systems, performance measures, and reporting structures. Correspondingly, there are no generally accepted worldwide auditing or reporting standards. There is, however, some guidance on reporting social, environmental, and economic performance provided by the GRI and other organizations, as discussed earlier. And, just as demands for disclosure of sustainability performance fostered the increase in reporting, they should also drive the scope and nature of the underlying assurance process.

The AA1000 Assurance Standard launched by AccountAbility in 2003 is based on an assessment of reports against three assurance principles:

- **Materiality.** Does the sustainability report provide an account covering all the areas of performance that stakeholders need to judge the organization's sustainability performance?
- **Completeness.** Is the information complete and accurate enough to assess and understand the organization's performance in all these areas?
- **Responsiveness.** Has the organization responded coherently and consistently to stakeholders' concerns and interests?

The AA1000 framework has several important attributes. It helps organizations define goals and targets, measure progress against targets, and audit and report performance. It also provides a means for others to judge the validity of reported performance. Guidelines established by industries and other organizations can also be used simultaneously. British Telecom, a leading provider of communications solutions and services, is one of the largest companies to apply the AA1000 Standard. However, it uses AA1000 in conjunction with the GRI guidelines and the Ten Principles of the United Nations Global Compact. Another advantage is its focus on continuous improvement. Companies using AA1000 gradually increase their level of assurance over time and the standard requires that companies indicate how they will meet future standards and expectations.[31]

ISO 14001 requires that organizations conduct periodic EMS audits to determine whether the EMS has been properly implemented, and the results of the audit are then reported to management. The ISO standards provide guidance on the general principles for conducting a social, environmental, and economic audit, criteria for selection and composition of audit teams, and the qualifications of internal and external auditors. In 2006, ISO added a guideline specifically addressing the accounting and verification of GHG emissions. The purpose of this standard is to promote consistency, transparency, and credibility in GHG quantification, monitoring, reporting, and verification.

A corporate sustainability reporting and verification system usually involves internal and external reporting and audits. Through extensive internal auditing processes, companies can identify areas of concern and improvement and gather information to aid in managerial decision-making. They can monitor processes and performance and report progress to relevant managers.

Internal sustainability audits

Since the 1970s, various researchers and companies have pioneered methods of "social auditing," a term that has come to mean various combinations of accounting for, reporting on, and verifying sustainability performance. A few companies in the 1970s did develop and implement well-developed models, but the systems were soon dropped. Social auditing did not fully develop in the 1970s because companies never adopted it as an integral tool for defining strategy, improving performance, or delivering value. Among the most prominent advocates of the social audit, Abt Associates, a consulting firm based in Cambridge, Massachusetts, produced a social balance sheet and income statement in 1973. The Abt statements divided benefits and costs, in dollars, by stakeholders, and then computed net social income. The company also worked with many prominent clients on the measurement and reporting of social, environmental, and economic impacts and production of both internal and external social reports.[32]

Currently, in most organizations, a social and environmental or sustainability internal audit program is well developed and routine. It is typically conducted by some combination of central staff from the sustainability department and staff from the facilities or business units, with wide variation in reporting responsibilities. Many companies send the results of the audits to the business unit managers who set the action plans and the schedules for reporting deficiencies. Others report to a central sustainability office (and senior management) which coordinates social, environmental, and economic improvements.

A report should be made to the head of sustainability, to a member of the senior management team, and to a member of the board of directors, as well as the business unit manager. In addition, the audit should be part of a more comprehensive program of evaluating the social, environmental, and economic performance of the business unit, the facility, the business unit manager, and other management and staff. It also should be part of a comprehensive performance evaluation system in the organization to provide the incentives necessary to motivate improved corporate sustainability performance.

Many companies have created internal auditing frameworks and checklists to record and evaluate social, environmental, and economic performance. Using these frameworks and checklists enhances audit reliability and also the comparability of the information, both over time and between units of the company.

Honda uses an Environmental Audit System which includes internal and external auditors (Fig.9.4). Internal auditing is carried out to confirm that factories are implementing the environmental management system correctly and to ensure that targets are being met. Honda has also established a Mutual Visit Environmental Audit Team. This audit is implemented by peer factories to confirm compliance with legal regulations and the progress made in achieving targets based on company policy. For example, Honda North America, Inc., a subsidiary of Honda Motor Co., Ltd., serves as auditor, helping to ensure that Honda's various subsidiary companies and its affiliated suppliers in the North America region are in compliance with all applicable environmental laws and regulations. At the same time, Honda North America uses a third-party evaluator to conduct environmental audits of participating dealers and recommend strategies for reducing their energy use.[33]

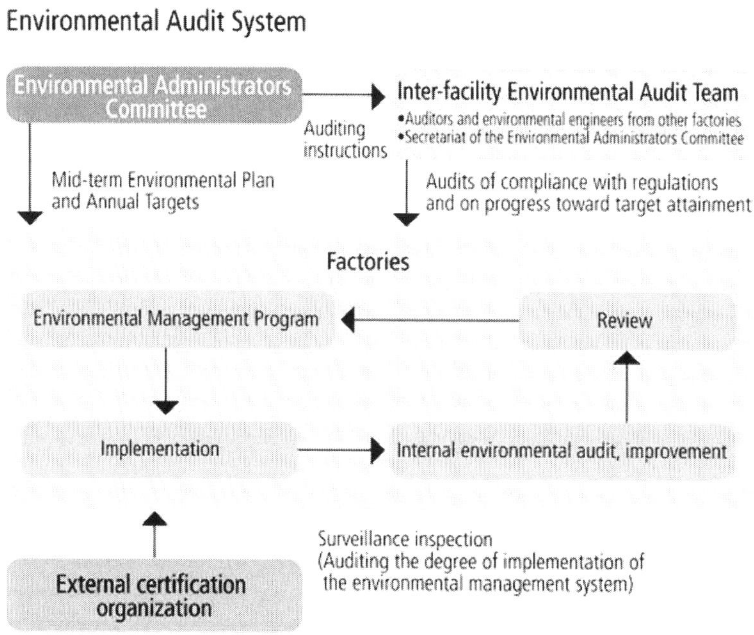

FIGURE 9.4 **Honda Environmental Audit System**

Source: Honda (2006) *Environmental Annual Report*

Corporate internal audits can be conducted for compliance with government regulations, corporate goals, procedures, and practices and to monitor, evaluate, and control company risks. However, some companies have expanded the internal audit role to proactively identify points in the organization's processes that impact environmental, social, and economic performance, identify the risks involved and measure the current or potential damage, and evaluate and suggest organizational changes to mitigate these risks. Social and environmental, or sustainability auditing practices vary widely among organizations depending on the objectives of the audit and the types of social, environmental, and economic risks faced. Among the types of audits are:

Compliance audit. The most common internal sustainability audit is the compliance audit. The compliance audit procedure includes a detailed, site-specific audit of current, past, and likely future operations.

Social and environmental management systems audit. As companies become more certain that they are in compliance with regulations, the audit emphasis shifts to sustainability management systems. To assess the many elements of the sustainability management systems, different types of performance indicators should be used. These might be selected from the sample indicators identified in Chapter 8 and customized to company needs.

Due diligence audit. Due diligence or transactional audits are conducted to assess the social, environmental, and economic risks and liabilities of land or facilities. These are typically conducted prior to a real-estate or business acquisition but can be completed at any time.

Treatment, storage, and disposal facility audit. Companies that produce hazardous waste material may contract with other companies to store, treat, or dispose of that material. Some companies conduct audits on the facilities they own and on facilities that handle hazardous waste material with which they contract.

Pollution prevention audit. Pollution prevention audits are designed to minimize waste at the source rather than at the "end of pipe." Companies conduct these audits because they recognize that eliminating or reducing the production of waste is usually much less expensive in total environmental and company costs than cleaning it up at the end of the production process.

Social and environmental liability accrual audit. These internal audits address the issues of reasonable, probable, and estimable in determining the social and environmental liabilities to be accrued for financial reporting.

Product audit. Some companies perform audits on specific products to determine whether more should be done to make them socially and environmentally friendly and to confirm that product and chemical restrictions are being met.[34]

Social and environmental, or sustainability auditing should be a proactive exercise that drives continuous and breakthrough improvement. When conducting audits, companies should:

- Reconsider strategy
- State objectives
- Pinpoint critical success factors
- Devise measures that gauge success among appropriate stakeholders

- Evaluate impacts on company stakeholders
- Work the measures into the remaining steps of the Corporate Sustainability Model to drive high performance

By following these steps, audits are conducted in the context of the overall strategy of the company and results can be integrated into the organization.

Internal sustainability audits can be critical elements of the management of processes, products, and projects to improve corporate sustainability (see Chapter 8). They can provide important information to aid in the evaluation of the impacts of both sustainability and financial performance. These internal sustainability audits also provide essential information to facilitate the external reporting and verification of sustainability processes and performance.

External sustainability audits

Earlier in this chapter, we discussed examples of the rapid rise in the quality and quantity of sustainability disclosures and attempts at standardizing these disclosures. Concurrently, companies have also found it desirable to obtain independent verification and attestation of progress toward improved sustainability management and performance. It is likely that stakeholder demands for increased external sustainability reports and audits will influence the number of corporations providing them. A report found that 59% of stakeholders want sustainability reports to be "verified by a professional assurance or verification body," with financial analysts and investors most strongly favoring verification statements.[35] Many consulting firms and accounting firms have begun performing external environmental audits and we describe some examples of audit reports below. The level of detail of the investigation and the external verification and assurance vary significantly.

British American Tobacco had its 2012 *Social Report* reviewed by Ernst & Young. Throughout the report, where appropriate, Ernst & Young provides assurance comments (Fig.9.5) to indicate areas that are supported by underlying evidence.

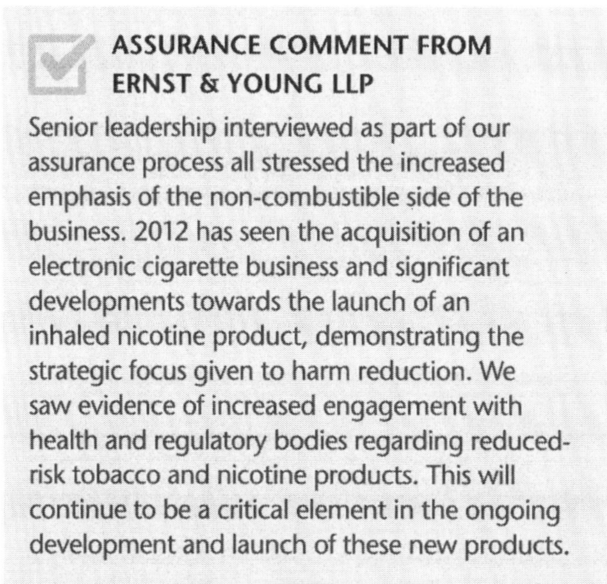
ASSURANCE COMMENT FROM ERNST & YOUNG LLP

Senior leadership interviewed as part of our assurance process all stressed the increased emphasis of the non-combustible side of the business. 2012 has seen the acquisition of an electronic cigarette business and significant developments towards the launch of an inhaled nicotine product, demonstrating the strategic focus given to harm reduction. We saw evidence of increased engagement with health and regulatory bodies regarding reduced-risk tobacco and nicotine products. This will continue to be a critical element in the ongoing development and launch of these new products.

FIGURE 9.5 **British American Tobacco report's assurance comments from Ernst & Young**

Source: Example from British American Tobacco (2012) *Sustainability: Why it Matters – Sustainability Summary 2012*

Verification should not seem to be just an extra step in the process; it should increase stakeholder trust in the reporting process. Therefore, any discrepancies and suggestions for improvement should be reported. Ernst & Young (along with the other major international accounting and auditing firms) has been involved in social and environmental, or sustainability audits. The company evaluated BP's *Sustainability Review 2012* on the basis of ISAE30001 (International Federation of the Accountants' International Standard for Assurance Engagements other than Audits or Reviews of Historical Financial Information) and to meet the requirements of assurance engagement as defined by AA1000AS (2008). To form its conclusions, Ernst & Young completed the following steps:

1. Interviewed a selection of BP executives and senior managers to understand the current status of safety, social, ethical, and environmental activities, and progress made during the reporting period

2. Reviewed selected group level documents relating to safety, social, ethical, and environmental aspects of BP's performance to understand progress made across the organization and test the coverage of topics within the report

3. Reviewed BP's approach to stakeholder engagement through interviews with employees with responsibility for managing engagement activities at group and local business level, and reviewed selected associated documentation
4. Carried out the following activities to review health, safety, and environment (HSE) and community investment data samples and processes:

   –Reviewed disaggregated HSE data reported by a sample of five businesses to assess whether the data had been collected, consolidated, and reported accurately

   –Reviewed and challenged supporting evidence from the sample of businesses

   –Tested whether HSE data had been collected, consolidated, and reported appropriately at group level

   –Reviewed community investment data at group level

5. Reviewed BP's processes for determining material issues to be included in the report
6. Reviewed the coverage of material issues within the report against the key issues raised by BP's stakeholder engagement activities, material issues

and areas of performance covered in external media reports, and sustainability reports of BP's peers, as well as the topics discussed by BP's SEEAC (safety, ethics and environment assurance committee)

7. Reviewed information or explanations about selected data, statements, and assertions regarding BP's sustainability performance[36]

Figure 9.6 provides Ernst & Young's observations and areas for improvement.

PwC (PricewaterhouseCoopers) is another of the major firms offering reporting and assurance of nonfinancial information. Their service has four main components:

Observations and areas for improvement

Our observations and areas for improvement will be raised in a report to BP management. Selected observations are provided below. These observations do not affect our conclusions on the Report set out above.

- Stakeholders continue to request more detail on BP's contribution to a low-carbon future, including its renewable energy strategy. The Sustainability Review sets out BP's alternative energy investment to date within the context of 'the energy future'. However, the contribution that these wind and biofuels assets currently make to a low-carbon future is less clear. For example, a more complete picture could be provided by comparing reported operational GHG emissions with an estimation of 'avoided' CO_2 emissions.
- BP has highlighted its renewed focus on diversity, with a new framework and associated goals for female representation in leadership positions. During our interviews we discussed the challenges for BP in delivering against these goals and whether interim milestones should be set. BP has introduced additional structures to support delivery and the reported increase in female leaders in the last three years shows that progress is being made. However, the fact that the same level of representation was first achieved in 2005 demonstrates the difficulty in sustaining this improvement.
- BP participates in a diverse range of joint ventures. BP acknowledges the importance of risks associated with non-operated joint ventures and the report highlights the pilot of a new group policy in this area. However, it does not explain how material these relationships are for BP, which account for an important part of BP's portfolio and overall value chain.
- We reviewed BP's GHG performance. BP explains that existing operations are required to incorporate energy use considerations into business planning but has also acknowledged that future upstream developments are likely to have higher CO_2 emissions. Interest in the steps that oil and gas companies are taking to reduce the intensity of emissions remains high. BP will need to continue providing a clear explanation of this relative performance in the context of ongoing changes to the business structure.
- For the second year, BP has reported on its direct economic contribution, including taxes paid and social investment spend. Whilst this provides a useful summary, stakeholders are increasingly looking for detailed performance information and BP should consider providing a more detailed breakdown for certain elements of this contribution, for example a regional split of spend with local suppliers.
- BP maintains a wide programme of stakeholder engagement. This takes place both during the course of running the business and specifically in relation to sustainability reporting. This Report has more explicitly addressed how BP is responding to some of the specific concerns that have been raised during this engagement process, for example in relation to the socio-economic impacts of the oil and gas industry.

Our independence

As auditors to BP p.l.c., Ernst & Young are required to comply with the requirements set out in the Auditing Practices Board's (APB) Ethical Standards for Auditors. Ernst & Young's independence policies apply to the firm, partners and professional staff. These policies prohibit any financial interests in our clients that would or might be seen to impair independence. Each year, partners and staff are required to confirm their compliance with the firm's policies.

We confirm annually to BP whether there have been any events including the provision of prohibited services that could impair our independence or objectivity. There were no such events or services in 2012.

Our assurance team

Our assurance team has been drawn from our global Climate Change and Sustainability Services Practice, which undertakes engagements similar to this with a number of significant UK and international businesses. The work has been led and reviewed by a Lead Sustainability Assurance Practitioner.

ERNST & YOUNG
Quality In Everything We Do

Ernst & Young LLP, London

20 March 2013

FIGURE 9.6 **BP assurance statement**

Source: Excerpt from BP (2012) *Sustainability Review: Building a Stronger, Safer BP*

1. **Reporting and communication planning and strategy.** Helps define the reporter's goals, audience, and the information that the readers will need. PwC helps select and develop performance measures to address stakeholder concerns for transparency and accountability
2. **Review and improvement of governance, systems, and reporting processes.** Helps companies review and establish governance structures, management and information systems
3. **Obtaining external assurance of nonfinancial information.** Evaluates and measures the quality of the company's information. Figure 9.7 displays their assurance process

4. **Reporting analysis and feedback.** Reviews reports and disclosures; helps company obtain feedback through surveys and focus groups

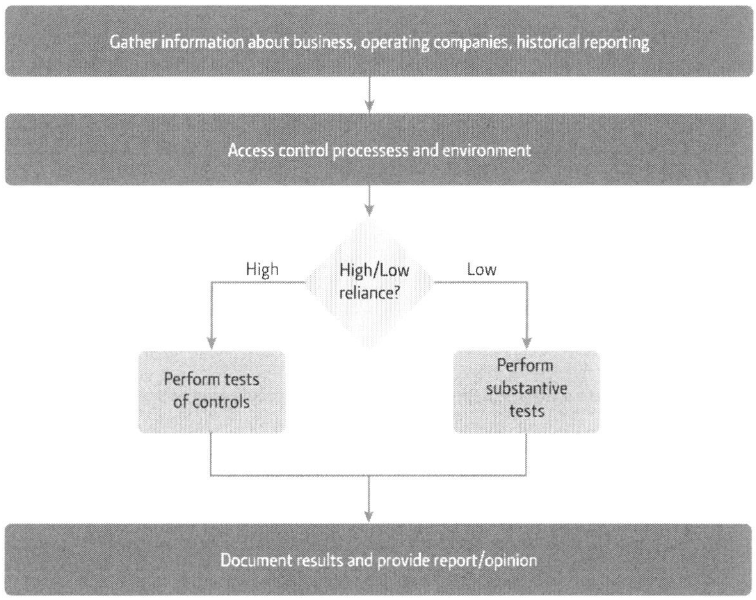

FIGURE 9.7 **PricewaterhouseCoopers reporting and assurance process**

Source: PricewaterhouseCoopers (2005) *Corporate Responsibility*

To ensure credibility of the report contents and reporting procedures, Samsung has received third party assurance for its 2011 *Sustainability Report* from PwC. This report was independently assured in accordance with the ISAE3000 and AA1000 Account-Ability Assurance Standard (AA1000AS Type II Assurance) (Fig.9.8).

Conclusion

Based on the results of the assurance work performed and the Assessment Criteria, our conclusion is as follows:

- On the AA1000APS principles;
  - Inclusivity: - Samsung Electronics has collected concerns and opinion through stakeholder communication channels that include customers, business partners, stockholders/investors, the government, local communities, employees, and NGOs
    - Nothing has come to our attention to suggest that material stakeholder groups were excluded in these channels.
  - Materiality: - Samsung Electronics has identified most relevant and significant sustainability issues through process for identifying material issues.
    - Nothing has come to our attention to suggest that material issues were omitted in this process.
  - Responsiveness: - Samsung Electronics has included in the Report its response to the material sustainability issues which are defined through process for identifying material issues.
    - Nothing has come to our attention to suggest that there is material deficiency in issue management system.
- Nothing has come to our attention that causes us to believe that Sustainability Data for the year ended December 31, 2010 are not fairly stated, in all material respects, in accordance with the Reporting Principles.

Responsiveness:

From our work, we have provided the following recommendations to the management.

- Reporting guideline, definition of key performance indicators ("KPI"), and data management control procedures should be improved to enhance consistency, accuracy and completeness of the sustainability data and information.
- Disclosure about the specific goals set up with reference to Samsung Electronics' sustainability strategy should be made together with related KPI to communicate with stakeholders more proactively by reporting progress toward achieving the goals.
- The methodology of determining reporting boundary should be developed and the reporting boundary should be expanded to all organizational level to communicate with all stakeholders on more accurate and complete sustainability performance data and to help management make more informed business decision.

Samil PricewaterhouseCoopers
May 31, 2011

Joonki Min
Partner, Assurance

Joonki Min

FIGURE 9.8 **PwC assurance statement for Samsung**

Source: Excerpt from Samsung (2011) *Sustainability Report*

While some companies employ these large accounting and auditing firms for external assurance, others use firms that specifically focus on sustainability. Coca-Cola commissioned FIRA Sustainability BV, Netherlands, to provide external assurance on its 2011/12 *Sustainability Report.* FIRA conducted the verification process in accordance with international assurance standards. Coca-Cola applied its own sustainability performance reporting criteria, derived from the GRI's

Sustainability Reporting Guidelines. FIRA reviewed Coca-Cola data and claims against Coca-Cola reporting criteria and the GRI, including explanatory notes related to disclosed performance information.

For years, the Dow Chemical Company believed that external assurance was important, but it did not hire an outside auditor to verify any data. By 2012, it was the fifth time that Dow had included assurance as a part of the *Annual Sustainability Report.* Dow has engaged Environmental Resources Management (ERM), one of the world's leading providers of environmental consulting services, to review its 2011 *Annual Sustainability Report* (see Fig.9.9).

Some observers have wondered whether, as with financial auditors, verifiers should act as both consultants and auditors and whether independence is jeopardized by the relationship. For both internal and external audits, companies should make sure that the independence of the audit is not compromised.

Other organizations and individual experts have become involved in verification of external sustainability reports. For years, Canon has had its sustainability reports reviewed by two stakeholder organizations, ASrIA (Association for Sustainable & Responsible Investment in Asia) and the Wuppertal Institute for Climate, Environment and Energy. They were invited to assess the appropriateness of the content, the quality of the treatment of individual topics, and the overall quality, balance, and relevance of the report. They were also invited to use the principles of the AA1000 Assurance Standard to inform their thinking. From 2008, Canon has received third-party opinions from an expert from the Wuppertal Institute. As a result, the expert is able to offer opinions from a medium-to long-term perspective on how well the information in Canon’s report meets expectations, the quality of performance it conveys, and its usefulness for substantial engagement. Canon reflects the expert’s suggestions in the report to the greatest extent possible (see Fig.9.10). This third-party comment, however, is the personal view of the expert and is not a verification of the report’s contents or data. The comment also does not imply any endorsement from the expert’s organization.

Ford Motor Company employed Ceres and a team of external stakeholders to review its *Sustainability Report.* The stakeholder team, selected by Ceres, was an independent group drawn primarily from the Ceres coalition, which has expertise in environmental, social, and governance issues. In reviewing the report, the team considered whether Ford adequately reported on its sustainability performance and key impacts, including goals, targets, systems, data, and initiatives. Through this review process, the stakeholder team provided feedback to the company, which was considered in the preparation of the final version of the report.

Some companies choose not to have any outside firm perform independent verification. Honda states that it has not obtained any external verification because no guidelines have been established for external verification, and the qualifications required of the verification organizations are not clear. Other companies have chosen to avoid external verification for other reasons, including cost.

Though there are no generally accepted international standards for the reporting or verification of sustainability performance or processes, auditors can verify the reliability and the fair representation of selected performance data. As shown in the above examples, this verification is done by reviewing management processes, interviewing employees, sample-testing key performance indicators, and reviewing other evidence to ensure compliance with all applicable laws and management directives. AA1000 and the GRI guidelines are also aiding auditors in verification and establishing benchmarks to compare reporting across companies.

Sustainability auditing and verification can create significant legal and operational benefits for organizations. The benefits include:

Independent Assurance Statement by ERM

EC EN LA HR SO PR

The Dow Chemical Company – 2011 Sustainability Report

Environmental Resources Management (ERM) was retained by The Dow Chemical Company (Dow) to provide an independent review and assurance statement for Dow's Annual Sustainability Report covering 2011 (the Report).

Scope of Work

The assurance process was conducted in accordance with AA1000AS (2008).

ERM was engaged to provide moderate level, Type 2 assurance, covering:

- evaluation of adherence to the AA1000AS (2008) principles of inclusivity, materiality and responsiveness (the Principles); and
- the reliability of specified sustainability performance information.

In addition, ERM reviewed the Report against the Global Reporting Initiative (GRI) G3.1 Sustainability Reporting Guidelines (2011) – Application Level A criteria.

The intended user of this statement is the reader of the Report.

Assurance Work Performed

ERM tested the effectiveness of processes used to adhere to the AA1000AS Principles through management interviews, the review of data and visits to selected operations.

ERM conducted interviews with personnel at The Dow Chemical Company's headquarters in Midland, Michigan and Dow AgroSciences' headquarters in Indianapolis, Indiana and visited manufacturing operations in Plaquemine, Louisiana and Dow Central Germany (Schkopau Operations). These sites were selected to represent a range of operations and material sustainability impacts and included a site located in a non-English speaking country.

During the interviews and site visits, ERM focused on:

- Understanding the Dow business context and its sustainability strategy;
- Reviewing the Report and other relevant supporting documents;
- Understanding the logistics of Dow's data collection and reporting processes and systems, including the Global Incident Reporting Database (GIRD), Global Emissions Inventory (GEI) and Global Asset Utilization Report (GAUR);
- Checking the accuracy of a sample of data from the sites visited;
- Reviewing the mechanisms for checking and verifying data at the site, business unit, and corporate level; and
- Reviewing the process for the management of change to data, including how errors are communicated to the data providers and corrective action checked to ensure accuracy in the reported data.

The review was conducted during May 2012 and included an assessment of Dow's data collection and verification processes and the Report's content as well as the development of this assurance statement.

Responsibility and Methodology

Dow was responsible for the preparation of the Report and this statement represents ERM's independent opinion. ERM's responsibility was to express our assurance conclusions within the agreed scope.

We delivered our work in accordance with ERM's assurance methodology, which is based on AA1000AS, the GRI Sustainability Reporting Standards and the International Standard on Assurance Engagements (ISAE) 3000. We planned and performed our work to obtain all the information and explanations that we believe were necessary to provide a basis for our assurance conclusions as to whether the reported information and data set out in our scope of work was appropriately reported (moderate assurance).

The accuracy and completeness of this information is subject to inherent limitations given their nature and methods for determining, calculating and estimating such data.

Evaluation of Alignment with AA1000AS (2008)

ERM evaluated Dow's alignment with the three AA1000AS principles of inclusivity, materiality and responsiveness. ERM determined through interviews with internal and external Dow stakeholders at both the corporate and site level that Dow has effectively included stakeholders, determined materiality of issues and responded to those issues.

Inclusivity

Dow has established robust processes for effective and quality stakeholder engagement. Discussions held as part of the assurance project with representatives of Dow's Sustainability External Advisory Council (SEAC), the business community in Midland, Michigan and the Community Advisory Panels (CAP) at the plants visited confirmed that Dow has provided a robust channel for engagement and dialogue with stakeholders and that Dow is responsive. Dow has also demonstrated their commitment to engaging their internal stakeholders through examples such as their annual Global Employee Opinion and Action Survey (GEOAS) and local sustainability committee such as that at Dow AgroSciences.

Materiality

Dow was assessed to have identified the material issues for the company and those issues that are taken into consideration when the organization or its stakeholders make decisions. Dow has a number of management processes for identifying issues that present material risks to the business. These include public affairs, enterprise risk management and GRI reporting processes. While we confirmed that Dow has identified all of the material issues, we make a recommendation that a more formal process be developed to organize this data and to provide future due process for the identification of changes to the material issues currently being managed.

Responsiveness

Dow was assessed to be responding to material issues, that is, they have developed and implemented policies, strategies and plans consistent with stakeholder and organizational interests and expectations. Dow's established goals for sustainability performance are aligned with the issues that are material to their business. The resources allocated to the goals and the quarterly tracking of progress towards these goals provides a measure of the responsiveness of the company.

ERM Conclusions

On the basis of the work undertaken, nothing came to our attention to suggest that the Report does not properly describe Dow's adherence to the principles, content and quality requirements of GRI G3.1 and AA1000 AS (2008) or the associated performance.

In terms of data accuracy, nothing came to our attention to suggest that data has not been properly collated from information reported from manufacturing sites. We are not aware of any errors that would materially affect the data reported.

During the review period, ERM identified a number of issues that were discussed and subsequently addressed by Dow in the final draft of the Report. These issues and the associated recommendations are included in detail in a separate assurance project report to Dow management.

ERM Competence

ERM operates strict conflict of interest checks and has confirmed our independence to work on this engagement with Dow. The members of the review team have not provided consulting services to Dow outside of the review of the Report. During 2011, ERM worked with Dow on unrelated consulting engagements. The members of the review team have not provided consulting services and were not involved in the preparation of any part of the Report. The review team has the required combination of education, experience, training, and skills for this engagement.

Environmental Resources Management
Chicago, USA
June 2012

ERM is an independent global provider of environmental, social and corporate responsibility consulting and assurance services. Over the past 4 years we have worked with over half of the world's 500 largest companies, in addition to numerous governments, international organisations and NGOs.

ERM®

2011 Annual Sustainability Report | 25

FIGURE 9.9 **Dow's independent assurance statement by ERM**

Source: Dow (2011) *Annual Sustainability Report*

Commentator Opinions and Canon's Response in Third-Party Dialogue

Main Suggestions	Main Reflections (New Content in the 2013 Report)	Pages Posted
The distinction between "Key Activities Report" and "Reporting in Accordance with the ISO 26000 Core Subjects" should be clarified.	As the report's opening explains, the "Key Activities Report" section of the report is organized according to the "Key Activities" listed in the Canon Group CSR Activity Policy and focuses on activities implemented in 2012 of high interest to stakeholders, while, mindful of international ISO 26000 standards, the "Reporting in Accordance with the ISO 26000 Core Subjects" section covers all CSR activity at Canon with a focus on our duty to disclose certain information and data to stakeholders.	p. 10
The GRI Sustainability Reporting Guidelines will be revised this fiscal year. In the future, I hope to see contents included which correspond to these revisions.	Canon has also been keeping an eye on the progress of revisions to the GRI guidelines as the organization works out policies of investor concern. We plan to hold internal investigations after revisions are complete so as to comply with new guidelines to the fullest extent possible.	–
It would be better if concrete figure were incorporated for the included objectives and actual results.	Due to repeated advice to this effect, we are disclosing a number of concrete figures in this report. Additionally, for items which are difficult to disclose, we plan to work to clarify related issues so as to comply as fully as possible in future reports as well.	pp. 25, 69, 75, 87, 99, 109, 115

FIGURE 9.10 **Canon independent third-party opinion**

Source: Canon (2013) *Sustainability Report*

- Ensuring compliance with applicable laws and regulations
- Ensuring compliance with management directives and procedures
- Proactively identifying areas of potential or actual noncompliance
- Minimizing the risk of civil and criminal liability to the corporation and to its employees
- Ensuring accurate certifications
- Ensuring accurate regulatory disclosures
- Raising employee consciousness about the importance of compliance

- Providing independent verification of a program, which some companies use as a public relations or marketing tool
- Assessing the potential impact of new or expected regulation
- Helping to standardize systems and measures in multiple facilities by providing a common framework for assessment[37]

Summary

The growth of social, environmental, and economic costs and corporate managers' recognition that they need to better manage corporate sustainability impacts have dramatically increased the demand for both internal and external sustainability reports. Improved internal audits are necessary to monitor and reduce the impacts, but external audits provide additional benefits. External audits:

- Increase stakeholder confidence in the quality of corporate sustainability controls, planning, and performance
- Provide senior management with an independent verification and analysis of the strengths and deficiencies of the sustainability program
- Provide additional confidence that hazards and violations will be minimized

Additionally, stakeholders want more verification of corporate sustainability. They want to understand corporate plans and processes to reduce social, environmental, and economic impacts. Shareholders and financial analysts want more information to better assess a company's future social, environmental, and economic liabilities. Managers need more information about these issues to develop a corporate sustainability strategy and manage impacts more effectively.

Companies and their stakeholders need to ensure that the flurry of activity created by external sustainability reporting and external environmental auditing is supported by actual company progress. External reporting is an opportunity for a company to tell the story of its performance. The external report should not, however, precede the integration of social, environmental, and economic considerations into product costing, capital investment decisions, company processes, product design, or performance evaluation.

In the final chapter we look at the significant benefits accruing to corporations and society by making sustainability work.

CHAPTER 10

The benefits of sustainability for corporations and society

Global companies are increasingly faced with difficult dilemmas. There is significant pressure to reduce costs in the supply chain, yet switching to lower-cost suppliers may increase social, environmental, and economic impacts, and reactions from various stakeholders, including employees, customers, regulators, and community activists, may have a detrimental effect on financial performance. Senior management often faces complex decisions about facility location that in simpler times could be made by examining differentials in labor, shipping, and raw material costs. Now social, environmental, economic, and political risk must become part of the calculus.

Business unit managers are regularly told by the CEO about the importance of sustainability, yet they receive daily pressure to increase short-term profitability. And their bonuses are typically based entirely on profits. Making the decisions (and these are often trade-offs) about achieving excellence in both sustainability *and* financial performance is a big challenge.

Though much has been written and discussed in both the academic and the business press about the motivations for sustainability and how to formulate a sustainability strategy, much less has been said about how to implement sustainability. Managers have often been frustrated by the challenges of execution in complex business organizations. Even the most socially concerned senior corporate and business unit managers find it difficult to simultaneously meet social, environmental, economic, and financial goals. In addition, senior environmental, community affairs, and sustainability executives are often frustrated by their inability to obtain the resources they need to execute programs that they are convinced create societal and organizational value.

This book has focused on how to implement sustainability in complex organizations. The question facing most senior general managers and most sustainability, community affairs, and EH&S managers is not *whether* to improve sustainability performance but *how* to do it in their global corporation given the strategies, structures, systems, culture, people, and pressures that already exist. Based on extensive research from field studies of companies, surveys, examples of company best practices, and other academic and company research and analysis, this book has offered guidance for successful implementation of sustainability to simultaneously improve corporate social, environmental, economic, and financial performance.

Make sustainability work

The Corporate Sustainability Model (Fig.1.4) describes the antecedents (drivers of success) and consequences (payoffs and measures of success) of investments in sustainability, and a way to analyze the social, environmental, and economic impacts of corporate products, services, processes, and other activities. This model is used to improve decision-making related to both targeted sustainability expenditures and other more general capital and operational investment decisions. It describes the critical role of management control and performance measurement in improving social, environmental, economic, *and* financial performance. It recognizes the importance of both the formal processes of strategy, structure, systems, performance measures, and rewards and the more informal systems of leadership, organizational culture, and people.

The model shows the cause-and-effect relationship between managerial actions and improvements in sustainability and financial performance. The three major sets of impacts, indicated by the numbered arrows, relate to: (1) the direct and specific financial costs and benefits of corporate actions; (2) the social, environmental, and economic (or sustainability) impacts of these corporate actions; and (3) the financial impacts that are a consequence of the sustainability performance and the related stakeholder reactions.

In spite of numerous inputs that act as constraints, managers have significant capability to affect corporate sustainability performance through leadership and the formulation and implementation of a sustainability strategy, structure, and systems. The output of these processes is the sustainability performance—that is, the effect of corporate activity on the social, environmental, and economic fabric of society. In addition to having an effect on society, these activities often affect corporate financial performance. Stakeholders (such as customers, employees, regulators, and consumer activists) can have various positive and negative reactions such as additional purchases, consumer protests, employee loyalty or resistance, and government regulations. These stakeholder reactions affect corporate profits and are a part of the business case that has been widely discussed in both academic and managerial circles.[1] They also often create valuable feedback on existing sustainability strategies and implementation. Remember that sustainability performance can be both an intermediate output and an ultimate outcome. So social, environmental, and economic impacts are important as companies attempt to minimize the impacts and at the same time to identify opportunities to simultaneously improve sustainability *and* financial performance.

A better understanding of the implications of decisions and specific actions can improve both sustainability and long-term financial performance. Some companies have already recognized the significant value that can be added by the identification and measurement of social, environmental, and economic impacts into business decisions, particularly for environmental expenditures. Though sustainability initiatives are admittedly often driven by regulatory requirements, an increasing number of companies are noticing that they frequently result in decreased operating costs and increased revenues. Recent research has shown a strong and positive link between successful sustainability strategy and corporate value. Sustainability can enhance businesses in several ways.[2] These are some of the documented payoffs of improved sustainability performance:

Financial payoffs

- Reduced operating costs (including lower litigation costs)
- Increased revenues
- Lower administrative costs
- Lower capital costs
- Stock market premiums

Customer-related payoffs

- Increased customer satisfaction
- Product innovation

- Market share increases
- Improved reputation
- New market opportunities

Operational payoffs

- Process innovation
- Productivity gains
- Reduced cycle times
- Improved resource yields
- Waste minimization

Organizational payoffs

- Employee satisfaction
- Improved stakeholder relationships
- Reduced regulatory intervention
- Reduced risk
- Increased learning[3]

Although executives increasingly recognize the importance of sustainability for fulfilling responsibilities to communities, increasing shareholder value, and improving social, environmental, economic, *and* financial performance, they have often not been able to implement it successfully. Many managers decide that they cannot develop the systems to effectively implement sustainability in their organizations. Implementing sustainability is particularly difficult because:

- The goal is to simultaneously achieve excellence in social, environmental, economic, *and* financial performance
- It is often unclear how to make trade-offs
- It is often unclear how stakeholders will respond
- Corporate and societal priorities often change
- The costs of implementing sustainability constantly change

Whereas in most other organizational changes the sole objective is to improve financial performance, sustainability has broadened the focus to simultaneously improve social, environmental, economic, *and* financial performance. Managers can find it hard to evaluate the trade-offs between sustainability and financial performance when excellence in both is expected. The social, environmental, and economic impacts of corporate activities also have effects that are often long-term and more difficult to measure than most of the impacts managers typically confront. However, through a mix of leadership, strategy, and "hard" and "soft" systems, sustainability can be implemented and measured successfully.

To integrate sustainability into day-to-day decision-making, companies need to make sustainability a central tenet of their strategy and exercise leadership to reinforce these objectives throughout the organization. However, for improved sustainability performance, strategy and leadership are only minimum enablers. Best-practice companies will have a strategy that includes sustainability and leaders who will show their commitment to sustainability by articulating trade-offs to managers and aligning the organization's strategy, structure, systems, people, and culture.

Companies also have a choice of hard or soft implementation systems. Hard systems are the formal systems that include structure, performance measurement and evaluation, and incentive systems used to motivate employee behavior. Performance measurement systems and rewards that include a broader set of performance metrics than financial performance alone encourage employees to include sustainability in their day-to-day decision-making. Soft systems are the informal systems such as organizational culture and people and they too can motivate behavior. A strong mission statement emphasizing the need for sustainability can convey to employees the importance of sustainability as a core corporate value. Some companies may prefer soft systems to hard systems in order to implement it; others may choose a mix of the two.

This book has presented many examples of company sustainability strategies, structures, and systems. General Electric's Ecomagination program or Nike's Considered Index® focus on innovative products to decrease social, environmental, and economic impact but is primarily focused on profit. Timberland's collaboration with City Year involves employees in sustainability. Novartis and Procter & Gamble use leadership commitment, people, and culture to drive sustainability throughout the organization. Canon uses life-cycle analysis to improve sustainability performance. Though why and how they implement it may vary significantly, most company leaders recognize the critical importance of stakeholder engagement and improved identification, measurement, and management of corporate sustainability performance.

Many of the companies presented throughout the book have faced scrutiny for their past environmental, economic, and social impacts, but most are making sincere efforts to improve their sustainability performance. The discussion around sustainability is no longer primarily focused on Patagonia, Ben & Jerry's, or The Body Shop. It now includes companies with huge social and environmental footprints such as General Electric, Walmart, and The Home Depot which are trying to face the significant challenges of simultaneously achieving excellence in financial *and* sustainability performance.

We have centered this book on how companies can integrate social, environmental, and economic impacts into management decisions and implement a corporate sustainability strategy. In Chapters 1–9, we discussed the components of corporate sustainability integration. They provided guidance on the best corporate practices and ways to implement a sustainability strategy. Here is a summary of the four steps that will help managers to get started or to progress if they have already embarked on the process.

Steps to sustainability strategy implementation

1. Make sustainability a central component of strategy
2. Be committed to sustainability and build additional organizational capacity. Actions are more difficult to specify so distributed leadership is more critical
3. Support with formal processes such as management control, performance measurement, and reward systems as appropriate. Support with informal processes such as mission, organizational culture, and people as appropriate
4. Use sustainability processes and systems to learn how to make the trade-offs and make the challenging managerial decisions. Integrate sustainability into all strategic decisions and then

introduce additional systems and rewards to formalize and support

Leadership and strategy are key components in improving sustainability and financial performance. The CEO communicates the importance of sustainability to the organization and establishes a culture for integrating sustainability into day-to-day decision-making. This communication usually begins with a strong mission statement that conveys the company's commitment to sustainability and encourages employees to consider sustainability as an important part of their responsibilities. Commitment to social, environmental, and economic concerns must be consistently communicated both in words and actions. In developing sustainability strategies, corporate executives will also have to consider the role of various voluntary and industry standards, government regulations, and social investors.

Organizational design affects the success of sustainability *and* financial performance and should consider the merits of centralized or decentralized sustainability units, outsourced activities, and collaborations with NGOs. Sustainability managers should have direct access to senior corporate officers. Sustainability departments should be charged with the development and implementation of corporate sustainability strategies and improved management of resources rather than only legal compliance. Integrating sustainability throughout the organization can also lead to a change in organizational culture where everyone views sustainability as important to long-term financial performance.

Management systems are critical to any successful implementation. This includes costing, capital investment, and risk management systems. To improve decision-making, companies should integrate accounting and financial analysis techniques including risk assessment into sustainability decisions. Current and future social, environmental, and economic impacts (costs and benefits) should be included in all corporate decisions including product costing, product design, and capital investments. This integration will permit improved analysis of choices among product improvements, process improvements, and capital improvements and greater understanding of uncertainties related to changing regulations and technology.

All employees must view sustainability performance as critical to the long-term financial success of the corporation. Incentives based totally on profits provide a signal that social, environmental, and economic performance is unimportant. Corporations should consider sustainability performance as a variable in the evaluation of total corporate performance and provide incentives for employees to suggest social, environmental, and economic improvements. These suggestions will ultimately lead to corporate profit improvements.

Measuring the payoffs of sustainability actions is difficult but critical. Measures are usually imprecise and data difficult and expensive to collect. However, the business case for sustainability can be made only by measuring sustainability performance. Measuring the processes and the results is key to evaluating effectiveness. Managers need to think broadly and consider both current and future impacts, as well as impacts on both the company and society. Various techniques, including revealed preference and stated preference methods, can aid companies in measuring their sustainability impacts. Additionally, companies can select from the metrics we have described in this book to measure the inputs, processes, outputs, and outcomes of sustainability investments.

A feedback system helps identify areas where products, processes, and performance can be improved. Measurement systems should provide information that management can act on to improve sustainability *and* financial performance. This information should be reported internally, not only to management but also to all employees so that adjustments can be made to improve performance. Internal reporting is also a means of conveying to employees that sustainability performance is important to the organization.

Don't forget external stakeholders. External reporting is an opportunity for companies to share information about its sustainability performance to stakeholders. And verification of sustainability reports will increase stakeholder confidence in the quality of the reporting.

Success at Henkel

Henkel, a German-based manufacturer of laundry and homecare products, cosmetics, toiletries, and adhesives, began integrating sustainability into its corporate strategy in the early 1990s. At that time, sustainability was a corporate priority, but the structure and systems to implement sustainability had not been developed. Henkel has now created a structure and system to incorporate sustainability into day-to-day decision-making.

The Henkel management board bears overall responsibility for sustainability strategy and compliance. Chaired by a management board member, and reflecting all areas of the company, the Sustainability Council steers global sustainability activities as a central decision-making body. Its members represent the business sectors and all corporate functions responsible for putting sustainability strategy into operational action.

To further integrate sustainability into the organization, Henkel uses a mix of soft (informal) and hard (formal) systems. First, it established a Code of Conduct with the most important corporate principles and behavioral rules. This is supplemented by guidelines for dealing with potential conflicts of interest. Further corporate standards address specific topics such as: compliance with competition and antitrust laws; safety, health, environment, and social standards; and public affairs. Hard systems at Henkel include integrated management systems and regular audits that are performed at facilities to measure progress toward achievement of sustainability goals. Facilities also conduct self-assessments on safety, environmental protection, and occupational health and safety. By using a combination of soft and hard systems, Henkel has created a culture that motivates employees to take sustainability seriously.

The company is ranked as a sustainability leader on several global and European sustainability assessments. From 2008 alone, it has decreased energy consumption by 30%, CO_2 emissions by 29%, and water consumption by 35%, while production volumes have been rising since 2010. Henkel believes that its focus on creating a culture of sustainability, manufacturing innovative products, and developing efficient processes has contributed to its overall growth and financial performance.[4]

Use the Corporate Sustainability Model to improve performance

To move toward a more advanced stage of sustainability integration and improve the decision-making process, the drivers of sustainability performance and the linkages between them must be measured. A clear understanding of the broad set of impacts that are caused by corporate activities and an understanding of these impacts on stakeholders will also aid managerial decisions.[5]

Furthermore, translating strategy into action requires appropriate systems, structures and measures that provide managers with both information about their current and past performance, and insight into their ability to improve their competitive position in the future. Only with such systems and measures can managers make day-to-day and long-term decisions while being aware of risks and opportunities. This will also help to define the strategy, communicate a clear agenda for expected sustainability performance, accelerate feedback and learning, and inspire loyalty among stakeholders. Indeed, it will provide managers with relevant information to quantify their efforts and evaluate their impacts on stakeholders and ultimate financial performance.[6] Improved sustainability performance will be produced by focusing on the following areas of management and leadership attention:

Understand the cause–effect relationship

A model for sustainability performance should accurately capture the range of corporate activities, the relevant effects of those activities, and define the cause–effect links that are crucial to the corporation's success. The first imperative is to learn more about how those relationships are currently functioning. The links are based in part on managers' experience and intuition. Employees, customers, and other stakeholders are a helpful source of hypotheses about links involving their own behavior and impacts. These links cannot be managed if they cannot be observed and measured (in commensurable units—preferably denominated in money—if at all possible), so managers should develop relevant indicators for each link from inputs to processes to outputs to outcomes, with measures at each end of each link. A well-designed, comprehensive, and accurate model defines the (presumed) links that must be observed—carefully and in detail—in order to facilitate improved performance.

Analyze and measure links

With accurate observational data replacing untested beliefs and assumptions, managers are in a position to conduct performance-enhancing analyses of their programs, projects, and activities. Precise measurement is challenging, but approximations are very useful. Since ignoring those impacts that are difficult to measure implicitly assigns a value of "zero," managers must be willing to accept approximations and predictions that point in the right direction. Using their explicit hypotheses about the effects of various actions, establishing measurable indicators of the actions and the effects, and analyzing the results, managers can systematically optimize their methods, approaches, and actions, and can find more efficient and effective allocations of time, effort, and funding across different activities.

Evaluate and learn from performance

By articulating explicit hypotheses about cause and effect and establishing measurable indicators at each end of the cause–effect links, managers have established conditions in which not only optimization but also systematic ongoing learning is possible (and even likely). Internal and external changes can lead to a change in the causal links and among the metrics. By forming comparisons—with prior performance, to a competitor's performance, or to performance in a different business unit—managers can determine those techniques, approaches, and actions that seem to produce the best results.

Use the links to produce alignment and drive action

Senior managers must consistently support the process of identifying and measuring causal relationships. Managers should communicate these relationships throughout the organizations to actively encourage their consideration in day-to-day decision-making.

Building and operating systems that communicate management objectives and the results of learning efforts to guide and align actions throughout the organization with the best current understanding of which activities create the most value is an essential step for enacting what has been discovered and learned. Effective systems of this kind both collect and promulgate organizational knowledge and ensure that it is effectively implemented.

Create opportunities for innovation

As discussed throughout the book, organizations are facing increased risks, from more sources with greater impact. Some current issues include:

- Poor working conditions and child labor
- Environmental emissions/climate change
- Joint-venture partner risk
- Unstable or corrupt governments
- Potentially dangerous products
- Nutrition and obesity
- Interrupted supply
- Unsafe supply

But these increased risks also create new opportunities for innovation to improve both sustainability *and* financial performance. So what can business leaders do to better integrate sustainability into operational and capital investment decisions? How can business leaders focus on both risk and opportunity in using innovation to increase both corporate profitability and sustainability?

The answers to these critical questions require: (1) more innovation and entrepreneurship from leaders in sustainability; and (2) more sensitivity to sustainability issues by innovation and R&D, business unit, and functional leaders. It requires companies to think not only about corporate social *responsibility* (CSR) but also corporate social *opportunity* (CSO).

Companies can become leaders in corporate sustainability by creating proactive strategies that create opportunities and increased profits rather than using reactive strategies that only respond to government regulations, industry standards, or consumer protests. The opportunity to gain competitive advantage through proactive sustainability strategies can be seen in companies such as General Electric and Toyota. Leadership companies view social, environmental, and economic responsiveness as an asset, producing increased revenues rather than a liability with the associated costs. They recognize that an investment in structures and systems to ensure strong social, environmental, and economic performance often pays dividends in terms of improved process and production quality, improved production efficiency and yields, lower risk, improved reputation, and increased profitability.[7]

Unfortunately, many companies forgo opportunities that might appear initially to be too risky but have not been formally analyzed. Risks can also present opportunities and provide significant possibilities for organizational innovation and new competitive advantage that can lead to improved sustainability *and* financial performance. Some companies may have superior organizational knowledge and capabilities which permit them to accept risk and respond to it effectively, while their competitors avoid potential opportunities because of their organization's assessment of these risks. Some organizations may be able to identify voids in the marketplace that provide opportunities for innovation that others may not see. Often it is the ability to identify and manage risks that others cannot that leads to innovation and market success. A company's ability to use tools to simultaneously perceive and assess risk and opportunity can enable it to manage offensively as an opportunity rather than defensively as a hazard.[8] The challenge for companies, then, is to develop strategies that anticipate the changing business landscape and use social, environmental, and economic pressures as a source for innovation.

Capturing opportunity: Toyota and the hybrid car

Aggressively seeking out opportunities for social, environmental, and economic improvements, with the explicit goal of investing in innovation, can produce significant advantage that the competition will not be able to easily or quickly match. Toyota is an example of how an organization can respond to social, environmental, and economic pressures through innovation while improving its financial performance.

Trying to envision what might transform its industry and threaten its market share in the future, Toyota's leaders convened a team to create the first great car of the 21st century in 1993, nearly a decade before that century arrived. Toyota's leadership pushed the team beyond the technological limits that it had previously worked within, and created a new equal-access system of communication and information-sharing to replace the traditional hierarchical model. It also brought engineers normally based at production plants to the planning floor to work out glitches at the blueprint stage, before the new car was being produced on the assembly line.

The Toyota Prius, an electric/gas hybrid, was introduced in Japan in 1997 and in the US market in 2001. The Prius has an average fuel efficiency of 28.9 miles per gallon. Realizing that the US market differs from others, the Prius was altered to appeal to American car buyers. The US version had more horsepower and cargo space than the Japanese model.[9] Toyota has sold more than 4 million hybrid cars worldwide since introducing them in 1997. Over the intervening years, Toyota has expanded its range of cars fitted with hybrid engines to include minivans, sedans, SUVs, and wagons, broadening the scope of hybrids. In addition, responding to the increasing environmental awareness of customers in recent years and the demand for fuel-efficient cars, Toyota has developed the compact and affordably-priced car, Aqua. The concept behind Aqua is to present a revolutionary hybrid compact car designed for 2020. It is a lighter and more compact, fun, and easy-to-drive car, boasting the best fuel efficiency at an affordable price. It has the same hybrid system as the Prius—the Toyota Hybrid System II (THS II)—but most parts, such as the smaller motor, have been redesigned to make them lighter and more compact.

As a result of a series of technological breakthroughs, manufacturing innovations, and careful marketing, Toyota's sales of hybrid cars account for about a half of all the hybrid car sales in North America. It has a strong reputation in producing environmentally-friendly vehicles.[10]

Stakeholder engagement plays an important role in seizing opportunities for innovation. It is important to evaluate stakeholder impacts and the level of trust or distrust from the perspective of external stakeholders (including activists, consumers, and suppliers), internal stakeholders (including employees and managers), and the senior and top management team. This evaluation will often highlight the differences between the real and perceived risk of company impacts and should reduce the likelihood of significant crises or surprises. Companies can respond by more effectively managing perception, reality, or both. Engaging with stakeholders can also help to identify issues that may become critical management concerns in the future. This is all part of an important process that is essential for both improved sustainability *and* financial performance. Effective stakeholder engagement not only improves trust and reputation but also presents opportunities for response to stakeholder concerns through innovative products.

Corporate executives need to recognize the opportunities for both technological innovation (products) and business model innovation (processes). A change to a product or service that a company offers in the marketplace, or the introduction of an entirely new product or service, is the most easily recognized type of innovation because consumers see the changes first-hand. In today's fast-changing market, consumers have come to expect significant and recurring technological innovation.

The bottom of the pyramid (BOP), discussed in Chapter 8, provides an excellent opportunity for innovation. For example, Procter & Gamble is providing sachets of ingredients used in large water treatment facilities to individual homes in developing countries to improve the quality of their stored water.[11] Investing in the BOP benefits companies because:

- It is a very large and ignored underserved market

- It causes companies to be innovative to sell at affordable prices
- The innovation can be transferred from developing countries to developed economies for increased profits

Alternatively, changes in product manufacturing and service delivery can result in products and services that are more socially, environmentally, and economically friendly. These business model changes are usually invisible to the consumer but often vital for reducing social, environmental, and economic impacts.[12] In Chapter 8, we presented many examples of how companies have used technological innovation and business model innovation to improve sustainability *and* financial performance.

And, for many leaders, this is where the opportunities lie. For example:

> **Bill Joy, co-founder Sun MicroSystems:** "The Next Big Thing—and the greatest creation of wealth today—is in the green area: not just in the US but also in the developing world—new fuels (ethanol, fuel cells, using biotech to make fuels), new green technologies. This will create the Googles and Microsofts of the new era."[13]

> **Ellen Kullman, Chair of the Board and CEO, DuPont:** "We are focused on addressing the key challenges of the future related to global population growth and looking for opportunities

to innovate sustainable solutions. We have identified three specific global areas that come with a growing population: feeding the world, reducing dependence on fossil fuels and protecting people and the environment."[14]

Muhtar Kent, CEO of Coca-Cola: "At The Coca-Cola Company, we believe the most profound and impactful innovations over the next decade and beyond will emerge at the intersection of sustainability and our vast global value chain—the suppliers, retailers, technologies, people, and infrastructure that bring our beverages to market every day around the world. For this reason, we're working to embed sustainability-minded innovations into every aspect of our business, from sourcing ingredients to increasing beverage options to aspiring to be water neutral and recovering packages for recycling."[15]

The objective is to think beyond the current business model. Companies that want to compete will need to explore new ideas, acquire competencies that focus on new markets for the organization or emerging industries or markets, and generate a new business model. Best-practice companies of the future will:

- Develop a strategy, in general and for sustainability, that relies on innovation
- Drive transformation in the organization, creating market-changing ideas and products

- Invest in technological and business model innovation

Leading companies have the power to influence how the rest of their industry will be judged and can benefit from advanced technological and business process innovation. Nike, for example, entered into a strategic partnership with DyeCoo Textile Systems, a Netherlands-based company that has developed and built the first commercially available waterless textile dyeing machines. By using recycled CO_2, the new technology eliminates the use of water in the textile dyeing process. Nike believes this technology has the potential to revolutionize textile manufacturing. It will have a particularly positive impact in Asia, where much of the world's textile dyeing occurs. The removal of water from the textile dyeing process also eliminates the risk of effluent discharge, a known environmental hazard.[16] Stakeholders, particularly customers through increased purchases, may reward these companies for their responsiveness. And the innovations to increase sustainability performance often increase innovation throughout the company, thereby providing benefits to both corporations and society.

A last word

The results of corporate decisions and strategies are being scrutinized more closely than ever before. Some companies have been ineffective in the development and implementation of a strategy for addressing environmental, social, and economic concerns or integrating these issues in day-to-day management decisions. It is a challenge.

To implement strategies generally, and sustainability strategies particularly, managers need to better understand the implications of their decisions and the actions they can take to produce improved performance. This requires a careful analysis of the key drivers of performance and a measurement of both the drivers and the causal linkages between them. It also requires a clear understanding of the broad set of impacts that are caused by corporate activities and to understand these impacts on a broad set of stakeholders.

The Corporate Sustainability Model provides a comprehensive approach for examining, measuring, and managing the drivers of corporate sustainability. It has been extensively tested and revised in both academic and managerial studies and implementations. Its use by managers can provide a clearer understanding of the impacts of the various past, pending, and future corporate decisions on both the corporation and society. It can aid managers in operationalizing a sustainability strategy and tying it to the specific actions that will improve both sustainability performance *and* financial performance. Through a careful identification and measurement of key performance drivers, the strategy implementation process is improved.

Though many think that sustainability is too difficult to measure, companies have found that, unless the impacts are measured, they are commonly ignored in the resource allocation process. Thus, sustainability managers do not receive the necessary resources for effective implementation, and senior managers do not make the improvements necessary to improve both financial *and* sustainability performance. The consequences are huge. So what do leading companies do to facilitate integration into day-to-day decision-making? Through a combination of a well-articulated and well-communicated sustainability strategy, senior-management commitment to a broad set of objectives, and use of a variety of management structures and systems, leading companies have been able to improve their sustainability performance.

To develop processes more effectively senior managers need to:

- Identify, measure, manage, monitor, and report corporate social, environmental, and economic impacts
- Integrate into operational, strategic, and resource allocation decisions
- Assist colleagues in managing the paradox of simultaneously improving social, environmental, economic, *and* financial performance
- Recognize that strategy, leadership, and implementation tools are all essential components

Without appropriate management systems, corporations may not reap the benefits associated with sustainability performance. The alignment of leadership, strategy, structure, management systems, and performance measures is essential for companies to both coordinate activities and motivate employees toward implementing a sustainability strategy. This must be viewed over a long time horizon so that both the leading and lagging indicators of performance can be examined.

By integrating the evaluation of sustainability performance into the corporate decision-making process, managers can make better operational and investment decisions. To do this, they need to measure inputs, processes, outputs, and outcomes and identify the causal relationships and the specific actions they can take. When this is done, companies find that they are more prepared for understating the long-term impacts and have better information for better managerial decisions. With this information, senior sustainability managers and senior corporate and business unit managers can help improve both society and their companies.

Endnotes

Introduction

[1] Greenhouse (2013) "Major Retailers Join Bangladesh Safety Plan."

[2] Brundtland (1985) *Our Common Future;* Elkington (2006) "The Triple Bottom Line." In their recent book, Ehrenfeld and Hoffman (2013) develop the definition of sustainability as "the possibility that humans and other life will flourish on the Earth forever."

[3] Many terms have been used to describe this concept of sustainability. Both academic and corporate use of the words and the concepts vary widely. In some companies or countries, it may be more common to use corporate social responsibility, corporate citizenship, sustainability, sustainable development, corporate accountability, stakeholder management, or a variety of other terms. In this book, we will at times use many of these terms to describe the focus of improving the social, environmental, and economic impacts of corporate products, services, processes, and activities.

[4] Hoffman and Woody (2008) *Climate Change: What's Your Business Strategy?*

[5] Cescau (2007) "Beyond Corporate Responsibility."

[6] Coca-Cola (2012) *2011/2012 Sustainability Report.*

[7] Casey (2007) "Eminence Green."

[8] Harvey (2005) "GE Looks Out for a Cleaner Profit;" General Electric (2013) *Our Global Impact.*

[9] Mackey *et al.* (2007) "Corporate Social Responsibility and Firm Performance: Investor Preferences and Corporate Strategies;" Smith and Tushman (2005) "Managing Strategic Contradictions: A Top Management Model for Managing Innovation Streams."

[10] Margolis and Walsh (2003) "Misery Loves Company: Rethinking Social Initiatives by Business."

[11] Aram (1989) "The Paradox of Interdependent Relations in the Field of Social Issues in Management."

[12] Epstein *et al.* (2013) "Managing Social, Environmental and Financial Performance Simultaneously."

[13] Epstein *et al.* (2009) *Managing Social, Environmental, and Financial Performance Simultaneously.*

[14] Epstein *et al.* (2009) *Managing Social, Environmental, and Financial Performance Simultaneously;* Epstein *et al.* (2013) "Managing Social, Environmental, and Financial Performance Simultaneously."

[15] Schnietz and Epstein (2005) "Exploring the Financial Value of a Reputation."

[16] CEMEX (2012) *Building the Cities of the Future: 2012 Sustainable Development Report.*

[17] Batelle Memorial Institute (2002) *Toward a Sustainable Cement Industry.*

[18] Radin (2003) "Chiquita Brands International, Inc."

[19] Entine (2007) "Chiquita Counts the Costs of Honesty."

Chapter 1

[1] Elkington (1998) *Cannibals with Forks.*

[2] Berman *et al.* (1999) "Does Stakeholder Orientation Matter? The Relationship Between Stakeholder Management Models and Firm Financial Performance."

[3] Dow (2011) *Annual Sustainability Report.*

[4] Epstein and Roy (2003) "Improving Sustainability Performance."

[5] For examples of governance processes and measurements, see Epstein and Roy (2003) "Designing a Board Evaluation System for Corporate Directors" and Epstein and Roy (2002) *Measuring and Improving the Performance of Corporate Boards.*

[6] Alcatel-Lucent (2012) *Sustainability Report.*

[7] Kol and Pinske (2006) "Stakeholder Mismanagement and Corporate Social Responsibility Crises."

[8] Hart and Sharma (2004) "Engaging Fringe Stakeholders for Competitive Imagination."

[9] Kanal (2010) "Lessons from Dell's Stakeholder Engagement Model."

[10] Walker and Marr (2001) *Stakeholder Power.*

[11] Wheeler and Sillanpää (1997) *The Stakeholder Corporation.*

[12] Epstein and Birchard (1999) *Counting What Counts.*

[13] This model was first introduced in Epstein and Roy (2001) "Sustainability in Action" and revised in Epstein (2006) "Improving Organizations and Society."

[14] Christmann and Taylor (2001) "Globalization and the Environment;" Wisner and Epstein (2005) "'Push' and 'Pull' Impacts of NAFTA on

Environmental Responsiveness and Performance in Mexican Industry."

[15] Tysiac (2013) "Conflict Minerals Rule Poses Compliance Challenge."

[16] Esty and Winston (2006) *Green to Gold.*

[17] Wisner *et al.* (2006) "Organizational Antecedents and Consequences of Environmental Performance."

[18] Epstein and Wisner (2005) "Managing and Controlling Environmental Performance."

[19] Hart and Ahuja (1996) "Does it Pay to be Green?;" Klassen and Whybark (1999) "The Impact of Environmental Technologies;" Wisner *et al.* (2006) "Organizational Antecedents and Consequences of Environmental Performance."

[20] Epstein (1996) *Measuring Corporate Environmental Performance.*

[21] Cone Inc. (2007) www.coneinc.com.

[22] Alcoa (2012) *Sustainability at a Glance.*

[23] Siemens (2012) *Siemens Sustainability Report 2012.*

[24] Gunther (2013) "Unilever's CEO Has a Green Thumb."

[25] McDonald's (2012) *Global Sustainability Highlights.*

[26] Standard-Friel (2004) "Proving the Win–Win Strategy of Cause-Related Marketing."

[27] Epstein and Wisner (2001) "Using a Balanced Scorecard to Implement Sustainability."

Chapter 2

[1] Epstein *et al.* (2010) "Why Nike Kicks Butt in Sustainability."

[2] Epstein and Roy (2004) "Improving the Performance of Corporate Boards."

[3] Santander (2005) *Sustainability Report.*

[4] Nelson *et al.* (2001) *Power to Change.*

[5] Social Enterprise Knowledge Network (2006) *Effective Management of Social Enterprises.*

[6] Kanter (2009) *Super Corp.*

[7] Adapted from Bevan *et al.* (2004) *Achieving High Performance.*

[8] Fujitsu Group (2012) *Fujitsu Group Sustainability Report.*

[9] Unilever (2012) *Progress Report – Unilever Sustainable Living Plan.*

[10] BHP Billiton (2012) *Sustainability Report.*

[11] Starbucks (2012) *Global Responsibility Report.*

[12] Nestlé (2012) *Nestlé in Society.*

[13] Freeman *et al.* (2007) *Managing for Stakeholders.*

[14] Collins (2005) *Good to Great and the Social Sectors.*

[15] Professional Accountants in Business Committee (2006) "Professional Accountants in Business: At the Heart of Sustainability?"

[16] Esty and Simmons (2011) *The Green to Gold Business Playbook;* Laszlo (2008) *Sustainable Value.*

[17] Epstein *et al.* (2009) *Managing Social, Environmental, and Financial Performance Simultaneously.*

[18] Gunther (2004) "Money and Morals at GE."

[19] Lye and Müller (2004) *The Changing Landscape of Liability.*

[20] Stern (2006) *Economics of Change: The Stern Review.*

[21] Carey (2004) "Global Warming."

[22] Corporate Average Fuel Economy (CAFE) are US regulations established by the Energy Policy and Conservation Act of 1975 and enacted by the US Congress in the same year. The intention was to improve the average fuel economy of cars and light trucks (trucks, vans and sport utility vehicles) sold in the US. CAFE

standards are expressed as mathematical functions depending on vehicle "footprint," a measure of vehicle size determined by multiplying the vehicle's wheelbase by its average track width. CAFE footprint requirements are set up such that a vehicle with a larger footprint has a lower fuel economy requirement than a vehicle with a smaller footprint. For example, the 2012 Honda Fit, with a footprint of 40 sq ft (3.7 m2), must achieve fuel economy (as measured for CAFE) of 36 miles per US gallon (6.5 l/100 km), equivalent to a published fuel economy of 27 miles per US gallon (8.7 1/100 km). Ford F-150, on the other hand, with its footprint of 65–75 sq ft (6.0–7.0 m2) must achieve CAFE fuel economy of 22 miles per US gallon (11 1/100 km), that is 17 miles per US gallon (14 1/100 km) published (for more, see National Highway Traffic Safety Administration (2010) *CAFE 2011–2016 Final Rule*).

[23] Ford Motor Company (2012) *Sustainability Report Summary.*

[24] Lash and Wellington (2007) "Competitive Advantage on a Warming Planet."

[25] Davila *et al.* (2013) *Making Innovation Work.*

[26] Russo (2010) *Companies on a Mission.*

[27] Toyota Motor Corporation (2012) *Sustainability Report.*

[28] Peloza *et al.* (2012) "Sustainability: How Stakeholder Perceptions Differ from Corporate Reality."

[29] Laszlo and Zhexembayeva (2013) *Embedded Sustainability.*

[30] Bertels (2010) *Embedding Sustainability in Organizational Culture.*

[31] Nidumolu *et al.* (2012) "Connecting Heart to Head: A Framework for Sustainable Growth."

[32] Epstein *et al.* (2009) *Managing Social, Environmental, and Financial Performance Simultaneously.*

[33] Porter and Kramer (2006) "Strategy and Society."

[34] These stages could also be seen as minimization versus optimization strategies, as described in Savitz and Weber (2006) *The Triple Bottom Line.*

[35] CEMEX (2012) *Building the Cities of the Future, 2012 Sustainable Development Report.*

[36] Epstein and Roy (1998) "Managing Corporate Environmental Performance."

[37] Stuart (2003) "Multinationals Should Face the Same Rules no Matter Where They Set up Shop."

[38] Hawkins (2006) *Corporate Social Responsibility.*

[39] Gunther (2005) "Cops of the Global Village."

[40] Hopkins (2007) *Corporate Social Responsibility and International Development.*

[41] Novartis (2007) "Implementing a Living Wage Globally."

[42] Wood *et al.* (2006) *Global Business Citizenship.*

[43] ISO (2011) "The ISO Survey of Management System Standard Certifications."

[44] Orenstein (1997) "Lucent Seeks Self-Regulation Nod."

[45] ASQ and Manpower Professional (2011) *Social Responsibility and the Quality Professional: The Implications of ISO 26000.*

[46] ISO (2010) *ISO 26000:2010: Guidance on Social Responsibility.*

[47] Lambert (2013) "Event Sustainability Management – ISO 20121 Passes 2012 Olympic Games Test."

[48] Social Accountability International (2008) *Social Accountability 8000.*

[49] International Trade Centre (2012) *Social Accountability International – SA8000.*

[50] United Nations Global Compact (2013) www.unglobalcompact.org/ParticipantsAndStakeholders/index.html.

[51] Hernandez (2004) "Institutionalising Global Standards of Responsible Corporate Citizenship: Assessing the Role of the UN Global Compact."

[52] United Nations (2000) *United Nations Millennium Declaration.*

[53] United Nations (2013) *The Millennium Development Goals Report 2013.*

[54] Sethi (2003) *Setting Global Standards.*

[55] Electronic Industry Citizenship Coalition (2012) *Electronic Industry Citizenship Coalition® Code of Conduct.*

[56] Sethi (2003) *Setting Global Standards.*

[57] Cramer (2006) *Corporate Social Responsibility and Globalisation.*

[58] Lydenberg (2005) *Corporations and the Public Interest.*

[59] Schnietz (2006) "The Purpose and History of Business Regulation."

[60] Glazebrook (2001) "How Australia's Top 500 Companies are Becoming Corporate Citizens."

[61] Porter and van der Linde (1995) "Green and Competitive: Ending the Stalemate."

[62] Wisner and Epstein (2005) "'Push' and 'Pull' Impacts of NAFTA on Environmental Responsiveness and Performance in Mexican Industry."

[63] Pinkham (2006) "Business and Government: Friends and Foes."

[64] Cramer and Pruzan-Jorgensen (2006) "Engaging Governments in Support of Corporate Social Responsibility in Global Supply Chains;" Zadek (2001) *The Civil Corporation.*

[65] Beloe *et al.* (2006) "Can Business Help Governments Change the System?"

[66] General Electric (2007) *Citizenship Report.*

[67] Freeman *et al.* (2000) *Environmentalism and the New Logic of Business.*

[68] Jaffe *et al.* (1995) "Environmental Regulation and the Competitiveness of US Manufacturing."

[69] Wood (2012) *The Role of Philanthropic Capital in Entrepreneurship.*

[70] Social Investment Forum (2006) *2005 Report of Socially Responsible Investing Trends in the United States.*

[71] Knoepfel (2001) "Dow Jones Sustainability Group Index."

[72] SIRAN (2006) www.siran.org/csr.php.

[73] Zadek (2005) "Responsibility Isn't a Blame Game."

[74] Vogel (2005) *The Market for Virtue.*

[75] Cogan (2003) *Corporate Governance and Climate Change.*

[76] Danone (2012) *Sustainability Report: Strategy and Performance.*

Chapter 3

[1] This section is drawn heavily from Epstein and Roy (2007) "Implementing a Corporate Environmental Strategy" and Epstein and Roy (1998) "Managing Corporate Environmental Performance."

[2] CEMEX (2012) *Building the Cities of the Future, 2012 Sustainable Development Report.*

[3] Mitsubishi Corporation (2010) *Sustainability Report.*

[4] Canon (2013) *Sustainability Report.*

[5] McDonald's (2012) *Global Sustainability Highlights.*

[6] McDonald's (2004) *Worldwide Corporate Responsibility Report.*

[7] Epstein *et al.* (2010) "Why Nike Kicks Butt in Sustainability."

[8] Hennes & Mauritz (2012) *Conscious Actions: Sustainability Report 2012.*

[9] Social Enterprise Knowledge Network (2006) *Effective Management of Social Enterprises.*

[10] MacLean (2004) "EHS Organizational Quality: A DuPont Case Study."

[11] The Covin Interview (2009) "Linda Fisher: Chief Sustainability Officer, DuPont. A Report from the Battlefield in the War on Climate Change."

[12] Waddock and Bodwell (2007) *Total Responsibility Management.*

[13] Logan *et al.* (1997) *Global Corporate Citizenship.*

[14] Ashridge Business School (2009) *Developing the Global Leader for Tomorrow.*

[15] Winston (2009) *Green Recovery.*

[16] Hart (2005) *Capitalism at the Crossroads.*

[17] Hewlett-Packard (2012) *Global Citizenship Report.*

[18] MacLean (2005) "Engagement at the Top."

[19] Epstein *et al.* (2010) "Why Nike Kicks Butt in Sustainability."

[20] Cogan (2003) *Corporate Governance and Climate Change.*

[21] Grayson and Hodges (2004) *Corporate Social Opportunity.*

[22] Blackburn (2007) *The Sustainability Handbook.*

[23] Deutsch (2007) "Companies Giving Green an Office."

[24] Austin *et al.* (2004) *Timberland.*

[25] Austin (2000) *The Collaboration Challenge.*

[26] Kotler and Lee (2005) *Corporate Social Responsibility.*

[27] Porter and Kramer (2002) "The Competitive Advantage of Corporate Philanthropy."

[28] The Boeing Company (2012) *Corporate Citizenship Report.*

[29] Logan *et al.* (1997) *Global Corporate Citizenship.*

[30] The Boeing Company (2012) *Corporate Citizenship Report.*

[31] Wisner (2004) "Keynote Speech."

[32] Goodwin (2003) "Workplace Volunteering is Good Business."

[33] Wisner (2004) "Keynote Speech."

[34] Epstein *et al.* (2009) *Managing Social, Environmental, and Financial Performance Simultaneously.*

[35] Roche (2012) *Annual Report.*

[36] Pracejus and Olsen (2004) "The Role of Brand/Cause Fit in the Effectiveness of Cause-Related Marketing Campaigns."

[37] Kotler and Lee (2005) *Corporate Social Responsibility.*

[38] Avon Products (2012) *Corporate Responsibility Report 2010–2011.*

[39] Starbucks (2012) *Global Responsibility Report.*

[40] Deutsch (2006) "Companies and Critics Try Collaboration."

[41] Vodafone Group (2013) *Sustainability Report 2012/2013.*

[42] Svendsen (1998) *The Stakeholder Strategy.*

[43] Novartis (2007) "Novartis Comunidad."

[44] This draws on the excellent work of Argenti (2004) "Collaborating with Activists" and Austin (2000) *The Collaboration Challenge.*

Chapter 4

[1] Bekefi and Epstein (2008) "Measuring and Managing Social and Political Risk."

[2] Finn *et al.* (2006) "Unrecognized Assets."

[3] Mitsubishi Corporation (2010) *Sustainability Report.*

[4] Alcoa (2003) *Sustainability Report.*

[5] De Beers (2005/6) *Report to Stakeholders.*

[6] Rangone (1998) "On the Applicability of Analytical Techniques..."

[7] Epstein and Roy (2006) "Measuring the Effectiveness of Corporate Boards and Directors."

[8] This section is drawn heavily from Epstein and Wisner (2002) "Measuring and Managing Social and Environmental Impact."

[9] Deegan (2003) *Environmental Management Accounting.*

[10] Schaltegger *et al.* (2003) *An Introduction to Corporate Environmental Management.*

[11] Fava and Smith (1998) "Integrating Financial and Environmental Information."

[12] Canon (2013) *Sustainability Report.*

[13] Stead and Stead (2004) *Sustainable Strategic Management.*

[14] Canon (2004). *Sustainability Report.*

[15] Berry and Rondinelli (1998) "Proactive Corporate Environmental Management."

[16] Holliday *et al.* (2002) *Walking the Talk.*

[17] This section is heavily drawn from Bekefi and Epstein (2006) *Integrating Social and Political Risk into Management Decision-Making.*

[18] Rio Tinto (2012) *Sustainable Development Report.*

[19] De Beers (2006) *HIV/Aids Case Study.*

[20] De Beers (2012) *Report to Society.*

[21] Alexander (1984) "A Calamity for Union Carbide."

[22] Dobrzynski (1984) "Union Carbide Fights for its Life."

[23] Jackson (1984) "Union Carbide's Good Name Takes a Beating."

[24] Antosh and Hyde (1999) "Chemical Giants Joining Forces."

[25] Bekefi and Epstein (2006) *Integrating Social and Political Risk into Management Decision-Making.*

[26] Coca-Cola (2012) *2011/2012 Sustainability Report.*

[27] *Professional Engineering* (1999) "Brent Spar Outcry Leaves Shell with a 60m Pound Bill."

[28] De Beers (2012) *Report to Society.*

[29] Equator Principles (2013) www.equator-principles.com/index.php/ep3/process-and-launch.

[30] Equator Principles Association (2013) *The Equator Principles – A Financial Industry Benchmark for Determining, Assessing and Managing Environmental and Social Risk in Projects.*

[31] HSBC Holdings (2012) *Sustainability Report.*

Chapter 5

[1] This chapter is drawn in part from Epstein and Wisner (2002) "Measuring and Managing Social and Environmental Impact;" Epstein and Birchard (1999) *Counting What Counts;* and Davila *et al.* (2006) *Making Innovation Work.*

[2] Colgate (2012) *2011/2012 Sustainability Report.*

[3] Jensen (2001) "Value Maximization, Stakeholder Theory and the Corporate Objective Function."

[4] Howell (1994) *Developing Comprehensive Performance Indicators.*

[5] Volvo Car Group (2012) *Sustainability Scorecard.*

[6] Kingfisher (2013) *Net Positive Report.*

[7] Nike (2012) *Nike Inc., FY10/11 – Sustainable Business Report Summary.*

[8] Ferracone (2011) "The Role of Environmental Sustainability in Executive Compensation."

[9] National Australia Bank (2012) *Report of the Directors.*

[10] Epstein and Roy (2006) "Measuring the Effectiveness of Corporate Boards and Directors."

[11] Warhurst (2002) *Environmental and Social Performance Indicators in Minerals Development.*

[12] Larkin (2003) *Strategic Reputation Risk Management.*

[13] Alcoa (2012) *Sustainability at a Glance.*

[14] Henkel (2012) *Sustainability Report.*

[15] Anglo-American (2004) *Annual Review.*

[16] Ferracone (2011) "The Role of Environmental Sustainability in Executive Compensation."

[17] UPS (2005) *Sustainability Report.*

[18] Werther (2006) *Strategic Corporate Social Responsibility.*

[19] Carey (2000) "A Free Market Cure for Global Warming."

[20] Gardner (2007) "Investors to Press Congress on Warming."

[21] Annala and Howe (2004) "Trading Solutions for Lowering Air Pollution."

[22] Chicago Climate Exchange (2007) www.chicagoclimatex.com/content.jsf?id=821.

[23] See, for example, Kaplan and Norton (1996) *The Balanced Scorecard;* Kaplan and Norton (2000) *The Strategy-Focused Organization.*

[24] Epstein and Wisner (2006) "Actions and Measures to Improve Sustainability."

[25] Epstein and Wisner (2001) "Good Neighbors."

[26] Fiksel (2003) "Revealing the Value of Sustainable Development."

[27] Holliday (2001) "Sustainable Growth, the DuPont Way."

[28] Epstein and Young (1999) "Greening with EVA."

Chapter 6

[1] J.P. Morgan and GIIN (2013) *Perspectives on Progress: The Impact Investor Survey.*

[2] Deutsch Bank (2012) *Corporate Responsibility Report.*

[3] Freeman (2003) *The Measurement of Environmental and Resource Values.*

[4] Shechter and Freeman (1994) "Nonuse Value."

[5] Goodstein (1999) *Economics and the Environment.*

[6] Kahneman *et al.* (1999) "Economic Preferences or Attitude Expressions?"

[7] Freeman (2003) *The Measurement of Environmental and Resource Values.*

[8] Trudel and Cotte (2009) "Does It Pay to be Good?"

[9] For more on insurance and the mitigation of risk, see Bekefi and Epstein (2006) *Integrating Social and Political Risk into Management Decision-Making.*

[10] Champ *et al.* (2003) *A Primer on Nonmarket Valuation.*

[11] Bekefi and Epstein (2006) *Integrating Social and Political Risk into Management Decision-Making.*

[12] Young (2005) *Determining the Economic Value of Water.*

[13] Layman *et al.* (1996) "Economic Valuation of the Chinook Salmon Sport Fishery of the Gulkana River, Alaska, Under Current and Alternate Management Plans."

[14] Boyle (2003) "Contingent Valuation in Practice"; Davis (1963) "Recreation Planning as an

Economic Problem"; Hoevenagel (1994) "An Assessment of the Contingent Valuation Model."

[15] Scott (1965) *The Valuation of Game Resources.*

[16] Arrow *et al.* (1993) *Report of the NOAA Panel on Contingent Valuation.*

[17] Epstein and Widener (2007) *Measuring Multiple Stakeholder Costs and Benefits.*

[18] BP (2012) *Sustainability Review: Building a Stronger, Safer BP.*

[19] Patt (2006) "Dealing with Uncertainty."

[20] Volkswagen Group (2012) *Sustainability Report.*

[21] Epstein and Rejc (2005) *Identifying, Measuring, and Managing Organizational Risks for Improved Performance.*

[22] Shell International Global Business Environment (2003) *Exploring the Future Scenarios.*

[23] Carbon Disclosure Project (2012) *Business Resilience in an Uncertain, Resource-Constrained World.*

[24] Birbeck (1999) *Forewarned is Forearmed.*

[25] Bekefi and Epstein (2006) *Integrating Social and Political Risk into Management Decision-Making.*

[26] AccountAbility *et al.* (2001) *The Sigma Project.*

[27] Vodafone Group (2013) *Sustainability Report 2012/2013.*

[28] US Environmental Protection Agency (1992) *Facility Pollution Prevention Guide.*

Chapter 7

[1] PricewaterhouseCoopers (2002) *Sustainability Survey Report.*

[2] Epstein *et al.* (2009) *Managing Social, Environmental, and Financial Performance Simultaneously.*

[3] This section is drawn from Epstein and Rejc (2005) *Evaluating Performance in Information Technology*.

[4] Epstein *et al.* (2013) "Managing Social, Environmental, and Financial Performance Simultaneously."

[5] Epstein and Roy (2003) "Sustainability in Action."

[6] Vergin and Qoronfleh (1998) "Corporate Reputation and the Stock Market."

[7] Schnietz and Epstein (2005) "Exploring the Financial Value of a Reputation for Corporate Social Responsibility Dduring a Crisis."

[8] Epstein and Roy (2003)

[9] For a more thorough discussion of using a balanced scorecard to measure board performance, see Epstein and Roy (2004) "How Does Your Board Rate?"

[10] Sberbank (2012) *Corporate Social Responsibility Report 2012.*

[11] For more on organizational trust, see Currall and Epstein (2003) "The Fragility of Organizational Trust."

[12] See, for example, Schnietz and Epstein (2005) "Exploring the Financial Value of a Reputation."

[13] Clifford (2013) "Some Retailers Say More about their Clothing's Origins."

[14] Bell and Morse (2003) *Measuring Sustainability.*

[15] BHP Billiton (2012) *Sustainability Report: Resourcing the Future–We Value Sustainability.*

[16] McElhaney (2008) *Just Good Business.*

[17] Svendsen and Laberge (2007) *Convening Stakeholder Networks.*

[18] BP (2012) *Sustainability Review: Building a Stronger, Safer BP.*

[19] Eccles *et al.* (2007) "Reputation and Its Risks."

[20] For further discussion of the reputation quotient, see Fombrun *et al.* (2000) "The Reputation Quotient."

[21] Fombrun (1996) *Reputation.*

[22] Sundaram (1998) *Perrier, Nestlé, and the Agnellis.*

[23] Interbrand (2012) *Best Global Brands 2012: The Definitive Guide to the 100 Best Global Brands.*

[24] This section is heavily drawn from Bekefi and Epstein (2006) *Integrating Social and Political Risk into Management Decision-Making.*

[25] Gunther (2004) "The Mosquito in the Tent."

[26] *Business Journal of Milwaukee* (2004) "Marsh & McLennan to Cut 3,000 Jobs;" *Boston Globe* (2004) "Cherkasky Says Marsh May Settle Spitzer's Lawsuit within a Month."

[27] This section is heavily drawn from Boardman *et al.* (1996) *Cost Benefit Analysis: Concepts and Practice* and Mitchell and Carson (1989) *Using Surveys to Value Public Goods.*

[28] Boyle (2003) "Introduction to Revealed Preference Methods."

[29] The Co-operative Bank (2001) *Partnership Report 2001: Our Impact.*

[30] Epstein and Widener (2007) *Measuring Multiple Stakeholder Costs and Benefits.*

Chapter 8

[1] Boguslaw (2002) "Have We Arrived? Only When the Integration of Stakeholder Interests Becomes 'Business as Usual'."

[2] This section is heavily drawn from Epstein and Roy (1997) "Using ISO 14000 for Improved Organizational Learning and Environmental Management."

[3] Leonard-Barton (1995) *Wellsprings of Knowledge.*

[4] Johnson & Johnson (2012) *Citizenship and Sustainability Report.*

[5] MillerCoors (2012) *Sustainability Report: Great Things on Tap.*

[6] Argyris and Schön (1978) *Organizational Learning.*

[7] Chronos (2013) www.sdchronos.org/ImmChronos/english/chronos_and_sustainability.html.

[8] Strom (2012) "Companies Pick up Used Packaging and Recycling's Cost."

[9] Power (2006) "Take it Back."

[10] Sony Ericsson (2011) *Sustainability Report.*

[11] Bristol-Myers Squibb (2010) *Sustainability Highlights.*

[12] Baxter (2012) *Sustainability Report.*

[13] Lehni (2000) *Eco-efficiency.*

[14] Holmes (2006) "Nike Goes for the Green."

[15] UPS (2012) *Corporate Sustainability Report.*

[16] Dell (2012) *Corporate Responsibility Report.*

[17] China National Petroleum Corporation (2012) *Annual Report.*

[18] Pfizer (2013) *Kalamazoo Cuts 'Wasted' Expenses.*

[19] Apple (2012) *Apple Facilities: Environmental Footprint Report – Fiscal 2012.*

[20] Gazprom (2012) *Environmental Report.*

[21] For a more thorough discussion of the cradle-to-cradle concept, see McDonough and Braungart (2002) *Cradle to Cradle.*

[22] Strom (2012) "Companies Pick Up Used Packaging, and Recycling's Cost."

[23] Hennes & Mauritz (2012) *Conscious Actions: Sustainability Report 2012.*

[24] Vodafone Group (2013) *Sustainability Report 2012/2013.*

[25] Baxter (2012) *Sustainability Report.*

[26] McDonough Braungart Design Chemistry (2013) www.mbdc.com/cradle-to-cradle/.

[27] Reinhardt (1999) "Bringing the Environment Down to Earth."

[28] Larkin (2003) *Strategic Reputation Risk Management.*

[29] Reinhardt (2000) *Down to Earth: Applying Business Principles to Environmental Management.*

[30] Oliva and Quinn (2003) "Interface's Evergreen Services Agreement."

[31] For a more thorough discussion on the concept of the bottom of the pyramid, see Prahalad (2004) *The Fortune at the Bottom of the Pyramid;* Hart and Christensen (2002) "The Great Leap"; Prahalad and Hammond (2002) "Serving the World's Poor, Profitably."

[32] Shah (2013) "Poverty Facts and Stats."

[33] Hart (2010) *Capitalism at the Crossroads.*

[34] Serrano (2008) *Amanco: Providing Irrigation Systems to the Rural Poor.*

[35] London and Hart (2011) *Next Generation Business Strategies for the Base of the Pyramid.*

[36] CEMEX (2012) *Building the Cities of the Future, 2012 Sustainable Development Report.*

[37] Roche (2012) *Annual Report.*

[38] For more general strategies, see London and Hart (2011) *Next Generations Business Strategies for the Base of the Pyramid.*

[39] Hammond *et al.* (2007) *The Next 4 Billion.*

[40] Welford *et al.* (1998) "Toward Sustainable Production and Consumption."

[41] Foxconn Technology Group (2011) *CSER 2011 Annual Report.*

[42] Toyota Motor Corporation (2012) *Sustainability Report.*

[43] Dell (2006) *Sustainability Report.*

[44] British American Tobacco (2004) *Social Report 2003/4.*

[45] Lippman (2002) "Environmental Responsibility."

[46] Nike (2012) *Nike Inc., FY10/11 – Sustainable Business Report Summary.*

[47] L'Oréal (2010) *Sustainable Development Report.*

[48] adidas Group (2012) *Sustainability Progress Report.*

[49] Walmart (2013) *Global Responsibility Report.*

[50] Cescau (2007) "Beyond Corporate Responsibility."

[51] This section is adapted from Epstein and Rejc Buhovac (2006) *The Reporting of Organizational*

Risks for Internal and External Decision Making and Bekefi and Epstein (2006) *Integrating Social and Political Risk into Management Decision-Making.*

Chapter 9

[1] Layoux and Soltis Martel (2013) "Sustainability Rising."

[2] Koehler and Hespenheide (2013) "Finding the Value in Environmental, Social, and Governance Performance."

[3] Ibid.

[4] Kolk (2005) "Sustainability Reporting."

[5] SIRAN (2006) "Socially Responsible Investment Analysts Find More Large US Companies Reporting on Social and Environmental Issues."

[6] Layoux and Soltis Martel (2013) "Sustainability Rising."

[7] General Mills (2010) *Corporate Social Responsibility.*

[8] Pleon (2005) *Accounting for Good.*

[9] Global Reporting Initiative (2013) *G4 Sustainability Reporting Guidelines.*

[10] NCIF (2012) *Collaborating to Harmonize Standardized Metrics for Impact Investors.*

[11] GIIRS (2010) *Sample Fund and Company Rating.*

[12] B-Corporation (2012) *Annual Report.*

[13] GOOD (2012) www.good.is/posts/b-corps-go-global-sistema-b-certifies-south-americansocial-enterprise.

[14] IIRC (2013) *Consultation Draft of the International Framework – Integrated Reporting.*

[15] Global Reporting Initiative (2013) *G4 Sustainability Reporting Guidelines.*

[16] IPIECA, API, and OGP (2010) *Oil and Gas Industry Guidance on Voluntary Sustainability Reporting.*

[17] This section is adapted from Epstein and Rejc Buhovac (2006) *The Reporting of Organizational Risks for Internal and External Decision Making.*

[18] London Stock Exchange (2010) *Investor Relations – A Practical Guide.*

[19] Lewis and Little (2004) *Fooling Investors and Fooling Themselves.*

[20] Kramer and Cooch (2006) *Investing for Impact.*

[21] Bayer (2012) *Sustainable Development Report.*

[22] Zadek and Merme (2003) "Redefining Materiality."

[23] Nike (2012) *Nike Inc., FY10/11 – Sustainable Business Report Summary.*

[24] adidas Group (2012) *Sustainability Progress Report: Performance Counts.*

[25] Dell (2012) *Corporate Responsibility Report: Powering the Possible.*

[26] This section is drawn from Epstein and Wisner (2001) "Increasing Corporate Accountability."

[27] Epstein and Palepu (1999) "What Financial Analysts Want."

[28] Baxter (2012) *Environmental Financial Statement.*

[29] Novo Nordisk (2012) *Annual Report.*

[30] KPMG (2005) *International Survey of Corporate Responsibility Reporting.*

[31] Leipziger (2003) *The Corporate Responsibility Code Book.*

[32] There have been more recent approaches to the measurement of social impact. Among the most prominent are the "blended value" approach developed by Jed Emerson and the social return on investment (SROI) developed by Emerson and REDF (a San Francisco-based venture philanthropy) and further developed by the New Economics Foundation. See Emerson (2003) "The Blended Value Proposition." A

recent review of innovative and coordinated webbased approaches to reporting performance, outcome, and impact measurement is provided by Kramer *et al.* (2009) *Breakthroughs in Shared Measurement and Social Impact.*

[33] Honda North America (2012) *North American Environmental Report.*

[34] CH_2M Hill (1993) *The Role of Internal Auditors in Environmental Issues.*

[35] Pleon (2005) *Accounting for Good.*

[36] BP (2012) *Sustainability Review: Building a Stronger, Safer BP.*

[37] Epstein and Wisner (2002) "Measuring and Managing Social and Environmental Impact."

Chapter 10

[1] Schnietz and Epstein (2005) "Exploring the Financial Value of a Reputation;" Epstein and Roy (2003) "Making the Business Case for Sustainability."

[2] Savitz and Weber (2006) The Triple Bottom Line; Bansal and Roth (2000) "Why Companies Go Green;" Elkington (2001) The Chrysalis Economy; Stanwick and Stanwick (1998) "The Relationship between Corporate Social Performance and Organizational Size, Financial Performance, and

Environmental Performance: An Empirical Examination."

[3] Adapted from Epstein and Wisner (2001) "Good Neighbors."

[4] Henkel (2012) *Sustainability Report.*

[5] This section is drawn from Epstein and Westbrook (2001) "Linking Actions to Profits in Strategic Decision Making" and Epstein and Leonard (2007) *The Value Proposition for Corporate Social Responsibility.*

[6] Epstein and Roy (2003) "Making the Business Case for Sustainability."

[7] Epstein and Wisner (2002) "Measuring and Managing Social and Environmental Impact."

[8] Bekefi *et al.* (2007) *Business Opportunities from Effective Risk Management.*

[9] Piasecki (2007) *World Inc.*

[10] Toyota Motor Corporation (2012) *Sustainability Report.*

[11] Churet (2005) *Business for Development.*

[12] Davila *et al.* (2013) *Making Innovation Work: How to Manage it, Measure it, and Profit from it.*

[13] Bartiromo (2006) "Green: The Next Big Thing."

[14] DuPont (2012) *Sustainability Progress Report.*

[15] Coca-Cola (2012) *2011/2012 Sustainability Report.*

[16] Nike (2012) "NIKE, Inc. Announces Strategic Partnership to Scale Waterless Dyeing Technologies."

Bibliography

AccountAbility, Forum for the Future, and BSI (2001) *The Sigma Project: Sustainability in Practice* (AccountAbility, Forum for the Future, and BSI).

adidas Group (2012) *Sustainability Progress Report: Performance Counts.*

Alcatel-Lucent (2012) *Sustainability Report: A Business Imperative.*

Alcoa (2003) *Sustainability Report.*

_____(2012) *Sustainability at a Glance.*

Alexander, C. (1984) "A Calamity for Union Carbide," *Time,* December 17, 1984.

Anglo-American (2004) *Annual Review.*

Annala, C., and H. Howe (2004) "Trading Solutions for Lowering Air Pollution: Trading of Greenhouse Gas Emissions Allowances Can Lower the Cost of Regulatory Compliance and Create Business Opportunities," *Strategic Finance,* September 2004.

Antosh, N., and J. Hyde (1999) "Chemical Giants Joining Forces: Dow to Buy Union Carbide," *Houston Chronicle,* August 5, 1999.

Apple (2012) *Apple Facilities: Environmental Footprint Report – Fiscal 2012.*

Aram, J.D. (1989) "The Paradox of Interdependent Relations in the Field of Social Issues in Management," *Academy of Management Review* 14.2: 266-83.

Argenti, P.A. (2004) "Collaborating with Activists: How Starbucks Works with NGOs," *California Management Review* 47.1: 91-116.

Argyris, C., and D. Schön (1978) *Organizational Learning: A Theory of Action Perspective* (Reading, MA: Addison-Wesley).

Arrow, K., R. Solow, P.R. Portney, E.E. Leamer, R. Radner, and H. Schuman (1993) *Report of the NOAA Panel on Contingent Valuation* (Washington, DC: Federal Register).

Ashridge Business School (2009) *Developing the Global Leader for Tomorrow.*

ASQ and Manpower Professional (2011) *Social Responsibility and the Quality Professional: The Implications of ISO 26000.*

Austin, J.E. (2000) *The Collaboration Challenge: How Nonprofits and Businesses Succeed through Strategic Alliances* (San Francisco, CA: Jossey-Bass).

_____H.B. Leonard, and J.W. Quinn (2004) *Timberland: Commerce and Justice* (Course Materials; Boston, MA: Harvard Business School).

Avon Products (2012) *Corporate Responsibility Report 2010–2011: The Beauty of Doing Good.*

B-Corporation (2012) *Annual Report.*

Ball, J. (2003) "New Market Shows Industry Moving on Global Warming," *Wall Street Journal,* January 16, 2003.

Bansal, P., and K. Roth (2000) "Why Companies Go Green: A Model of Ecological Responsiveness," *Academy of Management Journal* 43.4: 717-36.

Bartiromo, M. (2006) "Green: The Next Big Thing," *Bloomberg BusinessWeek Magazine,* May 28, 2006.

Batelle Memorial Institute (2002) *Toward a Sustainable Cement Industry: Summary Report* (Geneva: World Business Council for Sustainable Development).

Baxter (2012) *Sustainability Report.*

Baxter (2012) *Environmental Financial Statement.*

Bayer (2012) *Sustainable Development Report.*

Bekefi, T., and M.J. Epstein (2006) *Integrating Social and Political Risk into Management Decision-Making.*

CMA Canada/AICPA Management Accounting Guideline (www.cpa2biz.com/search/results.jsp, accessed September 6, 2007).

_____(2008) "Measuring and Managing Social and Political Risk," *Strategic Finance,* February 2008: 33-41.

_____and K. Yuthas (2007) *Business Opportunities from Effective Risk Management* (Working Paper; Houston, TX: Rice University).

Bell, S., and S. Morse (2003) *Measuring Sustainability: Learning from Doing* (London: Earthscan Publications).

Beloe, S., J. Elkington, and J. Thorpe (2006) "Can Business Help Governments Change the System?" in M.J. Epstein and K.O. Hanson (eds.), *The Accountable Corporation* (Westport, CT: Praeger Publishers, vol.4): 45-74.

Berman, S.L., A.C. Wicks, S. Kotha, and T.M. Jones (1999) "Does Stakeholder Orientation Matter? The Relationship Between Stakeholder Management Models and Firm Financial Performance," *Academy of Management Journal* 42.5: 488-506.

Berry, M.A., and D.A. Rondinelli (1998) "Proactive Corporate Environmental Management: A New Industrial Revolution," *Academy of Management Executive* 12.2: 38-50.

Bertels, S. (2010) *Embedding Sustainability in Organizational Culture: A How to Guide for Executives* (London, Ontario: Network for Business Sustainability).

Bevan, S., N. Isles, P. Emery, and T. Haskins (2004) *Achieving High Performance: CSR at the Heart of Business* (London: Work Foundation).

BHP Billiton (2012) *Sustainability Report: Resourcing the Future – We Value Sustainability.*

Birbeck, K. (1999) *Forewarned is Forearmed: Identification and Measurement in Integrated Risk Management* (Report 249-99; Ottawa: Conference Board of Canada).

Bishop, R.C., and T.A. Heberlein (1979) "Measuring Values of Extra-Market Goods: Are Indirect Measures Biased?" *American Journal of Agricultural Economics* 61: 926-30.

Black, E.L., T.A. Carnes, and V.J. Richardson (2000) "The Market Valuation of Corporate Reputation," *Corporate Reputation Review* 3.1: 31-42.

Blackburn, W.R. (2007) *The Sustainability Handbook: The Complete Management Guide to Achieving Social, Economic, and Environmental Responsibility* (Washington, DC: Environmental Law Institute).

Boardman, A.E., D.H. Greenberg, A.R. Vining, and D.L. Weimer (1996) *Cost Benefit Analysis: Concepts and Practice* (Upper Saddle River, NJ: Prentice Hall).

The Boeing Company (2012) *Corporate Citizenship Report.*

Boguslaw, J. (2002) "Have We Arrived? Only When the Integration of Stakeholder Interests Becomes 'Business as Usual'," *AccountAbility Quarterly,* December 2002: 52-58.

Boston Globe (2004) "Cherkasky Says Marsh May Settle Spitzer's Lawsuit within a Month: CEO is Seeking to Fix 40% Drop in Share Price," *Boston Globe,* November 23, 2004.

Boyle, K.J. (2003) "Contingent Valuation in Practice," in P.A. Champ, K.J. Boyle, and T.C. Brown (eds.), *A Primer on Nonmarket Valuation* (Boston, MA: Kluwer Academic Publishers): 111-70.

_____(2003) "Introduction to Revealed Preference Methods," in P.A. Champ, K.J. Boyle, and T.C. Brown (eds.), *A Primer on Nonmarket Valuation* (Boston, MA: Kluwer Academic Publishers): 259-68.

BP (2006) *Sustainability Report.*

_____(2012) *Sustainability Review: Building a Stronger, Safer BP.*

Bristol-Myers Squibb (2010) *Sustainability Highlights.*

British American Tobacco (2004) *Social Report 2003/4.*

British American Tobacco (2012) *Sustainability: Why it Matters – Sustainability Summary 2012.*

Brundtland, G. (1987) *Our Common Future: The World Commission on Environment and Development* (Oxford, UK: Oxford University Press).

Burke, L., and J.M. Logsdon (1996) "How Corporate Social Responsibility Pays Off," *Long Range Planning* 29.4: 495-502.

Business Journal of Milwaukee (2004) "Marsh & McLennan to Cut 3,000 Jobs," *Business Journal of Milwaukee,* November 9, 2004.

Bussey, J. (2007) "Terror Victims' Families Sue Chiquita Brands International," *Miami Herald,* June 15, 2007.

Canon (2004) *Sustainability Report.*

_____(2013) *Sustainability Report.*

Carbon Disclosure Project (2012) *Business Resilience in an Uncertain, Resource-Constrained World – CDP Global 500 Climate Change Report 2012.*

Carey, J. (2000) "A Free Market Cure for Global Warming," *BusinessWeek,* May 15, 2000: 167-69.

_____(2004) "Global Warming," *BusinessWeek,* August 16, 2004.

Casey, S. (2007) "Patagonia: How Founder Yvon Chouinard Took His Passion for the Outdoors and Turned it into a Truly Radical Business," *Fortune,* April 2, 2007.

CEMEX (2012) *Building the Cities of the Future: 2012 Sustainable Development Report.*

Cescau, P. (2007) "Beyond Corporate Responsibility: Social Innovation and Sustainable Development as Drivers of Business Growth," speech given at *INDEVOR Alumni Forum,* INSEAD, May 25, 2007.

CH_2M HILL (1993) *The Role of Internal Auditors in Environmental Issues* (Altamonte Springs, FL: Institute of Internal Auditors Research Foundation).

Champ, P.A., K.J. Boyle, and T.C. Brown (2003) *A Primer on Nonmarket Valuation* (Boston, MA: Kluwer Academic Publishers).

Chicago Climate Exchange (2007) www.chicagoclimatex.com/content.jsf?id=821, accessed September 5, 2007.

China National Petroleum Corporation (2012) *Annual Report.*

Christmann, P., and G. Taylor (2001) "Globalization and the Environment: Determinants of Corporation Self-Regulation in China," *Journal of International Business Studies* 32.3: 439-58.

Chronos (2013) www.sdchronos.org/ImmChronos/english/chronos_and_sustainability.html, accessed July 31, 2013.

Churet, C. (2005) *Business for Development: Business Solutions in Support of the Millennium Development Goals* (Geneva: World Business Council for Sustainable Development).

Clifford, S. (2013) "Some Retailers Say More about their Clothing's Origins," *The New York Times,* May 8, 2013.

Coca-Cola (2012) *2011/2012 Sustainability Report.*

Cogan, D.G. (2003) *Corporate Governance and Climate Change: Making the Connection* (Rockville, MD: Investor Responsibility Research Center for Ceres).

Colgate (2012) *Sustainability Report: Giving the World Reasons to Smile.*

Collins, J. (2005) *Good to Great and the Social Sectors: A Monograph to Accompany Good to Great* (New York: HarperCollins).

Cone Inc. (2007) www.coneinc.com, accessed September 5, 2007.

The Co-operative Bank (2001) *Partnership Report 2001: Our Impact.*

The Covin Interview (2009) "Linda Fisher: Chief Sustainability Officer, DuPont. A Report from the Battlefield in the War on Climate Change," *Fortune,* November 23, 2009.

Cramer, A., and P.M. Pruzan-Jorgensen (2006) "Engaging Governments in Support of Corporate Social Responsibility in Global Supply Chains," in M.J. Epstein and K.O. Hanson (eds.), *The Accountable Corporation* (Westport, CT: Praeger Publishers, vol.4): 75-122.

Cramer, J. (2006) *Corporate Social Responsibility and Globalisation: An Action Plan for Business* (Sheffield, UK: Greenleaf Publishing).

Currall, S., and M.J. Epstein (2003) "The Fragility of Organizational Trust: Lessons From the Rise and Fall of Enron," *Organizational Dynamics* 32.2.

Danone (2012) *Sustainability Report: Strategy and Performance.*

Davies, G., R. Chun, R. Vinhas Da Silva, and S. Roper (2003) *Corporate Reputation and Competitiveness* (New York: Routledge).

Davila, T., M.J. Epstein, and R. Shelton (2013) *Making Innovation Work: How to Manage it, Measure it, and Profit from it – Updated Edition* (Upper Saddle River, NJ: FT Press).

Davis, R.K. (1963) "Recreation Planning as an Economic Problem," *Natural Resources Journal* 3: 239-49.

De Beers (2005/6) *Report to Stakeholders.*

De Beers (2006) *HIV/Aids Case Study.*

De Beers (2012) *Report to Society.*

Deegan, C. (2003) *Environmental Management Accounting: An Introduction and Case Studies for Australia* (Sydney: Institute of Chartered Accountants in Australia).

Dell (2006) *Sustainability Report.*

_____(2012) *Corporate Responsibility Report: Powering the Possible.*

Deutsch, C.H. (2006) "Companies and Critics Try Collaboration," *The New York Times,* May 17, 2006.

_____(2007) “Companies Giving Green an Office,” *The New York Times,* July 3, 2007.

Deutsche Bank (2012) *Corporate Responsibility Report.*

Diageo (2006) *Corporate Citizenship Report.*

Dhaliwal, D., O. Zhen Li, and A. Tsang (2010) *Voluntary Nonfinancial Disclosure and the Cost of Equity Capital: The Initiation of Corporate Social Responsibility Reporting* (University of Arizona, The Chinese University of Hong Kong).

Dobrzynski, J.H. (1984) “Union Carbide Fights for its Life,” *BusinessWeek,* December 24, 1984.

Doppelt, B. (2003) *Leading Change Toward Sustainability: A Change-Management Guide for Business, Government and Civil Society* (Sheffield, UK: Greenleaf Publishing).

Dow (2011) *Annual Sustainability Report.*

Dowell, G., S. Hart, and B. Yeung (2000) “Do Corporate Global Environmental Standards Create or Destroy Value?” *Management Science* 46.8: 1059-74.

DuPont (2012) *Sustainability Progress Report.*

Eccles, R.G., and M.P. Krzus (2010) *One Report: Integrated Reporting for a Sustainable Strategy* (Hoboken, NY: John Wiley & Sons).

_____, S.C. Newquist, and R. Schatz (2007) "Reputation and its Risks," *Harvard Business Review,* February 2007: 104-14.

_____and G. Serafeim (2013) "Innovating for a Sustainable Strategy", *Harvard Business Review,* May 2013: 50-60.

Ehrenfeld, J.R. (2008) *Sustainability by Design: A Subversive Strategy for Transforming Our Consumer Culture* (New Haven, CT, and London: Yale University Press).

_____and A.J Hoffman (2013) *Flourishing: A Frank Conversation about Sustainability* (Stanford, CA: Stanford University Press).

Electronic Industry Citizenship Coalition (2012) *Electronic Industry Citizenship Coalition® Code of Conduct.*

Elkington, J. (1998) *Cannibals with Forks* (Gabriola Island, Canada: New Society Publishers).

_____(2001) *The Chrysalis Economy: How Citizen CEOs and Corporations Can Fuse Values and Value Creation* (Ashburton, UK: Capstone Publishing).

_____(2006) "The Triple Bottom Line," in M.J. Epstein and K.O. Hanson (eds.), *The Accountable Corporation* (Westport, CT: Praeger, vol.3): 97-110.

_____and P. Hartigan (2008) *The Power of Unreasonable People: How Social Entrepreneurs Create Markets that Change the World* (Boston, MA: Harvard Business Press).

Emerson, J. (2003) "The Blended Value Proposition: Integrating Social and Financial Returns," *California Management Review* 45.4.

Engardio, P. (2007) "Beyond the Green Corporation," *BusinessWeek,* January 29, 2007: 50-64.

Entine, J. (2007) "Chiquita Counts the Costs of Honesty," *Ethical Corporation,* May 7, 2007.

Epstein, M.J. (1996) *Measuring Corporate Environmental Performance: Best Practices for Costing and Managing an Effective Environmental Strategy* (Burr Ridge, IL: Institute of Management Accountants/Irwin Professional Publishing).

_____(2006) "Improving Organizations and Society: Performance Measurement and Management Control," *Studies in Managerial and Financial Accounting* 16: 3-19.

_____and B. Birchard (1999) *Counting What Counts: Turning Corporate Accountability to Competitive Advantage* (Reading, MA: Perseus Books).

_____and H.B. Leonard. (2007) *The Value Proposition for Corporate Social Responsibility: A Causal Linkage*

Approach (Working Paper; Houston, TX: Rice University).

_____and K. Palepu (1999) "What Financial Analysts Want," *Strategic Finance,* April 1999: 548-52.

_____and A. Rejc (2005) *Evaluating Performance in Information Technology* (Management Accounting Guideline; New York: AICPA; Toronto, CA: CMA).

_____and A. Rejc (2005) *Identifying, Measuring, and Managing Organizational Risks for Improved Performance* (Management Accounting Guideline; New York: AICPA; Toronto, CA: CMA).

_____and A. Rejc Buhovac (2006) *The Reporting of Organizational Risks for Internal and External Decision Making* (Management Accounting Guideline; New York: AICPA; Toronto, CA: CMA).

_____, A. Rejc Buhovac, and K. Yuthas (2009) *Managing Social, Environmental, and Financial Performance Simultaneously: What Can We Learn from Corporate Best Practices?* (Montvale, NY: Institute of Management Accountants).

_____, A. Rejc Buhovac, and K. Yuthas (2010) "Why Nike Kicks Butt in Sustainability," *Organizational Dynamics* 39.4: 535-56.

_____, A. Rejc Buhovac, and K. Yuthas (2013) "Managing Social, Environmental, and Financial

Performance Simultaneously," *Long Range Planning,* available online December 20, 2012.

_____and M.-J. Roy (1997) "Environmental Management to Improve Corporate Profitability," *Journal of Cost Management,* November/December 1997: 26-34.

_____and M.-J. Roy (1997) "Using ISO 14000 for Improved Organizational Learning and Environmental Management," *Environmental Quality Management,* Autumn 1997: 21-30.

_____and M.-J. Roy (1998) "Managing Corporate Environmental Performance: A Multinational Perspective," *European Management Journal* (June 1998): 286-89.

_____and M.-J. Roy (1998) *Understanding and Implementing ISO 14000* (Management Accounting Guideline; New York: AICPA; Toronto, CA: CMA).

_____and M.-J. Roy (2000) "Strategic Evaluation of Environmental Projects in SMEs," *Environmental Quality Management* 9.3: 37-47.

_____and M.-J. Roy (2001) "Sustainability in Action: Identifying and Measuring the Key Performance Drivers," *Long Range Planning* 34: 585-604.

_____and M.-J. Roy (2002) *Measuring and Improving the Performance of Corporate Boards* (Management

Accounting Guideline; New York: AICPA; Toronto, CA: CMA).

_____and M.-J. Roy (2003) "Designing a Board Evaluation System for Corporate Directors," *NACD Directors Monthly,* December 2003: 13-17.

_____and M.-J. Roy (2003) "Improving Sustainability Performance: Specifying, Implementing, and Measuring Key Principles," *Journal of General Management* 29.1: 15-31.

_____and M.-J. Roy (2003) "Making the Business Case for Sustainability: Linking Social and Environmental Actions to Financial Performance," *Journal of Corporate Citizenship* 9 (Spring 2003): 79-97.

_____and M.-J. Roy (2004) "How Does Your Board Rate?" *Strategic Finance,* February 2004.

_____and M.-J. Roy (2004) "Improving the Performance of Corporate Boards: Identifying and Measuring the Key Drivers of Success," *Journal of General Management* 29.3: 1-23.

_____and M.-J. Roy (2006) "Measuring the Effectiveness of Corporate Boards and Directors," in M.J. Epstein and K.O. Hanson (eds.), *The Accountable Corporation* (Westport, CT: Praeger Publishing, vol.1): 155-82.

_____and M.-J. Roy (2007) "Implementing a Corporate Environmental Strategy: Establishing Coordination and Control within Multinational Companies," *Business Strategy and the Environment* 16.6: 389-403.

_____and R.A. Westbrook (2001) "Linking Actions to Profits in Strategic Decision Making," *MIT Sloan Management Review,* Spring 2001.

_____and S. Widener (2007) *Measuring Multiple Stakeholder Costs and Benefits for Improved Decision-Making* (Working Paper; Houston, TX: Rice University).

_____and P.S. Wisner (2001) "Good Neighbors: Implementing Social and Environmental Strategies with the BSC," *Balanced Scorecard Report* 3.3.

_____and P.S. Wisner (2001) "Increasing Corporate Accountability: The External Disclosure of Balanced Scorecard Measures," *Balanced Scorecard Report* 3.1.

_____and P.S. Wisner (2001) "Using a Balanced Scorecard to Implement Sustainability," *Environmental Quality Management,* Winter 2001: 1-10.

_____and P.S. Wisner (2002) "Measuring and Managing Social and Environmental Impact," in *Handbook of Cost Management* (Boston, MA: Research Institute of America/Warren, Gorham & Lamont).

_____and P.S. Wisner (2005) "Managing and Controlling Environmental Performance: Evidence from Mexico," in J.Y. Lee and M.J. Epstein (eds.), *Advances in Management Accounting* (San Diego, CA: Elsevier): 115-37.

_____and P.S. Wisner (2006) "Actions and Measures to Improve Sustainability," in M.J. Epstein and K.O. Hanson (eds.), *The Accountable Corporation* (Westport, CT: Praeger, vol.3): 207-34.

_____and S.D. Young (1999) "Greening with EVA," *Strategic Finance,* January 1999: 45-49.

Equator Principles (2013) www.equator-principles.com /index.php/ep3/process-and-launch, July 30, 2013.

Equator Principles Association (2013) *The Equator Principles – A Financial Industry Benchmark for Determining, Assessing and Managing Environmental and Social Risk in Projects,* June 2013.

Esty, D.C., and A.S. Winston (2006) *Green to Gold: How Smart Companies Use Environmental Strategy to Innovate, Create Value, and Build Competitive Advantage* (New Haven, CT: Yale University Press).

_____and P.J. Simmons (2011) *The Green to Gold Business Playbook: How to Implement Sustainability Practices for Bottom-Line Results in Every Business Function* (Hoboken, NJ: John Wiley & Sons).

Fava, J.A., and J.K. Smith (1998) "Integrating Financial and Environmental Information for Better Decision Making," *Journal of Industrial Ecology* 2.1: 9-11.

Ferracone, R. (2011) "The Role of Environmental Sustainability in Executive Compensation," *Forbes,* April 26, 2011.

Finn, M., G.M. Rahl, and W. Rowe Jr. (2006) "Unrecognized Assets," *Strategy+Business* 44.

Fiksel, J. (2003) "Revealing the Value of Sustainable Development," *Corporate Strategy Today,* June 2003: 28-36.

Fombrun, C.J. (1996) *Reputation: Realizing Value from the Corporate Image* (Boston, MA: Harvard Business School Press).

_____N.A. Gardberg, and J.M. Sever (2000) "The Reputation Quotient: A Multi-Stakeholder Measure of Corporate Reputation," *Journal of Brand Management* 7.4: 241-55.

_____and M. Shanley (1990) "What's in a Name? Reputation Building and Corporate Strategy," *Academy of Management Journal* 33.2: 233-58.

Ford Motor Company (2012) *Sustainability Report Summary.*

Fortis (2006) *Corporate Social Responsibility Report.*

Foxconn Technology Group (2011) *CSER 2011 Annual Report.*

The Forum for Sustainable and Responsible Investment (US SIF) (2010) *Report on Socially Responsible Investing Trends in the United States.*

Freeman, A. M., III (2003) *The Measurement of Environmental and Resource Values: Theory and Methods* (Washington, DC: Resources for the Future Press, 2nd edn).

Freeman, R.E., J. Pierce, and R.H. Dodd (2000) *Environmentalism and the New Logic of Business: How Firms can be Profitable and Leave Our Children a Living Planet* (Oxford, UK: Oxford University Press).

_____J.S. Harrison, and A.C. Wicks (2007) *Managing for Stakeholders: Survival, Reputation, and Success* (New Haven, CT, and London: Yale University Press).

Fortune (2006) "Value Driven Leadership: Responsible Companies Committed to Tackling Global Societal Woes have Discovered they Gain Strategic Advantage," *Fortune,* November 13, 2006.

Foxconn Technology Group (2011) *CSER 2011 Annual Report.*

Fujitsu Group (2004) *Annual Report.*

_____(2012) *Fujitsu Group Sustainability Report – The Power of ICT for Sustainability and Beyond.*

Gardner, T. (2007) "Investors to Press Congress on Warming," *The New York Times,* March 17, 2007.

Gazprom (2012) *Environmental Report.*

General Electric (2007) *Citizenship Report.*

_____(2013) *Our Global Impact.*

General Mills (2010) *Corporate Social Responsibility.*

GIIRS (2010) *Sample Fund and Company Rating.*

Glazebrook, M. (2001) "How Australia's Top 500 Companies are Becoming Corporate Citizens," in J. Andriof and M. McIntosh (eds.), *Perspectives on Corporate Citizenship* (Sheffield, UK: Greenleaf Publishing): 152-65.

Global Reporting Initiative (2013) *G4 Sustainability Reporting Guidelines.*

GOOD (2012) www.good.is/posts/b-corps-go-global-sistema-b-certifies-south-american-socialenterprise.

Goodstein, E.E. (1999) *Economics and the Environment* (Upper Saddle River, NJ: Prentice Hall).

Goodwin, R.K. (2003) "Workplace Volunteering is Good Business," in "Volunteerism: What Makes America Great," *BusinessWeek* special advertising section by New Futures Media (May 26, 2003).

Grayson, D., and A. Hodges (2004) *Corporate Social Opportunity: Seven Steps to Make Corporate Social Responsibility Work for your Business* (Sheffield, UK: Greenleaf Publishing).

Greenhouse, S. (2013) "Major Retailers Join Bangladesh Safety Plan," *The New York Times,* May 13, 2013.

Gunther, M. (2004) "Money and Morals at GE," *Fortune,* November 15, 2004: 176-82.

______(2004) "The Mosquito in the Tent," *Fortune,* May 31, 2004: 158-65.

______(2006) "Saving Seafood: Wal-Mart Has Unsentimental Business Reasons for Promoting Sustainable Fishing Practices," *Fortune,* July 31, 2006.

______(2007) "The End of Garbage," *Fortune,* March 19, 2007.

______(2013) "Unilever's CEO Has a Green Thumb: Paul Polman Embraces Sustainability, and the Consumer Products Colossus Grows Like Crazy," *Fortune,* June 10, 2013.

Hammond, A.L., W.J. Kramer, and R.S. Katz (2007) *The Next 4 Billion: Market Size and Business Strategy at the Base of the Pyramid* (Washington, DC: World Resources Institute).

Hart, S.L. (2005) *Capitalism at the Crossroads: The Unlimited Business Opportunities in Solving the World's Most Difficult Problems* (Upper Saddle River, NJ: Wharton School Publishing).

_____(2010) *Capitalism at the Crossroads: Next Generation Business Strategies for a Post-Crisis World* (Upper Saddle River, NJ: Pearson Education, 3rd edn).

_____and G. Ahuja (1996) "Does it Pay to be Green? An Empirical Examination of the Relationship between Emission Reduction and Corporation Performance," *Business Strategy and the Environment* 5: 30-37.

_____and C.M. Christensen (2002) "The Great Leap: Driving Innovation from the Base of the Pyramid," *Sloan Management Review,* Fall 2002: 51-56.

_____and T. London (2005) "Developing Native Capability: What Multinational Corporations Can Learn from the Base of the Pyramid," *Stanford Social Innovation Review,* Summer 2005: 28-33.

_____and S. Sharma (2004) "Engaging Fringe Stakeholders for Competitive Imagination," *Academy of Management Executive* 18.1: 7-18.

Harvey, F. (2005) "GE Looks out for a Cleaner Profit," *Financial Times,* July 1, 2005: 13.

Hawken, P., A. Lovins, and L.H. Lovins (1999) *Natural Capitalism: Creating the Next Industrial Revolution* (Boston, MA: Little, Brown).

Hawkins, D.E. (2006) *Corporate Social Responsibility: Balancing Tomorrow's Sustainability and Today's Profitability* (New York: Palgrave Macmillan).

Heal, G. (2008) *When Principles Pay: Corporate Social Responsibility and the Bottom Line* (New York: Columbia University Press).

Henkel (2012) *Sustainability Report.*

Hennes & Mauritz (2012) *Conscious Actions: Sustainability Report 2012.*

Hernandez, M.I. (2004) "Institutionalising Global Standards of Responsible Corporate Citizenship: Assessing the Role of the UN Global Compact," in M. McIntosh, S. Waddock, and G. Kell (eds.), *Learning to Talk: Corporate Citizenship and the Development of the UN Global Compact* (Sheffield, UK: Greenleaf Publishing): 114-28.

Hewlett-Packard (2012) *Global Citizenship Report.*

Hoevenagel, R. (1994) "An Assessment of the Contingent Valuation Method," in R. Pethig (ed.),

Valuing the Environment: Methodological and Measurement Issue s (Boston, MA: Kluwer Academic Publishers): 195-228.

Hoffman, A.J., and Woody, J.G. (2008) *Climate Change: What's Your Business Strategy?* (Boston, MA: Harvard Business Press).

Holliday, C. (2001) "Sustainable Growth, the DuPont Way," *Harvard Business Review,* September 2001: 129-32.

_____S. Schmidheiny, and P. Watts (2002) *Walking the Talk: The Business Case for Sustainable Development* (Sheffield, UK: Greenleaf Publishing).

Holmes, S. (2006) "Nike Goes for the Green," *BusinessWeek,* September 25, 2006.

Honda (2006) *Environmental Annual Report.*

Honda North America (2012) *North American Environmental Report.*

Hopkins, M. (2003) *The Planetary Bargain: Corporate Social Responsibility Matters* (London: Earthscan Publications).

Howell, R.A. (1994) *Developing Comprehensive Performance Indicators* (Hamilton, Canada: Society of Management Accountants of Canada, vol.31).

HSBC Holdings (2012) *Sustainability Report.*

IIRC (2013) *Consultation Draft of the International <IR> Framework – Integrated Reporting.*

International Trade Centre (2012) *Social Accountability International – SA8000.*

IPIECA, API, and OGP (2010) *Oil and Gas Industry Guidance on Voluntary Sustainability Reporting,* 2nd edn.

ISO (2010) *ISO 26000:2010: Guidance on Social Responsibility.* www.iso.org/iso/catalogue_detail?csnumber=42546.

_____(2011) "The ISO Survey of Management Systems Standard Certifications," www.iso.org/iso/home/news_index/news_archive/news.htm?refid=Ref1686.

Interbrand (2012) *Best Global Brands 2012: The Definitive Guide to the 100 Best Global Brands.*

Jackson, S. (1984) "Union Carbide's Good Name Takes a Beating," *BusinessWeek,* December 31, 1984.

Jaffe, A.B., S.R. Peterson, P.R. Portney, and R.N. Stavins (1995) "Environmental Regulation and the Competitiveness of US Manufacturing: What the Evidence Tells Us," *Journal of Economic Literature* 33.1: 132-63.

Jensen, M.C. (2001) "Value Maximization, Stakeholder Theory and the Corporate Objective Function," *Journal of Applied Corporate Finance* 14.3: 8-21.

Johnson & Johnson (2012) *Citizenship and Sustainability Report.*

Kahneman, D., I. Ritov, and D. Schkade (1999) "Economic Preferences or Attitude Expressions? An Analysis of Dollar Responses to Public Issues," *Journal of Risk and Uncertainty* 19: 203-35.

Kanal, V. (2010) "Lessons from Dell's Stakeholder Engagement Model," http://www.triplepundit.com/2010/12/lessons-dells-stakeholder-engagement-model/.

Kanter, R.M (2009) *Super Corp: How Vanguard Companies Create Innovation, Profits, and Social Good* (New York, NY: Crown Business).

Kaplan, R.S., and D.P. Norton (1996) *The Balanced Scorecard: Translating Strategy into Action* (Boston, MA: Harvard Business School Press).

______(2000) *The Strategy-Focused Organization: How Balanced Scorecard Companies Thrive in the New Business Environment* (Boston, MA: Harvard Business School Press).

Kingfisher (2013) *Net Positive Report.*

Klassen, R.D., and D.C. Whybark (1999) "The Impact of Environmental Technologies on Manufacturing Performance," *Academy of Management Journal* 42.6: 599-615.

Knoepfel, I. (2001) "Dow Jones Sustainability Group Index: A Global Benchmark for Corporate Sustainability," *Corporate Environmental Strategy* 8.1: 6-15.

Kodak (2004) *Health, Safety, and Environment Report.*

_____(2013) www.kodak.com/ek/US/en/Global_Sustainability/Health_Safety_and_Environment/HSE_Support_Center/Product_End_of_Life_Management/Recycling_Chemical.htm.

Koehler, D.A., and E. Hespenheide (2012) "Drivers of Long-term Business Value: Stakeholders, Stats and Strategy" (www.deloitte.com/view/en_US/us/Services/additional-services/deloittesustainability/a4521cabe6cc7310VgnVCM3000001c56f00aRCRD.htm).

_____(2013) "Finding the Value in Environmental, Social, and Governance Performance," *Deloitte Review* 12: 99-111.

Kolk, A. (2005) "Sustainability Reporting," *VBA Journal* 3.

_____and J. Pinske (2006) "Stakeholder Mismanagement and Corporate Social Responsibility Crises," *European Management Journal* 24.1: 59-72.

Kotler, P., and N. Lee (2005) *Corporate Social Responsibility: Doing the Most Good for Your Company and Your Cause* (Hoboken, NJ: John Wiley).

KPMG (2002) *International Survey of Corporate Sustainability Reporting* (Amsterdam: KPMG).

_____(2005) *International Survey of Corporate Responsibility Reporting* (Amsterdam: KPMG).

Kramer, M., and S. Cooch (2006) *Investing for Impact: Managing and Measuring Proactive Social Investments* (Boston, MA: Foundation Strategy Group).

_____M. Parkhurst, and L. Vaidyanathan (2009) *Breakthroughs in Shared Measurement and Social Impact* (FSG Social Impact Advisor).

Lambert (2013) "Event Sustainability Management – ISO 20121 Passes 2012 Olympic Games Test," *ISO News,* January 9, 2013.

Larkin, J. (2003) *Strategic Reputation Risk Management* (New York: Palgrave Macmillan).

Lash, J., and F. Wellington (2007) "Competitive Advantage on a Warming Planet," *Harvard Business Review,* March 2007: 94-102.

Laszlo, C. (2003) *The Sustainable Company: How to Create Lasting Value through Social and Environmental Performance* (Washington, DC: Island Press).

_____(2008) *Sustainable Value: How the World's Leading Companies are Doing Well by Doing Good* (Stanford, CA: Stanford University Press).

_____and N. Zhexembayeva (2013) *Embedded Sustainability: The Next Big Competitive Advantage* (Stanford, CA: Stanford University Press).

Layman, R.C., J.R. Boyce, and K.R. Criddle (1996) "Economic Valuation of the Chinook Salmon Sport Fishery of the Gulkana River, Alaska, Under Current and Alternate Management Plans," *Land Economics* 72.1: 112-28.

Layoux, A.R., and C. Soltis Martel (2013) "Sustainability Rising," *NACD Directory,* March 14, 2013.

Lehni, M. (2000) *Eco-Efficiency: Creating More Value with Less Impact* (Geneva: World Business Council for Sustainable Development).

Leipziger, D. (2003) *The Corporate Responsibility Code Book* (Sheffield, UK: Greenleaf Publishing).

Leonard-Barton, D. (1995) *Wellsprings of Knowledge: Building and Sustaining the Sources of Innovation* (Boston, MA: Harvard Business School Press).

Lewis, S., and T. Little (2004) *Fooling Investors and Fooling Themselves: How Aggressive Corporate Accounting and Asset Management Tactics Can Lead to Environmental Accounting Fraud* (Oakland, CA: Rose Foundation for Communities and the Environment).

Lippman, S. (2002) "Environmental Responsibility: Are Corporations Buying It? A Look at Corporate Green Purchasing," *Investing for a Better World* 17.2 (Summer 2002).

L'Oréal (2005) *Sustainability Report.*

_____(2010) *Sustainable Development Report.*

Logan, D., D. Roy, and L. Regelbrugge (1997) *Global Corporate Citizenship: Rationale and Strategies* (Washington, DC: Hitachi Foundation).

London Stock Exchange (2010) *Investor Relations – A Practical Guide.*

London, T., and S.L. Hart (2004) "Reinventing Strategies for Emerging Markets: Beyond the Transnational Model," *Journal of International Business Studies* 35: 350-70.

_____(2011) *Next Generation Business Strategies for the Base of the Pyramid: New Approaches for Building Mutual Value* (Upper Saddle River, NJ: Pearson Education).

Lovins, A., L.H. Lovins, and P. Hawken (1999) "A Road Map for Natural Capitalism," *Harvard Business Review*, May–June 1999.

Lydenberg, S. (2005) *Corporations and the Public Interest: Guiding the Invisible Hand* (San Francisco, CA: Berrett-Koehler Publishers).

Lye, G., and F. Müller (2004) *The Changing Landscape of Liability: A Director's Guide to Trends in Corporate Environmental, Social and Economic Liability* (London: SustainAbility).

Mackey, A., T.B. Mackey, and J.B. Barney (2007) "Corporate Social Responsibility and Firm Performance: Investor Preferences and Corporate Strategies," *Academy of Management Review* 32.3: 817-35.

MacLean, R. (2004) "EHS Organizational Quality: A DuPont Case Study," *Environmental Quality Management*, Winter 2004: 19-28.

_____(2005) "Engagement at the Top," *Environmental Protection* 16.8.

Margolis, J.D., and J. Walsh (2003) "Misery Loves Company: Rethinking Social Initiatives by Business," *Administrative Science Quarterly* 48.2: 268-305.

Makover, J. (2009) *Strategies for the Green Economy: Opportunities and Challenges in the New World of Business* (New York: McGraw-Hill).

McDonald's (2004) *Worldwide Corporate Responsibility Report.*

______(2012) *Global Sustainability Highlights.*

McDonough, W., and M. Braungart (2002) *Cradle to Cradle: Remaking the Way we Make Things* (New York: North Point Press).

McDonough Braungart Design Chemistry (2013) www.mbdc.com/cradle-to-cradle/, accessed July 31, 2013.

McElhaney, K.A. (2008) *Just Good Business: The Strategic Guide to Aligning Corporate Responsibility and Brand* (San Francisco, CA: Berrett-Koehler Publishers).

Meehan, M. (2010) "Tying Executive Pay to Sustainability Performance," *Forbes.com,* October 25, 2010.

MillerCoors (2012) *Sustainability Report: Great Things on Tap.*

Mitchell, R.C., and R.T. Carson (1989) *Using Surveys to Value Public Goods: The Contingent Valuation Method* (Washington, DC: Resources for the Future Press).

Mitsubishi Corporation (2010) *Sustainability Report.*

National Australia Bank (2012) *Report of the Directors.*

National Highway Traffic Safety Administration (2010) *CAFE 2011–2016 Final Rule.*

NCIF (2012) *Collaborating to Harmonize Standardized Metrics for Impact Investors.*

Nelson, J., P. Zollinger, and A. Singh (2001) *Power to Change: Mobilising Board Leadership to Deliver Sustainable Value to Markets and Society* (London: SustainAbility).

Nestlé (2012) *Nestlé in Society: Creating Shared Value and Meeting Our Commitments – Full Report.*

Nidumolu, R., K. Kramer, and A.J. Zeitz (2012) "Connecting Heart to Head: A Framework for Sustainable Growth," *The EcoInnovator,* February 26, 2012.

Nike (2004) *Corporate Responsibility Report.*

_____(2012) *Nike Inc., FY10/11 – Sustainable Business Report Summary.*

_____(2012) "NIKE, Inc. Announces Strategic Partnership to Scale Waterless Dyeing Technologies."

Novartis (2007) "Implementing a Living Wage Globally: The Novartis Approach" (www.novartis.com/about-novartis/corporate-citizenship/citizenship-in-action/case-studies.shtml, accessed September 5, 2007).

_____(2007) "Novartis Comunidad: Creating a Permanent Local Medicine Donation Program in Argentina" (www.novartis.com/about-novartis/corporate-citizenship/citizenship-in-action/case-studies.shtml, September 5, 2007).

Novo Nordisk (2012) *Annual Report.*

Nunn, M. (2003) "Building Social Capital," in "Volunteerism: What Makes America Great," *BusinessWeek* special advertising section by New Futures Media, May 26, 2003.

Oliva, R., and J. Quinn (2003) "Interface's Evergreen Services Agreement," *Harvard Business School Case* 9-603-112, June 2003.

Orenstein, B.W. (1997) "Lucent Seeks Self-regulation Nod," *Eastern PA Business Journal,* January 1997.

Patt, A. (2006) "Dealing with Uncertainty," in A.E. Farrell and J. Jäger (eds.), *Assessments of Regional and Global Environmental Risks: Designing Processes for the Effective Use of Science in Decisionmaking* (Washington, DC: Resources for the Future Press).

Peloza, J., M. Loock, J. Cerruti, and M. Muyot (2012) "Sustainability: How Stakeholder Perceptions Differ from Corporate Reality", *University of California, Berkeley,* Fall 2012: 74-97.

Peters, G. (1999) *Waltzing with the Raptors: A Practical Roadmap to Protecting your Company's Reputation* (New York: John Wiley).

Pfizer (2013) *Kalamazoo Cuts 'Wasted' Expenses.*

Piasecki, B. (2007) *World Inc.* (Naperville, IL: Sourcebooks).

Pinkham, D.G. (2006) "Business and Government: Friends and Foes," in M.J. Epstein and K.O. Hanson (eds.), *The Accountable Corporation* (Westport, CT: Praeger Publishers, vol.4): 23-32.

Pleon (2005) *Accounting for Good: The Global Stakeholder Report 2005* (Amsterdam/Bonn: Pleon Kohtes Klewes).

Porter, M.E. (1990) *The Competitive Advantage of Nations* (New York: Free Press).

_____and M.R. Kramer (2002) "The Competitive Advantage of Corporate Philanthropy," *Harvard Business Review,* December 2002: 56-69.

_____and M.R. Kramer (2006) "Strategy and Society: The Link between Competitive Advantage and Corporate Social Responsibility," *Harvard Business Review,* December 2006: 78-92.

_____and C. van der Linde (1995) "Green and Competitive: Ending the Stalemate," *Harvard Business Review,* September–October 1995: 120-34.

Power, S. (2006) "Take it Back: Where Do Cars Go When They Die? In Europe, They Have Little Choice," *Wall Street Journal,* April 17, 2006: R6.

Pracejus, J.W., and G.D. Olsen (2004) "The Role of Brand/Cause Fit in the Effectiveness of Cause-Related Marketing Campaigns," *Journal of Business Research* 57, 635-40.

Prahalad, C.K. (2004) *The Fortune at the Bottom of the Pyramid: Eradicating Poverty through Profits* (Upper Saddle River, NJ: Wharton School Publishing).

_____and A. Hammond (2002) "Serving the World's Poor, Profitably," *Harvard Business Review,* September 2002: 48-57.

PricewaterhouseCoopers (2002) *Sustainability Survey Report,* August 2002.

_____(2005) *Corporate Responsibility: Strategy, Management and Value: How PwC can Help.*

Professional Accountants in Business Committee (2006) "Professional Accountants in Business: At the Heart of Sustainability?" (www.ifac.org/store).

Professional Engineering (1999) "Brent Spar Outcry Leaves Shell with a 60m Pound Bill," *Professional Engineering* 12.16: 9.

PUMA (2011) *PUMA's Environmental Profit and Loss Account for the Year Ended 31 December 2010.* Radin, T.J. (2003) "Chiquita Brands International, Inc: Values-Based Management and Corporate Responsibility in Latin America," in L.P. Hartman, D.G. Arnold, and R.E. Wokutch (eds.), *Rising above Sweatshops: Innovative Approaches to Global Labor Challenges* (Westport, CT: Praeger Publishers): 353-84.

Rangone, A. (1998) "On the Applicability of Analytical Techniques for the Selection of AMTs in Small–Medium Sized Firms," *Small Business Economics* 10: 293-304.

Reinhardt, F.L. (1999) "Bringing the Environment Down to Earth," *Harvard Business Review,* July–August 1999: 149-57.

_____(2000) *Down to Earth: Applying Business Principles to Environmental Management* (Boston, MA: Harvard Business School Press).

Rio Tinto (2012) *Sustainable Development Report.*

Roche (2012) *Annual Report.*

Rummler, G.A., and A.P. Brache (1995) *Improving Performance: How to Manage the White Space on the Organization Chart* (San Francisco, CA: Jossey-Bass).

Russo, M.V. (2010) *Companies on a Mission: Entrepreneurial Strategies for Growing Sustainably, Responsibly, and Profitably* (Stanford, CA: Stanford University Press).

SAAS (2013) http://www.saasaccreditation.org/certfacilitieslist.htm

Samsung (2011) *Sustainability Report.*

Santander (2005) *Sustainability Report.*

Savitz, A., and K. Weber (2006) *The Triple Bottom Line; How Today's Best-Run Companies are Achieving Economic, Social, and Environmental Success – and How You Can Too* (San Francisco, CA: Jossey-Bass).

Sberbank (2012) *Corporate Social Responsibility Report 2012.*

Schaltegger, S., R. Burritt, and H. Petersen (2003) *An Introduction to Corporate Environmental Management: Striving for Sustainability* (Sheffield, UK: Greenleaf Publishing).

Schnietz, K.E. (2006) "The Purpose and History of Business Regulation," in M.J. Epstein and K.O. Hanson (eds.), *The Accountable Corporation* (Westport, CT: Praeger Publishers, vol.4): 3-22.

_____and M.J. Epstein (2004) "Does a Reputation for Corporate Social Responsibility Pay Off? Investor Reaction to the 1999 Seattle WTO Meeting," in J. Hooker (ed.), *International Corporate Responsibility: Exploring the Issues* (Pittsburgh, PA: Carnegie Mellon University Press).

_____and M.J. Epstein (2005) "Exploring the Financial Value of a Reputation for Corporate Social Responsibility during a Crisis," *Corporate Reputation Review* 7.4: 327-45.

Scott, A. (1965) *The Valuation of Game Resources: Some Theoretical Aspects* (Canadian Fisheries Report 4; Ottawa: Department of Fisheries).

Serrano, L. (2008) *Amanco: Providing Irrigation Systems to the Rural Poor* (Case study; UNDP: Growing Inclusive Markets).

Sethi, S.P. (2003) *Setting Global Standards: Guidelines for Creating Codes of Conduct in Multinational Corporations* (Hoboken, NJ: John Wiley).

Shah, A. (2013) "Poverty Facts and Stats," *Global Issues,* January 7, 2013

Sharma, S. (2000) "Managerial Interpretations and Organizational Context as Predictors of Corporate Choice of Environmental Strategy," *Academy of Management Journal* 43.4: 681-97.

Shechter, M., and S. Freeman (1994) "Nonuse Value: Reflection on the Definition and Measurement," in R. Pethig (ed.), *Valuing the Environment: Methodological and Measurement Issues* (Boston, MA: Kluwer Academic Publishers): 171-94.

Shell (2005) *Sustainability Report.*

Shell International Global Business Environment (2003) *Exploring the Future Scenarios: An Explorer's Guide* (Shell International).

Siemens (2012) *Sustainability Report 2012: Driving Sustainability.*

SIRAN (2006) "Socially Responsible Investment Analysts Find More Large US Companies Reporting on Social and Environmental Issues" (www.siran.org/pdfs/csrreportingpr2006.pdf, accessed September 5, 2007).

Smith, W.K., and M.L. Tushman (2005) "Managing Strategic Contradictions: A Top Management Model for Managing Innovation Streams," *Organization Science* 16.5: 522-36.

Social Accountability International (2008) *Social Accountability 8000.*

Social Enterprise Knowledge Network (2006) *Effective Management of Social Enterprises: Lessons from*

Businesses and Civil Society in Iberoamerica (Cambridge, MA: Harvard University Press).

Social Investment Forum (2006) *2005 Report of Socially Responsible Investing Trends in the United States* (Washington, DC: Social Investment Forum).

Sony Ericsson (2011) *Sustainability Report.*

Standard-Friel, J. (2004) "Proving the Win–Win Strategy of Cause-Related Marketing" (www.onphilanthropy.com, accessed September 5, 2007).

Stanwick, P.A., and S.D. Stanwick (1998) "The Relationship between Corporate Social Performance and Organizational Size, Financial Performance, and Environmental Performance: An Empirical Examination," *Journal of Business Ethics* 17: 195-204.

Starbucks (2012) *Global Responsibility Report.*

Stead, W.E., and J.G. Stead (2004) *Sustainable Strategic Management* (New York: M.E. Sharpe).

Stern, N. (2006) *Economics of Change: The Stern Review* (Cambridge, UK: Cambridge University Press).

Stevens, T.H., R. Belkner, D. Dennis, D. Kittredge, and C. Willis (2000) "Comparison of Contingent Valuation and Conjoint Analysis in Ecosystem Management," *Ecological Economics* 32: 63-74.

Strom, S. (2012) "Companies Pick up Used Packaging and Recycling's Cost," *The New York Times,* March 23, 2012.

Stuart, L. (2003) "Multinationals Should Face the Same Rules no Matter Where They Set up Shop," *The Guardian,* August 11, 2003 (business.guardian.co.uk/story/0,3604,1016103,00.html).

Sundaram, A.K. (1998) *Perrier, Nestlé, and the Agnellis* (Thunderbird Case Study; Glendale, AZ: Thunderbird School of Global Management).

Svendsen, A. (1998) *The Stakeholder Strategy: Profiting from Collaborative Business Relationships* (San Francisco, CA: Berrett-Koehler Publishers).

_____and M. Laberge (2007) *Convening Stakeholder Networks: A New Way of Engaging* (Chestnut Hill, MA: Center for Corporate Citizenship).

Time Warner (2007) "Green Building Design" (www.timewarner.com/corp/citizenship/environment/green_design/index.html, accessed September 5, 2007).

Toyota Motor Corporation (2012) *Sustainability Report.*

_____(2012) www.toyota-global.com/company/vision_philosophy/guiding_principles.html.

Trudel, R., and J. Cotte (2009) "Does It Pay to be Good?" *MIT Sloan Management Review,* Winter 2009, 61-68.

Tysiac, K. (2013) "Conflict Minerals Rule Poses Compliance Challenge: SEC Regulation Forces Deep Dive into Supply Chains," *Journal of Accountancy,* April 2013.

Unilever (2012) *Progress Report – Unilever Sustainable Living Plan.*

United Nations (2000) *United Nations Millennium Declaration.*

_____(2013) *The Millennium Development Goals Report 2013.*

United Nations Global Compact (undated) "The Ten Principles" (www.unglobalcompact.org/AboutTheGC/TheTenPrinciples/index.html, accessed September 5, 2007).

_____(2013) www.unglobalcompact.org/ParticipantsAndStakeholders/index.html, accessed July 3, 2013.

US Environmental Protection Agency (1992) *Facility Pollution Prevention Guide* (Washington, DC: EPA).

_____(1999) "An Introduction to Environmental Accounting as a Business Management Tool," in M.V.

Russo (ed.), *Environmental Management: Readings and Cases* (Boston, MA: Houghton Mifflin).

UPS (2005) *Sustainability Report.*

_____(2012) *Corporate Sustainability Report: More on What Matters.*

Vergin, R.C., and M.W. Qoronfleh (1998) "Corporate Reputation and the Stock Market," *Business Horizons,* January–February 1998: 19-26.

Vodafone Group (2013) *Sustainability Report 2012/2013.*

Vogel, D. (2005) *The Market for Virtue: The Potential and Limits of Corporate Social Responsibility* (Washington, DC: Brookings Institution Press).

Volkswagen Group (2012) *Sustainability Report.*

Volvo Car Group (2012) *Sustainability Report.*

Waddock, S., and C. Bodwell (2007) *Total Responsibility Management: The Manual* (Sheffield, UK: Greenleaf Publishing).

Walker, S.F., and J.W. Marr (2001) *Stakeholder Power: A Winning Plan for Building Stakeholder Commitment and Driving Corporate Growth* (Reading, MA: Perseus Books).

Walmart (2013) *Global Responsibility Report.*

Warhurst, A. (2002) *Environmental and Social Performance Indicators in Minerals Development* (Warwick, UK: Warwick Business School).

Webley, S., and E. More (2003) *Does Business Ethics Pay? Ethics and Financial Performance* (London: Institute of Business Ethics).

Weiser, J., M. Kahane, S. Rochlin, and J. Landis (2006) *Untapped: Creating Value in Underserved Markets* (San Francisco, CA: Berrett-Koehler Publishers).

Welford, R., W. Young, and B. Ytterhus (1998) "Toward Sustainable Production and Consumption: A Conceptual Framework," in N.J. Roome (ed.), *Sustainability Strategies for Industry* (Washington, DC: Island Press): 75-98.

Werther, W.B., Jr. (2006) *Strategic Corporate Social Responsibility: Stakeholders in a Global Environment* (Thousand Oaks, CA: Sage Publications).

Wheeler, D., and M. Sillanpää (1997) *The Stakeholder Corporation: A Blueprint for Maximizing Stakeholder Value* (Washington, DC: Pitman Publishing).

Winston, A.S. (2009) *Green Recovery: Get Lean, Get Smart, and Emerge from the Downturn on Top* (Boston, MA: Harvard Business Press).

Wisner, P.S. (2004) "Keynote Speech," *Corporate Volunteerism Awards,* Phoenix, AZ, April 22, 2004.

_____and M.J. Epstein (2005) "'Push' and 'Pull' Impacts of NAFTA on Environmental Responsiveness and Performance in Mexican Industry," *Management International Review* 45.3: 327-47.

_____M.J. Epstein, and R.P. Bagozzi (2006) "Organizational Antecedents and Consequences of Environmental Performance," in M. Freedman and B. Jaggi (eds.), *Environmental Accounting: Commitment or Propaganda* (San Diego, CA: Elsevier): 143-48.

Wood, S. (2012) *The Role of Philanthropic Capital in Entrepreneurship: An Empirical Analysis of Financial Vehicles at the Nonprofit/For-profit Boundary of Science and Engineering* (Boston: MA: MIT).

Wood, D.J., J.M. Logsdon, P.G. Lewellyn, and K. Davenport (2006) *Global Business Citizenship: A Transformative Framework for Ethics and Sustainable Capitalism* (New York: M.E. Sharpe).

Yip, G.S. (1989) "Global Strategy ... in a World of Nations?" *Sloan Management Review,* Fall 1989: 29-41.

Young, R.A. (2005) *Determining the Economic Value of Water: Concepts and Methods* (Washington, DC: Resources for the Future Press).

Zadek, S. (2001) *The Civil Corporation: The New Economy of Corporate Citizenship* (London/Stirling, VA: Earthscan Publications).

_____(2005) "Responsibility Isn't a Blame Game," *Fortune,* October 3, 2005.

_____and M. Merme (2003) "Redefining Materiality: Making Reporting More Relevant," *Account-Ability Quarterly,* September 2003.

Index

A

C

D

E

H

I

M

O

P